Artificial Intelligence for Multimedia Information Processing

Advances in artificial intelligence (AI), widespread mobile devices, internet technologies, multimedia data sources, and information processing have led to the emergence of multimedia processing. Multimedia processing is the application of signal processing tools to multimedia data—text, audio, images, and video—to allow the interpretation of these data, particularly in urban and smart city environments. This book discusses the new standards of multimedia and information processing from several technological perspectives, including analytics empowered by AI, streaming on the intelligent edge, multimedia edge caching and AI, services for edge AI, and hardware and devices for multimedia on edge intelligence.

FEATURES

- Covers a wide spectrum of enabling technologies for AI and machine learning for multimedia and information processing
- Includes many applications using AI, from robotics and driverless cars to environmental, human health, and remote sensing
- Presents an overview of the fundamentals of AI and multimedia processing: imaging, signal, and speech
- Explains new models and architectures for multimedia streaming, services, and caching for AI
- Discusses the emerging paradigms of the deployment of hardware and devices for multimedia on edge intelligence
- Gives recommendations for future research in multimedia and AI

This book is written for engineers and graduate students in image and signal processing, information processing, environmental engineering, medical and public health, etc., who are interested in machine learning, deep learning, and multimedia processing.

Artificial Intelligence for Multimedia Information Processing

Tools and Applications

Edited by
Xavier Savarimuthu, S.J.,
Sivakannan Subramani, and Alex Noel Joseph Raj

CRC Press
Taylor & Francis Group
Boca Raton London New York

CRC Press is an imprint of the
Taylor & Francis Group, an **informa** business

First edition published 2024
by CRC Press
2385 NW Executive Center Drive, Suite 320, Boca Raton FL 33431

and by CRC Press
4 Park Square, Milton Park, Abingdon, Oxon, OX14 4RN

CRC Press is an imprint of Taylor & Francis Group, LLC

ISBN: 978-1-032-52147-3 (hbk)
ISBN: 978-1-032-52148-0 (pbk)
ISBN: 978-1-003-40543-6 (ebk)

DOI: 10.1201/9781003405436

Typeset in Times
by Apex CoVantage, LLC

*We dedicate this volume to our parents, who stand as
a testament to our insatiable quest for knowledge.
Their boundless encouragement, unwavering support,
and enduring love have been the guiding lights
throughout our journey in completing this book.*

*We would also like to dedicate this book to all those
who believe in the transformative power of ideas that
inspires the next generation of thinkers, creators, and
innovators. The researchers and practitioners who
tirelessly push the boundaries of possibility, inspiring us
all to reach new heights. This book is dedicated to you.*

Contents

SECTION I   AI-Multimedia Information Processing

SECTION II   AI and Its Applications in Public Health

SECTION III AI and Its Applications in Environmental Science

Foreword

Artificial intelligence (AI), with its intricacies of machine learning, deep learning, and its convergence with other cutting-edge technologies is carving its unique imprints in the annals of human progress. We humans are not merely observers of technological progress, but active participants in a narrative that continues to unfold.

Recent developments in Large language models, thanks to their resemblance to what an artificial general intelligence might look like in the future, has somewhat obscured the unprecedented developments in machine learning and deep learning that are taking place, and threatens to leave unattended a vast number of industrial and domestic applications of AI that are becoming regular engineering practice. Thus, as the former have made their way to newspaper headlines, machine learning and vision applications are becoming the industry standard. As a consequence, there is a clear need for a thorough review of these technologies, considering their evolution, their current status, and the future application opportunities they offer, to bring them back to their well-earned front page of technological advancement.

This book, *Artificial Intelligence for Multimedia Information Processing: Tools and Applications*, clearly fulfills this need, by offering a comprehensive narrative of AI's epoch-making journey. This volume is more than a retrospective; it is a gaze into the future, inviting you to be part of the ongoing narrative of transformative intelligence created by AI.

Beyond mere theoretical underpinnings, this volume goes to showcase the practical applications of AI across diverse domains. It explores how AI is transforming multimedia information processing, public health and healthcare, environmental science and sustainability, automobile and transportation industry (self-driving cars, intelligent traffic management systems, and optimized logistics solutions), computer vision, material science, and various other industries, providing tangible examples of its impact on our society.

This wonderful edition has six sections and seventeen well-edited chapters. Chapter by chapter, readers will embark on a journey through the milestones of AI's evolution, from its nascent stages to the leading-edge intelligent systems that permeate our daily lives. As we traverse the pages of this insightful exploration, we are invited to witness the profound impact of AI across various domains.

I commend the authors for their commitment to documenting and interpreting the epoch-making journey of AI. Their insights, drawn from the intersection of theory and practice, provide a distinct perspective that will undoubtedly inspire and inform readers across a spectrum of backgrounds. This book that you hold in your hands is an enlightening exploration into the diverse realms where AI unfurls its transformative capabilities. It will equip readers with the knowledge and insights necessary to navigate this rapidly changing landscape, enabling them to contribute to the responsible and beneficial application of AI for the well-being of humanity.

As we embark on this intellectual journey, may the insights presented within these pages deepen our appreciation for the multifaceted nature of AI and inspire us to explore the boundless possibilities that lie ahead.

My best wishes.

Fabio Gómez-Estern, PhD
Rector
Universidad Loyola
Campus Sevilla

Preface

This book, titled *Artificial Intelligence for Multimedia Information Processing: Tools and Applications*, is a comprehensive exploration of the multifaceted relationship between AI and multimedia, offering a deep dive into the tools and applications that drive this dynamic synergy. It shines as a beacon in this transformative landscape, guiding readers through the dynamic fusion of AI and multimedia technologies. In an era characterized by an unprecedented increase in multimedia data, the intersection of artificial intelligence (AI) and multimedia information processing has become a focal point of innovation, transforming the landscape of technology and human interaction. At the academic level, an understanding of what AI is really capable of delivering is being sought, often with a view toward learning about it so that its unbridled power can be harnessed, for the greater good. *Artificial Intelligence for Multimedia Information Processing: Tools and Applications* is a step in this direction, the right direction.

This volume serves as a roadmap for researchers, practitioners, and enthusiasts eager to navigate the intricate terrain of AI applied to multimedia information processing. It spans a broad spectrum of tools and applications that leverage the power of AI to unlock new possibilities in understanding and harnessing the potential of multimedia information. From image and video analysis to natural language processing and beyond, the contributions within this volume encapsulate the cutting-edge advancements that have emerged at the crossroads of AI and multimedia. Our goal is to provide a holistic perspective, bridging theoretical foundations with practical insights, researched case studies, and real-world applications.

The chapters within this volume cover a wide spectrum of topics, ranging from fundamental concepts in AI and multimedia processing to cutting-edge applications that span diverse domains. Readers will find discussions on computer vision, natural language processing, machine learning, deep learning, automobile applications, environmental resource monitoring, and applications of AI to public health concerns, each tailored to the specific challenges and opportunities posed by multimedia data. The book additionally delves into the ethical considerations and societal implications of deploying AI in multimedia contexts, fostering a nuanced understanding of the responsible use of these technologies.

The contributors to this book are leading experts in their respective fields, bringing a wealth of knowledge and experience to the table. Their insights illuminate the current state of the art and offer valuable perspectives on the future trajectory of AI in multimedia information processing.

We invite you to embark on a journey through the exciting landscape of *Artificial Intelligence for Multimedia Information Processing*, where theoretical foundations meet innovative applications, and where the power of AI converges with the richness of multimedia data to shape the future of technology and human experience.

Xavier Savarimuthu, S.J.
Sivakannan Subramani
Alex Noel Joseph Raj

Acknowledgments

We would like to express our gratitude to Irma Shagla Britton, Senior Editor, and Chelsea Reeves, Senior Editorial Assistant, CRC Press, Taylor & Francis Group, for their able guidance and ready support through all stages of this project.

Dr. Fabio Gómez-Estern, Rector, Universidad Loyola, Sevilla, Spain, is very gratefully acknowledged for his brilliant exposition in the foreword that bestows certitude to the book's objective. Mr. Suvobrata Ganguly, Editor, Core Sector Communique, is thanked for providing his valuable editorial advice. We also offer our heartfelt appreciation to all the contributors for their work on this project. Their insights illuminate the challenges and opportunities inherent in the intertwining of AI and multimedia, providing readers with a comprehensive view of the state-of-the-art methodologies and their real-world applications.

X.S. (Xavier Savarimuthu, S.J.) would like to thank the Society of Jesus (Jesuit Order) for constantly encouraging and providing support of the author's research. The St. Xavier's College Jaipur also played a pivotal role by providing the necessary infrastructure. X.S. would also like to thank his coeditors Sivakannan Subramani and Alex Noel Joseph Raj, for their ready participation in this endeavour, notwithstanding their many research and teaching commitments. Grateful acknowledgment is due to Aleksandar Zecevic, Associate Dean, Santa Clara University, California; Rev. Fr. Arul Sivan, S.J.; Rev. Dr. Ignacimuthu, S.J.; Rev. Dr. Joseph Varghese, CMI; Rev. Dr. Sebasti L Raj, S.J.; Rev. Dr. Xavier Vedam, S.J.; Rev. Fr. Johnson Padiyara, S.J.; Rev. Fr. Dejus J R and my confidantes Ms. Margaret King and Dr. Pradip Kumar Kar, for their unflinching support and encouragement throughout his career; to Prof. Allan H. Smith, University of California, Berkeley, for inspiration during the formative years as a budding environmental scientist; and to Prof. Protap Chakravarti, Former Director, Geological Survey of India, and an internationally renowned geoscientist, for contributing to his development as a scientist and researcher.

Sivakannan would like to express profound gratitude to my parents, Father Subramani K. and Mother Selvi Subramani, for their unwavering support. Special thanks to Fr. Xavier Savarimuthu, S.J., for his guidance. I appreciate Fr. Victor Lobo, S.J., Vice Chancellor, St. Joseph's University, Bangalore, and Fr. Denzil Lobo, S.J., Dean, School of Information Technology, St. Joseph's University, Bangalore, for their leadership. Thanks to my entire family for their constant encouragement. Your support has been instrumental in my academic journey.

Alex Noel Joseph Raj would like to extend heartfelt gratitude to my parents for their unwavering support. Special thanks to my wife, Valentina Jesumani, for her love and encouragement. I appreciate the academic environment at Shantou University, China, for enriching my research. Your collective support has been instrumental in the completion of this book, and I am sincerely grateful for each contribution.

Xavier Savarimuthu, S.J.
Sivakannan Subramani
Alex Noel Joseph Raj

Editors

Xavier Savarimuthu, S.J., is the Principal of St. Xavier's College Jaipur, Rajasthan, India. He has spent more than two decades in the fields of scientific research, teaching, research, and consultancy in Jesuit higher education institutions like St. Xavier's College (Vice Principal), Kolkata, and St. Joseph's University (Research Director) in Bengaluru. He has taught at Santa Clara University, California, and Saint Joseph's University, Philadelphia, where he held the endowed Donald MacLean Jesuit Chair.

Dr. Savarimuthu's academic prowess is reflected in the fact that he was a research assistant and a Fogarty trainee in the University of California, Berkeley's Arsenic Research Program for his doctoral research. He was invited to deliver lectures in Stockholm (Sweden) and Manila (the Philippines). A few more feathers in his cap were added when he delivered invited lectures at the University of Oxford and at the United Nations climate summits while attending the famous COP 21 in Paris (France) and COP 23 in Bonn (Germany). He has presented and published extensively on arsenic pollution in West Bengal, both in India and abroad. His latest article, "The Earth We Want Requires Reconciliation", was published in four languages: French, Spanish, Latin, and English. He has authored a Cambridge University Press textbook titled *Fundamentals of Environmental Studies*. His second edited volume is *Go Green for Sustainability*, published by CRC Press/Taylor & Francis Group in Boca Raton, Florida, and Oxford, UK. *Artificial Intelligence for Multimedia Information Processing: Tools and Applications* is his third book in his lifetime achievement of focusing on current pertinent issues.

Sivakannan Subramani is Assistant Professor at St. Joseph's University, Bangalore. He earned a PhD in medical image processing at the Medical Image Processing Laboratory, Department of Electronics and Instrumentation Engineering, Annamalai University, India, in 2018. He earned a master's of engineering and a bachelor's of engineering at Anna University, Chennai. His primary interests are medical image processing, deep learning, machine learning, data analytics, and ASIC/FPGA. He has more than ten years of academic and seven years of research experience, primarily working on predictive analytics and deep learning, with a main focus on solving real-life problems. He has completed public-funded research projects for the Indian Space Research Organization (ISRO), Defense Research and Development Organization (DRDO), and other client projects as project co-investigator. He has published research articles in various international journals and presented his research at several international conferences. He is a reviewer for several international scientific journals and a member of the Data Science Foundation, Data Science Association, Institute of Electrical and Electronics Engineers, and Institute of Electronics and Telecommunication Engineers, among others.

Alex Noel Joseph Raj earned a BE in electrical engineering at Madras University, India, in 2001, an ME in applied electronics at Anna University in 2005, and a PhD in engineering at the University of Warwick, Coventry, UK, in 2009. From October 2009 to September 2011, he was with Valeport Ltd Totnes, UK, as a design engineer. From March 2013 to March 2017, he was with the Department of Embedded Technology, School of Electronics Engineering, VIT University, Vellore, India, as a professor. Since March 2017, he has been with the Department of Electronic Engineering, College of Engineering, Shantou University, China. His research interests include machine learning, signal and image processing, and FPGA implementations. He specializes in image processing and has industrial and teaching experience in machine learning, deep networks signal and medical image processing, FPGA system design, MATLAB®, Simulink, machine vision systems, sonar systems, and embedded systems. He is constantly looking for dynamic interns and postdocs to join his research team at Shantou University China. His major technical skills are VHDL, MATLAB®, Simulink, C and Embedded C. He has published research articles in various international journals and has presented his research at several international conferences.

Contributors

Prasenjeet Acharjee
Western Digital
Bangalore, India

Lohith Ashwa
Department of Advanced
    Computing
St. Joseph's University
Bangalore, Karnataka, India

B. Beulah Aswini
Centre for Advanced systems
Defence Research and Development
    Organisation
Hyderabad, Telangana, India

L. R. Aravind Babu
Department of Computer and
    Information Science
Annamalai University
Tamil Nadu, India

Sridevi Balu
Department of Electronics and
    Communication Engineering
Velammal College of Engineering
    and Technology
Madurai, Tamil Nadu, India

Ram Shenoy Basti
Department of Radio-diagnosis and
    Imaging
Father Muller Medical College
    Hospital
Mangaluru, Karnataka, India

Mahua Basu
Department of Commerce
St. Xavier's (Autonomous) College
Kolkata, India

S. Beatrice
Department of Computer Engineering
Xavier Institute of Engineering
Mahim, Mumbai, India

Namratha Betala
Department of Advanced Computing
St. Joseph's University
Bangalore, India

S. Daksa
Department of Environmental Science
St. Joseph's University
Bangalore, India

Pranav Elayadam
Department of Environmental Science
St. Joseph's University
Bangalore, India

Densil Raj V. Francis
Department of Computer and
    Information Science
Annamalai University
Tamil Nadu, India

R. Geetha
Saveetha School of Engineering
Saveetha University
Chennai, Tamil Nadu, India

Fiona Ghosh
Department of Advanced Computing
St. Joseph's University
Bangalore, India

Pawni Gupta
Department of Advanced Computing
St. Joseph's University
Bangalore, India

Mohammed Jabeer
Department of Software Technology
St. Aloysius Institute of
    Management and Information
    Technology
Mangaluru, Karnataka, India

Pramod Kumar Jha
Centre for Advanced Systems
Defence Research and Development
    Organisation
Hyderabad, Telangana, India

Sivasankari Jothiraj
Department of Electronics and
    Communication Engineering
Velammal College of Engineering and
    Technology
Madurai, Tamil Nadu, India

D. Kavitha
Department of Electronics and
    Communication Engineering
CMR Institute of Technology,
    Bengaluru
Karnataka, India

Nagarajan Nagarani
Department of Electronics and
    Communication Engineering
Velammal College of Engineering
    and Technology
Madurai, Tamil Nadu, India

Niranjana Nair
Department of Environmental
    Science
St. Joseph's University
Bangalore, India

Bhawani Narayan
Department of Chemistry
St. Joseph's University
Bengaluru, Karnataka, India

Mausumi Das Nath
Department of Commerce
St. Xavier's (Autonomous) College
Kolkata, India

Rengarajan Neelaveni
Department of Electronics and
    Communication Engineering
MNM Jain Engineering College
Chennai, Tamil Nadu, India

Saizel Pathania
Department of Advanced Computing
St. Joseph's University
Bangalore, India

John Rose, S.J.
Department of Information Technology
Xavier Institute of Engineering
Mahim, Mumbai, India

S. Ruban
Department of Software Technology
St. Aloysius Institute of Management
    and Information Technology
Mangaluru, Karnataka, India

Khushi Sabarad
d'Alchemy
Bangalore, India

Xavier Savarimuthu, S.J.
St. Xavier's College
Jaipur, Rajasthan, India

Vinod Seshadri
Tiger Analytics
Austin, Texas, United States

A. Shahela
Centre for Advanced Systems
Defence Research and Development
    Organisation
Hyderabad, Telangana, India

A. Sharwin
Laminaar Aviation Infotech Pvt Ltd
Bangalore, India

Sanjanavhas Srinivasan
Department of Environmental Science
St. Joseph's University
Bangalore, India

J. Stella
Department of Information Technology
Xavier Institute of Engineering
Mahim, Mumbai, India

Sivakannan Subramani
Department of Advanced Computing
St. Joseph's University
Bangalore, India

Gandu Suchitra
Centre for Advanced Systems
Defence Research and Development
    Organisation
Hyderabad, Telangana, India

Introduction

The profound impact of artificial intelligence (AI) on multimedia information processing stands at the forefront of a pivotal juncture in human history. As our globally interconnected world grapples with the challenges posed by an ever-evolving technological era, the imperative to devise innovative solutions has never been more pressing. This comprehensive volume, titled *Artificial Intelligence for Multimedia Information Processing: Tools and Applications,* emerges as a beacon of interdisciplinary exploration, echoing the urgency reflected in contemporary discourse on environmental sustainability.

Although firmly rooted in the domains of computer science, engineering, and information technology, the complexities of multimedia information processing transcend disciplinary boundaries. Acknowledging this, our compilation intentionally incorporates perspectives from diverse fields, such as public health, environmental science, automobiles, and material science. This convergence of various disciplines with AI aims to address the multifaceted issues inherent in the realm of multimedia information processing, fostering a holistic and integrated approach.

Esteemed experts, researchers, practitioners, and intellectuals contribute to this volume, delving into the intricacies of AI applications in multimedia contexts. These applications span a wide spectrum, from the analysis of images and videos to natural language processing. What makes this collaborative effort unique is its inclusivity, encompassing contributions from both established and emerging authors. This inclusivity is vital for fostering a comprehensive dialogue and essential for ethical and enduring advancements in artificial intelligence.

The visionary insights presented in this volume aspire to catalyze a deliberate and concerted effort among researchers, practitioners, and regulatory bodies. The objective is to pave the way for a future in which artificial intelligence not only transforms the landscape of multimedia information processing but also contributes to a sustainable and harmonious coexistence with the evolving tapestry of our technological landscape. The collective dedication advocated in this volume serves as a call to action, emphasizing that it is only through collaborative efforts that we can sculpt a future where artificial intelligence becomes a tremendous positive force for change.

In the first section of the book, Chapter 1 by Sivasankari, Sridevi, Neelaveni, and Nagarani discusses challenges in the preservation of historical monuments. It emphasizes the importance of efficient methods for detecting surface damage in historical buildings. The current reliance on manual inspection has drawbacks, including time consumption and potential misjudgments. The authors propose a solution that suggests using an unmanned aerial vehicle (UAV) image acquisition system coupled with an automatic damage detection method using mask region-based convolutional neural network (Mask R-CNN) with a Dense-Net structure. Industry 4.0 technologies such as the Internet of Things (IoT), AI, and big data have aided the manufacturing industry to progress significantly. However, this progress comes with cybersecurity challenges, leading to financial losses and reputation damage to various organizations. In response, organizations are adopting AI-driven digital forensics, employing machine learning (ML) algorithms to sift through data and detect anomalies. Digital forensics proves essential in understanding attack origins, responding to incidents, ensuring compliance, and safeguarding intellectual property. Chapter 2 explores specific ML methods tailored to diverse digital forensics aspects, addressing challenges such as complexity and data volume. Authors Shahela, Suchitra, Aswini, and Jha stress the ongoing significance of prioritizing digital forensics in the ever-evolving industry landscape. AI in healthcare, particularly in processing and analyzing multimedia medical information, has a transformative potential. In Chapter 3, authors Geetha and Kavitha explore reversible data embedding techniques, focusing on high-capacity reversible data hiding (RDH) schemes while preserving image quality. They discuss the intersection of RDH with machine learning and analyze computational complexities and reversibility issues. The chapter concludes with a performance comparison of existing RDH methods, providing insights into their efficacy and limitations in secure data concealment. Chemistry's vital role in emerging sciences like materials chemistry, chemical biology, and nanotechnology demands deep expertise and intuition. However, its complexity often leads to unpredictable outcomes, requiring significant time and resources. Artificial intelligence (AI) offers a promising solution, providing predictability and reproducibility. Authored by Bhawani and Vinod, Chapter 4 aims to demystify AI for chemists, highlighting its potential to enhance research outcomes while identifying areas for improvement in the field.

The second section of the book covers the potential of AI in processing medical and public health data. In Chapter 5, Sivakannan and Sharwin explore the use of artificial intelligence in healthcare, specifically in predicting anticancer peptide (ACP) activity from genomic data. It discusses MLACP 2.0, a machine learning-based anticancer peptide predictor, which surpasses current tools. Utilizing diverse feature encodings and a convolutional neural network, MLACP 2.0 simplifies ACP identification, aiding hypothesis-driven experiments and potentially improving cancer treatment outcomes. Further, in Chapter 6, an AI-based diagnostic system designed to predict Bi-Rads scores for detecting breast cancer from mammograms is discussed by Ruban, Jabeer, and Basti. They highlight the growing impact of AI in healthcare, particularly in cancer diagnosis through medical imaging. With breast cancer rates rising, early detection is crucial, and mammography's limitations have led to false positives and negatives. The study showcases how radiologists can leverage AI, specifically deep learning algorithms such as CNN, to improve the precision of cancer detection. The AI model developed demonstrates a 90% overall precision in predicting Bi-Rad scores, providing radiologists with a valuable tool for accurate diagnoses.

Kavitha and Geetha's work (Chapter 7) delve into the prenatal diagnosis of congenital heart defects (CHDs), a complex area with eighteen distinct types, some critical. They focus on reviewing existing work in this domain, particularly in automating the diagnosis of prenatal CHD anomalies. While cardiac anomaly screening using ultrasound is challenging, the approach involves a transversal application of image processing methods and AI algorithms. Technological advancements enable the use of computational intelligence techniques, surpassing traditional approaches. The survey work explores optimization methods, including pre-processing, segmentation, and edge detection algorithms. It also examines machine learning and deep learning algorithms actively contributing to prenatal CHD diagnosis. The study evaluates current approaches, highlighting their benefits and drawbacks, while also envisioning potential future applications in this critical field. Aortic stenosis, prevalent in aging populations, demands early detection to prevent severe complications like heart failure. While aortic valve replacement is standard, recent medical advances offer alternatives for high-risk patients. However, accurate diagnosis remains challenging, with traditional methods limited. Chapter 8 by Francis and Babu aims to utilize European Society of Cardiology (ESC)/European Association for Cardio-Thoracic Surgery (EACTS)/American College of Cardiology (ACC) guidelines and artificial intelligence (AI), machine learning (ML), and deep learning (DL) approaches for early identification of at-risk patients, addressing critical diagnostic gaps. Chapter 9 examines the intersection of environmental protection and health in the context of sustainable development. Authors Mahua and Mausumi explore how AI and ML techniques can address challenges such as waste management and marine research, contributing to the achievement of the Sustainable Development Goals (SDGs), a set of 17 global objectives aimed at promoting prosperity while protecting the planet by 2030. Cerebral palsy is a debilitating condition affecting children's movement, where early diagnosis and intervention are crucial for effective therapy. However, accurately assessing physical activity levels, especially in children with mobility challenges, remains challenging. Motion sensors with accelerometers are now standard for precise measurement. Chapter 10 by authors Beatrice, Stella, and John employs classical machine learning algorithms to predict cerebral palsy from visual impairment, offering a comprehensive evaluation to enhance diagnostic accuracy and therapeutic interventions.

AI can also play as crucial a role as reducing mortality due to sudden cardiac death (SCD). SCD is the unexpected loss of heart function and it is one of the fundamental drivers of death around the world, accounting for approximately 4.5 million deaths per year. Since it occurs relatively often in younger people, its socioeconomic impact is much higher than the impact of other major health issues such as cerebrovascular disease. So, it is important to monitor and predict SCD. In Chapter 11, Lohith, Sivakannan, and Savarimuthu discuss a system to predict SCD from electrocardiograph signals using ML approaches.

As described by Acharjee in Chapter 12, AI plays a pivotal role in environmental science by leveraging advanced algorithms and data analysis to model complex ecosystems, predict environmental changes, and optimize resource management for sustainable and efficient solutions. In Chapter 13, Saizel, Daksa, Sanjanavhas, and Savarimuthu address the diminishing availability of water, a crucial resource for all life forms, and the role of wastewater treatment as a solution. It explores the integration of AI and ML applications in wastewater treatment, utilizing algorithms to analyze various parameters. The chapter discusses the use of technologies such as IoT, sensors, and systems, with a focus on AI models such as artificial neural networks (ANNs), random forests (RFs), radial basis function (RBF) kernel, and adaptive neuro-fuzzy inference systems (ANFIS). These models aid in data analysis and predicting the optimal adsorbent for wastewater treatment. Notably, ANN and ANFIS demonstrate high accuracy with R^2 values of 0.9991 and 0.9997, respectively. The chapter also emphasizes the adsorption process for removing pollutants and critiques traditional wastewater treatment methods. It highlights limitations, including data availability and certain methodological challenges in the adsorption process analysis, while discussing avenues for improvement. In Chapter 14, authors Fiona, Niranjana, Pranav, and Savarimuthu merge bioacoustics and machine learning to classify animal species based on vocalizations. Various algorithms, including convolutional neural networks, linear discriminant analysis (LDA), decision trees, and support vector machines (SVMs), are compared, with SVM and LDA showing the highest accuracy rates. This integration promises to expand biodiversity data coverage, enhancing conservation efforts significantly.

Prasenjeet, in Chapter 15, presents the potential benefits of AI in driverless cars, such as enhanced road safety and reduced traffic congestion. The chapter covers the application of AI technologies such as computer vision, machine learning, and natural language processing, enabling these vehicles to perceive their surroundings, make decisions, and interact with passengers effectively. Despite the promising advancements, challenges such as regulatory hurdles and public acceptance are discussed, emphasizing the ongoing need for research and development in this rapidly evolving field.

Chapter 16 examines feature-based image stitching methods in computer vision, focusing on algorithms like scale-invariant feature transform (SIFT), speeded up robust features (SURF), and features from accelerated segment test (FAST). It offers a comparative analysis, guiding readers in selecting suitable approaches for their applications. Recent advancements, including convolutional neural networks, are discussed for enhanced performance. Authored by Pawni, Namratha, and Sivakannan, it's a valuable resource for researchers and practitioners in computer vision.

AI in material science plays a crucial role by leveraging advanced predictive models to forecast various material parameters, accelerating the discovery and optimization of materials with specific properties, ultimately enhancing efficiency and innovation in material design and development. The increasing global demand for concrete has driven the need for innovative variants. Researchers have focused on predicting concrete strength using robust models, aiming for advancements in preset standards. While traditional models in statistical and empirical studies prove time-inefficient and inconsistent in complex scenarios, ML models such as decision trees, support vector machines, and ANNs are explored as alternatives. In Chapter 17, Khushi evaluates the applications of each model, analyzing their performance to identify knowledge gaps, provide practical recommendations, and propose avenues for future research.

In the forthcoming pages, we shall embark upon a comprehensive exploration of the intricate intersection between AI and multimedia information processing. Navigating through the tools and applications that define this evolving discipline, we extend an invitation for you to partake in unravelling the myriad possibilities that AI presents across domains. Together, let us traverse the realms of innovation and discovery, bearing witness to the profound influence of AI on the continually expanding tapestry of multimedia information processing.

Section I

AI-Multimedia Information Processing

1 Monitoring Ancient Buildings Using UAV and Instant Image Segmentation Using Masked R-CNN

Sivasankari Jothiraj, Sridevi Balu,
Neelaveni Rengarajan, and Nagarani Nagarajan

1.1 INTRODUCTION

The history of any country speaks of its importance via its archaeological sources such as monuments, ancient structures, and other artefacts. One way of recognizing a country's history can be by protecting noteworthy buildings and structures. They may represent a particular design trend, or they may reflect an important period in the city's history. Ancient structures serve as witnesses to a city's cultural and social history, giving people a sense of place and connection to the past (Bacco et al. 2020). Famous structures typically express something important to the people who reside in or go to a city. While some ancient structures were built to last, more recent ones typically have a life expectancy of 30 to 40 years. Holding significant buildings and making improvements to bring them up to date with requirements and codes can make economic sense.

In addition to adding character to the area, restoring historic structures to their original appearance may also help attract investors and tourists, if the buildings are of sufficient importance. A remarkable but abandoned mechanical building, for example, could be converted into a small shop or a mixed-use development, giving the structure and the entire neighbourhood a modern makeover (Akbar et al. 2019). Significant ancient buildings are shown in Figure 1.1.

1.2 AESTHETIC SIGNIFICANCE OF HISTORIC BUILDINGS

Most of the time, unique and profitable materials such as heart pine, marble, or old brick were used to build older structures. They might have odd glasswork, copper lining, ornamental veneers, or details and accents that are no longer there. Because of their unique character and personality, many people believe that older buildings are

DOI: 10.1201/9781003405436-2

"

FIGURE 1.1   Some ancient architectural buildings in India.

more interesting than modern ones. Preservation of old buildings also helps preserve traditional methods of craftsmanship.

One of the primary structural elements in both historic and contemporary buildings is brick masonry. There are still many historic stone buildings, which is evidence that their lifespan can be greatly increased with proper preservation. Historical masonry constructions are frequently susceptible to seismic excitation, necessitating extensive damage assessments to provide appropriate restoration plans when necessary. Beams, frequent fatigue stress and cyclic loading on concrete surfaces results in cracks or other damage on the surface, usually at the microscopic level. The structure's fissures cause localised reductions in stiffness and material discontinuities (Shim et al. 2023). Noteworthy designs should be completely evaluated to identify any harm at a beginning phase or after a significant catastrophe to safeguard them. Nonetheless, this technique is staggeringly tedious, slow, and costly when you figure the quantity of worker hours expected to deal with the information gathered, both in the field and in the workplace. Additionally, as the quality of the inspection process depends significantly on the abilities and physical health of the inspector, as well as the fact that fatigue or inexperience might easily result in underreported damage, it is subject to interpretation (Li et al. 2022). Manual inspection might raise safety issues because some areas of the structures are difficult to access and have restricted access. As an option to manual examination, assessment using vision-based analysis and observation of common foundations is becoming more well known. For some time, scientists have been inspired by computer vision for identification of breaks. Vision-based break identification is an example of a safer evaluation strategy (Kim et al. 2017). It very well may be particularly helpful for antiquated structures where severe restrictions are in place and, surprisingly, basic interventions, such as the arrangement of break rulers, are not allowed by the preservation specialists (Kim and Cho 2020).

FIGURE 1.2   Unmanned aerial vehicle.

1.3   SIGNIFICANCE OF UNMANNED AERIAL VEHICLES

One of the greatest issues in development administration is the ability to access rooftops, tall dividers and other hard-to-reach open air spaces. Accessible strategies such as using a rope and platform are hazardous, costly and time consuming. This may restrain the capacity to see and assemble point by point data around certain parts of the building. Conventional ways of getting to certain parts of a building are costly and frequently hazardous (Shim et al. 2023). In any case, modern unmanned airborne vehicles are changing everything. Rambles of the type shown in Figure 1.2 are presently utilized for an assortment of individual and commercial applications and have become a secure and cost-effective choice for building observation (Sankarasrinivasan et al. 2015). Ramble-based reviews dispense with the dangers and high costs associated with manual or conventional building reviews. Rambles can effectively get to nearly any surface and take point by point photographs, recordings and warm pictures. In addition, the drone takes a fraction of the time and there is no need to buy and introduce costly devices to access the buildings for inspection.

Unmanned aerial vehicles (UAVs) or drones permit reviewers to perform investigations securely without having to send somebody up a step or pay for devices like a platform. This requires an administrator on the ground and a ramble modified to discover the specified design of the ceiling. This decreases security dangers for specialists and individuals around or inside the building. By taking hundreds of pictures of places inconceivable for humans to reach, UAVs can identify issues that a professional would have missed by conventional tools or the naked eye. Hence, the most complex rooftops and structures can be reviewed (Wang and Shu 2022).

Existing techniques for detecting cracks and other damage are either destructive or nondestructive. Deficiencies in the surface condition of a structure are evaluated utilising instruments for visual examination and surveying. The kind, quantity, width, and length of cracks on a structure's surface indicate the amount of initial degradation. In place of human inspection procedures, automatic crack detection has been developed for quick and precise surface defect analysis. The habitual fracture identification feature of nondestructive testing is highly effective. Degradation can't be accurately assessed through manual inspection.

In nondestructive testing, image-based crack detection is becoming more and more common. The resulting images present additional difficulties for its adaptive nature due to the change in shape and size of the cracks, as well as other noise, such as shadows, imperfections, and concrete cracks. There are four groups of these technologies: the morphological approach, the integrated algorithm, the percolation-based method, and the practical technique are all examples.

1.4 RELATED WORK

With Yolov3, (Villanueva et al. 2022) wanted to create a deep learning model that could use an Android app to identify moderate, severe, or very severe cracks in reinforced concrete structures. The Android application's overall accuracy is 93.33%, and its kappa value is 97. The Android application and the deep learning model were able to accurately calculate this, allowing them to identify the crack and classify it as such.

For the purpose of segmenting common roads and pavement cracks, Zawad et al. (2021) proposed a brand-new unsupervised image processing technique. The technique initial cuts the pictures into $M \times N$ sub-pictures and afterward eliminates the sub-pictures without breaks in view of a limit factor. The investigation is then performed on just a few halfway sub-pictures, which fundamentally lessens the estimation time.

Yu et al. (2020) proposed a pyramid organization with an enhanced dissect setting for recognizing breaks from asphalt pictures, planning and analyzing another setting to increase case pyramid organization. CCapFPN uses vector pictures to portray the undeniable level, inward, and significant highlights of breaks. By fostering an element pyramid design, CCapFPN can consolidate various levels and sizes of case capabilities to give a semantically vigorous component portrayal and high goal, for exact break discovery.

DeepCrack, an end-to-end trained deep convolutional neural network for automatic crack detection that learns advanced features to represent cracks, was proposed by Zou et al. (2019). To capture line structures, this method combines previously known multilevel deep convolution features in hierarchical convolution steps.

Maningo et al. (2020) developed a crack detection system that can analyse the physical properties of damages and required to map the wall surface. The model is used to classify and define cracks using a Faster region-based convolutional neural network (R-CNN) machine learning architecture.

A semi-supervised method for detecting pavement cracks was proposed by Li et al. (2020). To begin, the training algorithm can be applied to images of unlabelled road surfaces. To compensate for the absence of annotation in the image, our model is able to generate a tracking signal for unlabelled pavement images.

Using an advanced deep learning model known as You Look Only Once-version3 (YOLO-v3), Kumar et al. (2021) propose a real-time multi-drone damage detection system for tall civil structures. An open-source hex copter drone based on Pixhawk hardware standards is used to implement the proposed method, which makes use of Jetson-TX2 as a hardware platform to run YOLO-v3.

A deep learning model and image processing were used by Ali et al. (2020) to create an automatic inspection system for crack detection. With the aid of surface images, CNN transfer learning models are used to learn the internal characteristics of cracks, enabling the automatic classification of objects into cracked and non-cracked classes.

Image segmentation–based crack detection was also proposed by Ogawa et al. (2019), in a method that combines filtering and the Gaussian mixture model. Exploratory outcomes show that their proposed strategy beats break identification techniques with regard to both handling time and precision.

Altabey et al. (2022) proposed an image processing in crack detection against complicated backgrounds and developed image segmentation model related on semantic segmentation.

A road crack detection algorithm that combines a light-homogenization algorithm and a deep CNN method was proposed by Xu and Xu (2019). The Dice comparability coefficient for crack distinguishing proof expanded to 0.7937 because of examinations exhibiting the adequacy of their proposed strategy in working on the precision of break acknowledgment in genuine street pictures.

Girshick (2015) proposed R-CNN to handle the complication of object detection. An exhaustive search method was used with sliding window technique in R-CNN. The problem with R-CNN is the memory storage requirement and the time needed to train the model with a minimum of 2,000 samples. To handle the above problem Fast R-CNN was proposed, which combines the classification layer, regions of interest (RoI) pooling, and ConvNet. The main advantage is that it uses a SoftMax layer to replace support vector machines (SVMs) saving on memory storage. The main disadvantage is that Fast R-CNN omits the optional search region during detection.

1.5 PROPOSED METHODOLOGY

Infrastructure management needs to be safeguarded, which is a costly task. Cracks are a key element in managing infrastructure. More cracks in infrastructure cause severe damage and shorten the life of historic buildings. Several computer-vision applications are of interest in detecting cracks, which are common line structures. He et al. (2017) proposed Mask R-CNN to do object detection and semantic segmentation (deals with identifying object as a single class) instantly. It consists of two parts, namely backbone and head architectures.

Spalling and efflorescence are two types of damage that can be detected in ancient buildings using the UAV) associated with the Mask R-CNN model via DenseNet architecture as a backbone, as proposed in this work. So "L" slices of layers, one between each layer and the next layer can be utilized. More number of layers increases the complexity as they become deeper and hence the loss of information may occur due to the distance between input and output layers. Mask R-CNN is an extended version of conventional CNN. Figure 1.3 depicts the general architecture of rapid image analysis-based defect detection. The damage detection method begins by taking an image of the structure using a drone-controlled aerial vehicle. To enhance the image processing, the collected images are processed using a variety of methods, such as segmentation. Crack feature extraction is separated according to their propagation direction, depth, and breadth.

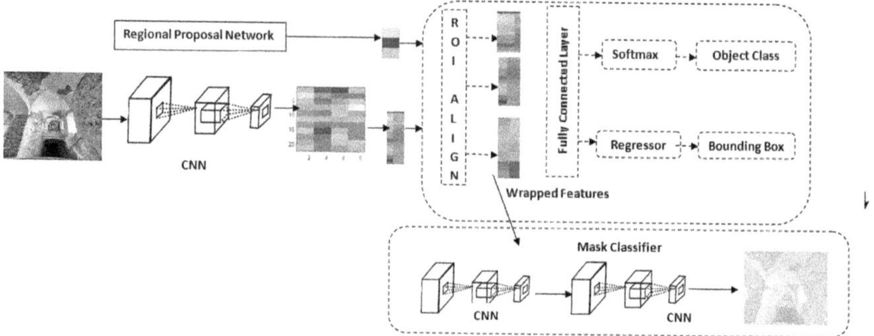

FIGURE 1.3 Mask R-CNN architecture.

In DensNet, each layer is given input from each and every layers. In each layer, the element guides of the multitude of previous layers are not added yet connected and utilized as data sources. As a result, DenseNets don't need as many parameters as a standard CNN, which makes it possible to reuse features because redundant feature maps are thrown out. Thus, the element guides of every previous layer, x0, . . . , xl-1, are taken care of into the lth layer as info:

$$X_l = H_l\left(\left[x_0, x_1, \ldots\ldots x_{l\text{-}1}\right]\right), \text{ exhibits the feature maps from all layers.} \tag{1.1}$$

1.5.1 DENSEBLOCKS

Mapping of feature maps plays an important role in denseNet architecture, string function cannot be used. However, the central part of CNN is layer down sampling, which minimizes feature map's dimension to achieve superior computational speed. To make this conceivable, thick organizations are separated into thick blocks, where the size of element maps stays consistent inside a block, but the quantity of channels between them is variable. Interblock layers called transition layers reduce the number of channels by half in comparison to the number of channels that are currently being used. For each layer, Hl in the above condition is characterized as a mix capability that utilizes three successive tasks: bunch standardization (BN), redressed straight unit (ReLU), and convolution (Conv). DenseNet architecture is shown in Figure 1.4.

1.5.2 CONNECTIVITY

Mask prediction is done using a fully connected network in Mask R-CNN. This ConvNet provides an m*m mask of the input RoI,1*1 convolutions are used to compress the channels to 256 and scaling of mask to infer the input image. For this fully connected network to anticipate mask, the input must be generated. The RoI Align function is to turn the region proposal network's various-sized feature maps into fixed-size feature maps.

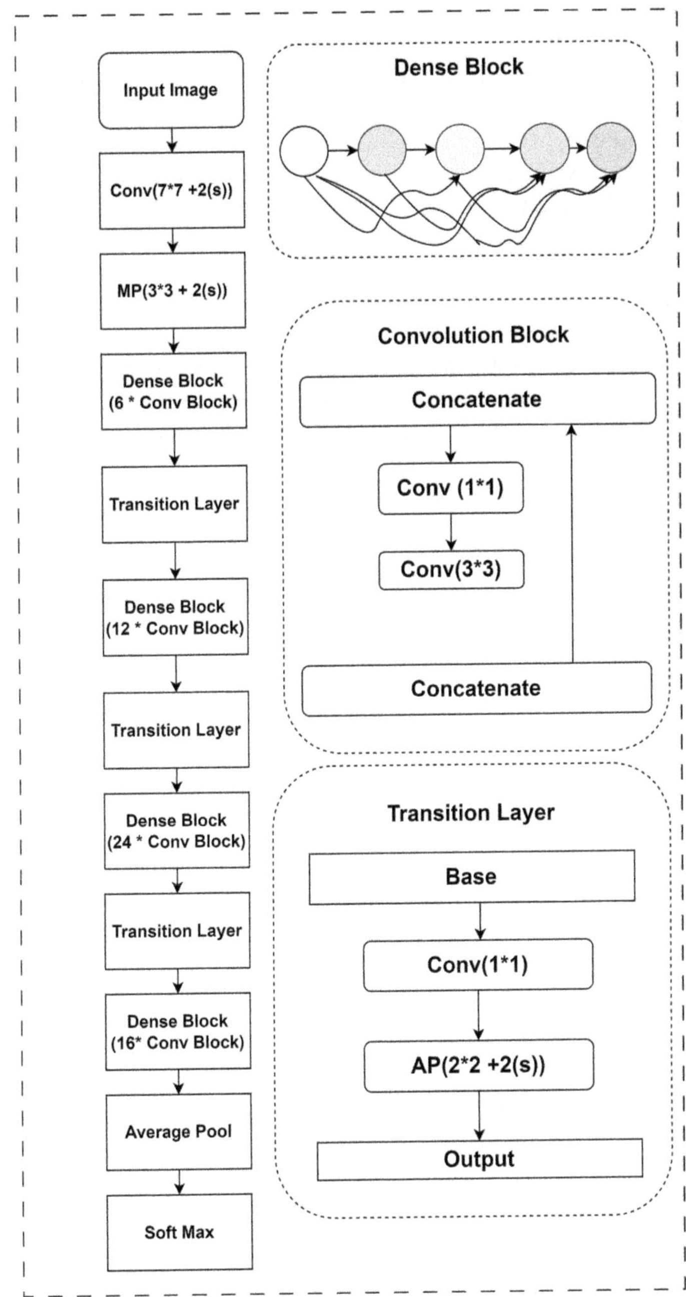

FIGURE 1.4 DenseNet architecture and components.

1.5.3 GROWTH RATE

The features can be regarded as the network's global state. After passing through every dense layer, the feature map increases in size as each layer adds "K" objects to the global space. The growth rate of the network, indicated by the "K" parameter, controls how much information is concatenated to every layer of the entire network. 'K' feature maps are generated from L layers using H1 function, number of input layer channels, k0. DenseNets can have very narrow layers, which is different from current network architectures. Its corresponding architecture specifications are shown in Table 1.1.

$$K_l = k_0 + k *_{(l-1)} \tag{1.2}$$

Although each layer only generates k output feature-maps, there might be a large number of inputs, especially for higher layers. To maximize computing speed and efficiency, a bottleneck layer with a single-dimension convolutional layer has been added as before each 3×3 convolution.

1.6 MASK REGION-BASED CONVOLUTIONAL NEURAL NETWORK

Mask R-CNN (He et al. 2017) is a high-tech deep learning model that is capable of object detection and instance segmentation. Developed by a group of researchers at Facebook AI Research (FAIR), Mask R-CNN has become one of the most popular and widely used models for computer vision tasks, particularly in the field of image understanding and scene analysis. In this chapter, we will delve into the details of Mask R-CNN.

Object detection involves locating and classifying objects within an image or a video. It has numerous applications, including autonomous vehicles, video surveillance, robotics, medical imaging, and augmented reality, among others. Traditional object detection approaches used hand-crafted features and machine learning algorithms, but these methods often had limitations of accuracy and scalability. The advent of deep learning revolutionized object detection, and Mask R-CNN has emerged as a leading model in this area. Mask R-CNN extends Faster R-CNN by adding a third branch for instance segmentation, which produces pixel-level object masks in addition to object bounding boxes and class labels.

1.7 COMBINING MASK R-CNN WITH DENSENET AS A BACKBONE

The drone-captured image is given to preprocessing; this is an important step in preparing the input data for Mask R-CNN to ensure that the model receives clean and properly formatted data for effective training and inference. The typical preprocessing steps before feeding the data into Mask R-CNN include image resizing and data augmentation.

Data augmentation techniques are nothing but rotation, scaling, flipping, and changing brightness/contrast to augment the training dataset and increase its

TABLE 1.1

DenseNet-121 Architecture

Layers	Output Size	DenseNet-169	DenseNet-121	DenseNet-201	DenseNet-264
Convolution	112×112		7×7 Convolution, Stride 2		
Pooling	56×56		3×3 max pool, Stride 2		
Dense Block (1)	56×56	$(1 \quad x13 \quad x3)x6$	$(1 \quad x13 \quad x3)x6$	$(1 \quad x13 \quad x3)x6$	$(1 \quad x13 \quad x3)x6$
Transition Layer (1)	56×56		1×1 Convolution		
	28×28		2×2 average pool, Stride 2		
Dense Block (2)	28×28	$(1 \quad x13 \quad x3)x12$	$(1 \quad x13 \quad x3)x12$	$(1 \quad x13 \quad x3)x12$	$(1 \quad x13 \quad x3)x12$
Transition Layer (2)	28×28		1×1 Convolution		
	14×14		2×2 average pool, Stride 2		
Dense Block (3)	14×14	$(1 \quad x13 \quad x3)x24$	$(1 \quad x13 \quad x3)x32$	$(1 \quad x13 \quad x3)x48$	$(1 \quad x13 \quad x3)x64$
Transition Layer (3)	14×14		7×7 Convolution, Stride 2		
	7×7		3×3 max pool, Stride 2		
Dense Block (4)	7×7	$(1 \quad x13 \quad x3)x16$	$(1 \quad x13 \quad x3)x32$	$(1 \quad x13 \quad x3)x32$	$(1 \quad x13 \quad x3)x48$
Classification	1×1		7×7 global average pool		
Layer			1000D fully connected, soft max		

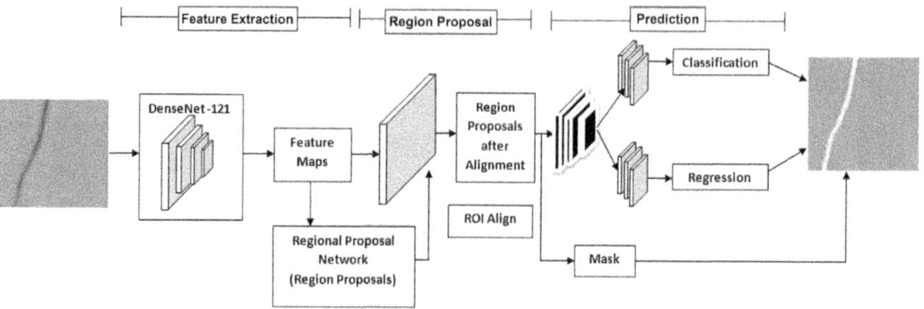

FIGURE 1.5 Overall architecture.

diversity. Data augmentation helps in improving the model's ability to generalize to different variations of the input images and enhances the model's robustness.

The Mask R-CNN includes the backbone network, region proposal network (RPN), mask representation and RoIAlign. As said earlier the preprocessed image is given to the DenseNet backbone where it produces feature maps. Each value in the feature map represents a learned feature at a specific spatial location in the input image. The values in the feature map are learned during the training process and represent abstract representations of the input image, capturing local and global patterns, textures, edges and other visual cues relevant for the task at hand, such as object detection or instance segmentation.

The feature map output from the DenseNet backbone is then used as input to other components in Mask R-CNN, such as the RPN and RoIAlign. The RPN generates region proposals, which are candidate bounding box proposals for potential object instances in the image. The RPN operates on the feature maps obtained from the backbone network and predicts objectless scores and bounding box regression offsets for candidate RoIs. The output of RPN and DenseNet is given to input of RoIAlign. The overall architecture of the proposed methodology employing Mask R-CNN is depicted in Figure 1.5.

1.7.1 RoIALIGN

RoIAlign is a key component in Mask R-CNN that allows for accurate pixel-level alignment of RoIs with the corresponding feature maps. It ensures that the masks and bounding box regression are accurately applied to the RoIs at the pixel level, which is crucial for precise instance segmentation. RoIAlign replaces the RoI Pooling layer used in Faster R-CNN and avoids quantization errors.

1.7.2 OBJECT CLASSIFICATION BRANCH

This branch is responsible for predicting the class labels of the objects within the RoIs. It takes the RoIs obtained from the RPN or RoIAlign as input and predicts the probability of each RoI containing an object and the corresponding object class using fully connected layers and softmax activation.

1.7.3 BOUNDING BOX REGRESSION BRANCH

This branch is responsible for predicting the refined bounding box coordinates for each RoI. It takes the RoIs obtained from the RPN or RoIAlign as input and predicts the offset values for adjusting the coordinates of the RoIs to tightly fit the objects using fully connected layers.

1.7.4 MASK PREDICTION BRANCH

This is the additional branch in Mask R-CNN that predicts the binary masks for each RoI, which represent the pixel-level segmentation of the objects. It takes the RoIs obtained from the RPN or RoIAlign as input and predicts a binary mask for each RoI using convolutional layers with sigmoid activation.

1.7.5 LOSS FUNCTIONS

Mask R-CNN uses various loss functions during training to optimize the model. These include the objective classification loss, the bounding box regression loss, and the mask prediction loss, which are used to train the RPN, object classification, and mask prediction branches, respectively. The losses are computed based on the predicted outputs and ground truth annotations.

$$L = Lcls + Lbox + Lmask \tag{1.3}$$

Overall performance of the proposed algorithm is calculated using the following formula.

Accuracy: The proportion of correctly classified instances to the total number of instances in the dataset.

$$Accuracy = \frac{TP}{TP + TN + FP + FN} \tag{1.4}$$

where TP is the number of true positives (correctly predicted positive instances), TN is the number of true negatives (correctly predicted negative instances), FP is the number of false positives (incorrectly predicted positive instances), and FN is the number of false negatives (incorrectly predicted negative instances).

Precision: The proportion of correctly predicted positive instances to the total number of instances predicted as positive. $\quad Precision = \dfrac{TP}{TP + FP} \quad \tag{1.5}$

Recall: The proportion of correctly predicted positive instances to the total number of positive instances in the dataset. $\quad Recall = \dfrac{TP}{TP + FN} \quad \tag{1.6}$

TABLE 1.2
Dataset Information

Attribute	Value
Total Images	3,886
Defected Images	757
Nondefected Images	3,139
Resolution	5,184 × 3,456
Ground Truths	Crack and noncrack

F1 score: The harmonic mean of precision and recall.

$$\text{F1 score} = 2 * \frac{\text{Precision} * \text{Recall}}{\text{Precision} + \text{Recall}} \tag{1.7}$$

Specificity: It measures the proportion of true negative instances correctly identified by the model.

$$\text{Specificity} = \frac{\text{TN}}{\text{TN} + \text{FP}} \tag{1.8}$$

Sensitivity: It measures the proportion of true positive instances correctly identified by the model.

$$\text{Sensitivity} = \frac{\text{TP}}{\text{TP} + \text{FN}} \tag{1.9}$$

The Mask R-CNN allows for both object detection and instance segmentation, while the DenseNet framework is a dense CNN architecture that allows for efficient use of parameters and better feature reuse. The proposed model achieved the highest accuracy of 91.4% and also had the highest sensitivity, specificity, and F1 score among all the compared models.

First, the "Historical _ **Building _ Crack _ 2019**" dataset was taken from the IEEE Xplore® paper "Using Hybrid Filter-Wrapper Feature Selection with Multi-Objective Improved-Salp Optimization for Crack Severity Recognition" (Elhariri et al. 2020). More information is given in Table 1.2. Using that set of images, we pretrained the DenseNet backbone; details have been predicted in Table 1.2.

More existing techniques via deep learning networks such as CNN, R-CNN were trained with different architectures like AlexNet, ResNet, GoogleNet, and the proposed Mask R-CNN model enhanced with DenseNet, and were compared in terms of their performance in crack detection. AlexNet is a popular deep learning model that was introduced in 2012. It has a shallow neural network design with three fully connected layers, a softmax layer, and five convolutional layers. In our study, AlexNet achieved an accuracy of 93.2%, which is lower compared to the other three models.

TABLE 1.3
Performance Metrics

	Accuracy			Sensitivity			Specificity			F1 Score		
Method	Mask R-CNN	R-CNN	CNN	Mask R-CNN	R-CNN	CNN	Mask R-CNN	R-CNN	CNN	Mask R-CNN	R-CNN	CNN
AlexNet	93.2	92.68	91.86	94.3	93.97	93.21	93.2	92.68	91.95	94.4	93.85	93.05
ResNet	94.6	93.24	92.14	95.1	94.65	94.02	95.89	95.01	94.28	94.9	94.07	93.86
Google-Net	96.7	95.35	93.98	95.8	94.95	94.02	95.97	95.23	94.93	96.43	95.98	95.27
Proposed	91.4	97.25	95.86	97.42	95.68	94.89	97.88	96.69	96.06	97.48	96.96	95.69

ResNet, in contrast, is a more complex neural network architecture that was introduced in 2015. It consists of residual blocks that allow for deeper network architectures without the problem of vanishing gradients. In this study, ResNet achieved an accuracy of 94.6%, which is higher compared to AlexNet but lower than the proposed model.

GoogleNet, also known as Inception-v1, is another deep learning model that was introduced in 2014 which allows multiple parallel convolutional layers with different kernel sizes, for better representation of the input data. In this study, GoogleNet achieved an accuracy of 96.7%, which is higher compared to both AlexNet and ResNet, but still lower compared to the proposed model.

The proposed model in this study uses the Mask R-CNN with DenseNet framework, which is a deep learning architecture for instant image segmentation. All the above architectures are well known and their performance evaluated for parameters such as accuracy, sensitivity, specificity and F1 score. The outcomes demonstrated the suggested model performed better than the other three models, with accuracy of about 91.4%, sensitivity of 97.42%, specificity of 97.88%, and an F1 score of 97.48%, as shown in Table 1.3.

The Mask R-CNN can do instance segmentation, while the DenseNet framework is a dense CNN architecture that allows for efficient use of parameters and better feature reuse. The proposed model reached high accuracy of about 91.4% and also had the highest sensitivity, specificity, and F1 score among all the compared models. The image analysis with masked R-CNN has been tested for certain sample images, shown in Figure 1.6

Overall, the results show that our methodology is highly efficient for detecting cracks in historic buildings using the DenseNet deep-learning backbone. The significant improvement in overall performance is that this methodology is a valuable tool for crack detection in historic buildings, which can help in their preservation and maintenance.

1.8 SUMMARY

This study presents a novel approach for crack/damage identification in ancient buildings using the deep learning architecture of Mask-RCNN and DenseNet framework. The proposed approach showed high accuracy, sensitivity, specificity, and F1 score

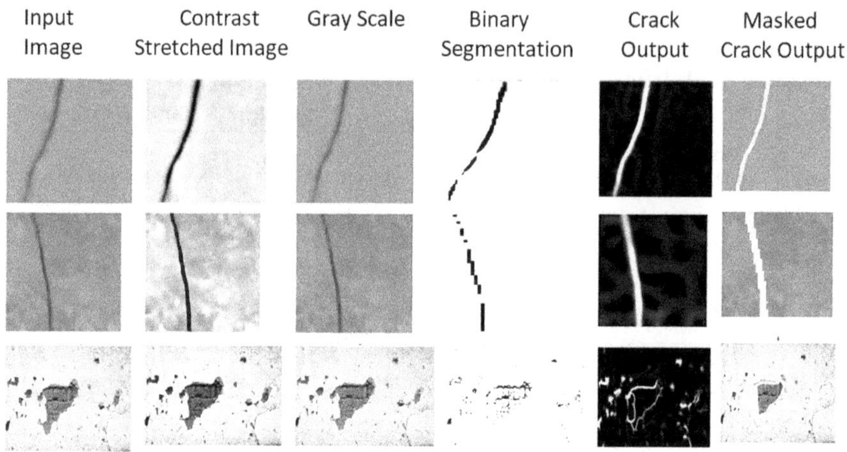

FIGURE 1.6 Evaluation results.

with the achieved evaluation measures of 91.4%, 97.42%, 97.88%, and 97.48%, respectively, which exceeded the performance of existing crack detection methods.

The output of this study shows that the proposed approach can effectively detect cracks of different sizes and shapes in historic buildings, which is critical for the preservation and maintenance of cultural heritage. In addition, the proposed approach is flexible and can be extended to other applications related to building and infrastructure inspection. The high accuracy and efficiency of the proposed approach can potentially diminish the time and cost of manual inspection of historic buildings, making it a valuable tool for building and infrastructure management. The proposed approach allows for the timely maintenance and repair of historic buildings, ensuring their longevity and sustainability.

Although the proposed approach has shown impressive performance, future research can focus on further optimization of the approach to more accurately and efficiently detect various structural faults in buildings and infrastructures, such as deformation, rust, and erosion. The proposed approach has the potential to change the way structural defects in historic buildings are identified and addressed, contributing to the conservation and sustainability of cultural heritage sites.

REFERENCES

Akbar, M.A., Qidwai, U. and Jahanshahi, M.R. 2019. An evaluation of image-based structural health monitoring using integrated unmanned aerial vehicle platform, Structural Control Health Monitoring, 26, E2276.

Ali, S.B., Wate, R., Kujur, S., Singh, A. and Kumar, S. 2020. Wall crack detection using transfer learning-based CNN models. In IEEE 17th India Council International Conference (INDICON), New Delhi, India.

Altabey, W.A., Kouritem, S.A., Abouheaf, M.I. and Nahas, N. 2022. Research in image processing for pipeline crack detection applications. In International Conference on Electrical, Computer, Communications and Mechatronics Engineering (ICECCME), Maldives.

Bacco, M., Barsocchi, P., Cassará, P., Germanese, D., Gotta, A. and Leone, G.R. 2020. Monitoring ancient buildings: Real deployment of an IoT system enhanced by UAVs and virtual reality. IEEE Access, 8, 50131–50148. https://doi.org/10.1109/ACCESS.2020. 2980359.

Elhariri, E., El-Bendary, N. and Taie, S.A. 2020. Using hybrid filter-wrapper feature selection with multi-objective improved-salp optimization for crack severity recognition. IEEE Access, 8. https://doi.org/10.1109/ACCESS.2020.2991968.

Girshick, R. 2015. Fast R-CNN. In 2015 IEEE International Conference on Computer Vision (ICCV). Santiago, Chile (pp. 1440–1448). https://doi.org/10.1109/ICCV.2015.169.

He, K., Gkioxari, G., Dollár, P. and Girshick, R. 2017. Mask R-CNN. In 2017 IEEE International Conference on Computer Vision (ICCV). Venice, Italy (pp. 2980–2988). https:// doi.org/10.1109/ICCV.2017.322.

Kim, B. and Cho, S. 2020. Automated multiple concrete damage detection using instance segmentation deep learning model. Applied Sciences, 10(22), 8001. https://doi.org/10.3390/ app10228001.

Kim, H., Lee, J., Ahn, E., Cho, S., Shin, M. and Sim, S.-H. 2017. Concrete crack identification using a UAV incorporating hybrid image processing. Sensors, 17, 2052.

Kumar, P., Batchu, S., Swamy, N.S. and Kota, S.R. 2021. Real-time concrete damage detection using deep learning for high rise structures. IEEE Access, 9.

Li, G., Wan, J., He, S., Liu, Q. and Ma, B. 2020. Semi-supervised semantic segmentation using adversarial learning for pavement crack detection. IEEE Access, 8.

Li, L., Chen, J., Su, X. and Nawaz, A. 2022. Advanced-technological UAVs-based enhanced reconstruction of edges for building models. Buildings, 12(8), 1241. https://doi. org/10.3390/buildings12081248

Maningo, J.M.Z., Bandala, A.A., Bedruz, R., Dadios, E.P., Lacuna, R.J.N., Manalo, A.B.O., Perez, P.L.E. and Sia, N.P.C. 2020. Crack detection with 2D wall mapping for building safety inspection. In IEEE Region 10 Conference (TENCON), Osaka, Japan (pp. 702–707). https://doi.org/10.1109/TENCON50793.2020.9293727.

Ogawa, S., Matsushima, K. and Takahashi, O. 2019. Crack detection based on gaussian mixture model using image filtering. In 2019 International Symposium on Electrical and Electronics Engineering (ISEE), Ho Chi Minh City, Vietnam.

Sankarasrinivasan, S., Balasubramanian, E., Karthik, K., Chandrasekar, U. and Gupta, R. 2015. Health monitoring of civil structures with integrated UAV and image processing system. Procedia Computer Science, 54, 508–515.

Shim, S., Lee, S.W., Cho, G.C., Kim, J. and Kang, S.M. 2023. Remote robotic system for 3D measurement of concrete damage in tunnel with ground vehicle and manipulator. Computer-Aided Civil and Infrastructure Engineering, 38(15), 2180–2201. https:// doi.org/10.1111/mice.12982.

Villanueva, A., Balba, J.K.B., Beceril, C.D., Belza, J.L.G., Tagle, R.I.P., Venal, M.C.A. and Rosales, M.M. 2022. Crack detection and classification for reinforced concrete structures using deep learning. In 2nd International Conference on Intelligent Technologies (CONIT), Hubli, India (pp. 1–6). https://doi.org/10.1109/CONIT55038.2022.9848129.

Wang, D. and Shu, H. 2022. Accuracy analysis of three-dimensional modeling of a multi-level UAV without control points. Buildings, 12(5), 592. https://doi.org/10.3390/ buildings12050592

Xu, D. and Xu, G. 2019. Crack identification algorithm based on MASK Dodging principle and deep learning. In Chinese Automation Congress (CAC), Hangzhou, China.

Yu, Y., Guan, H., Li, D., Zhang, Y., Jin, S. and Yu, C. 2020. CCapFPN: A context-augmented capsule feature pyramid network for pavement crack detection. IEEE Transactions on Intelligent Transportation Systems, 23(4), 3324–3335.

Zawad, M., Shahriar, R., Zawad, M., Shahriar, F., Rahman, M., and Priyom, S.N. 2021. A comparative review of image processing based crack detection techniques on civil engineering structures. Journal of Soft Computing in Civil Engineering, 5(3), 58–74.

Zou, Q., Zhang, Z., Li, Q., Qi, X., Wang, Q. and Wang, S. 2019. DeepCrack: Learning hierarchical convolutional features for crack detection. IEEE Transactions on Image Processing, 28(3), 1498–1512.

2 AI-Assisted Digital Forensics for Securing Industry 4.0 Assets

A. Shahela, Suchitra Gandu, B. Beulah Aswini, and Pramod Kumar Jha

2.1 ORIGINATION OF DIGITAL TRANSFORMATION 4.0

The manufacturing sector has seen a tremendous transformation as a result of Industry 4.0 technologies. These advanced technologies offer many benefits, such as improved efficiency and productivity, but they also pose new cybersecurity challenges. The interconnectedness of these technologies means that a cyberattack in one area of the system can quickly spread and cause damage across the entire network. Furthermore, it is challenging to quickly identify and respond to cyberattacks due to the complicated nature of these systems.

2.1.1 CHALLENGES OF CYBERSECURITY IN INDUSTRY 4.0

One of the biggest challenges of cybersecurity is the large number of connected devices. The Internet of Things (IoT) has enabled manufacturers to connect a vast number of devices, sensors, and machines, which can communicate with each other and share data. However, the more devices that are connected, the greater the attack surface, making it easier for cybercriminals to find vulnerabilities and exploit them.

Another challenge is the complexity of the systems. These systems are highly integrated, with data flowing between multiple devices and platforms. This makes it difficult to identify and isolate the source of an attack. For example, hackers can use AI tools to hack passwords within a minute, which is a huge challenge for an organization. A cyberattack on a single sensor could affect the entire production line, leading to downtime, financial losses, and damage to the company's reputation.

As more organizations adopt Industry 4.0 technologies, the risks of cyber threats increase significantly. Cybercriminals can infiltrate an organization's system through various means, including malware attacks, social engineering, and exploiting vulnerabilities in the system. In the examination of such situations, digital forensics is crucial since it aids in determining the origin of the assault, the degree of damage caused, and the steps required to prevent a repeat. It also involves collecting, analyzing, and preserving digital evidence, which can be used to identify the source and extent of a cyberattack and develop remediation strategies.

DOI: 10.1201/9781003405436-3

TABLE 2.1

Cybercrime Statistics across the World

Incident	Target System	Monetary Loss
The 2022 Colonial Pipeline Ransomware Attack	Oil and gas pipeline	$4.4 million
The 2021 Kaseya Data Breach	Managed service providers	$70 million
The 2020 SolarWinds Hack	Government agencies, businesses	Billions of dollars
The 2018 WannaCry Ransomware Attack	Computers worldwide	Billions of dollars
The 2017 Equifax Data Breach	Customer credit reporting data	$700 million

In addition, digital forensics can help organizations to ensure compliance with regulations. Industry 4.0 technologies have introduced new regulations and compliance requirements, which can be difficult to navigate without the proper tools and expertise. Digital forensics can assist in compliance efforts by identifying areas of noncompliance and providing remediation recommendations.

2.1.2 GLOBAL CYBERCRIME STATISTICS

Table 2.1 shows some examples of cyber breaches, focusing on the target system and monetary loss. The table demonstrates the increased cybercrime across the world.

It is important to note that these are just a few examples of the many cyber breach incidents that have occurred in the world. The amount of monetary loss from these incidents is difficult to estimate, as it can include factors such as lost productivity and reputational damage. However, it is clear that these incidents had a significant financial impact on the organizations that were targeted.

2.2 DIGITAL FORENSICS FOR INCIDENT RESPONSE

In the event of a cyberattack, digital forensics plays a significant role in the incident response process. The first step in this process is to identify the scope of the attack and gather as much information as possible about the incident. This is accomplished by the gathering and examination of digital data, such as user activity records, network activity, and system logs. The process flow chart (Figure 2.1) illustrates how this evidence can be utilized to determine the origin, mode of operation, and degree of damage of an attack.

The methodology for incident response includes the subsequent stages described in the following.

2.2.1 IDENTIFYING THE SCOPE OF THE ATTACK

Identifying the scope of the attack refers to determining the extent to which a system, network, or organization has been compromised due to a cybersecurity incident. This

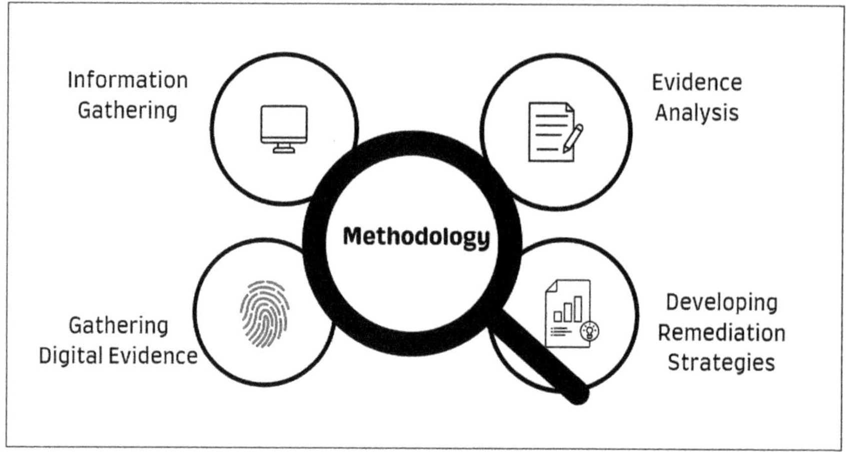

FIGURE 2.1   Process flow diagram of digital forensics.

involves analyzing digital evidence to understand the full extent of the attack and the potential impact on the organization. To determine the scope of the attack, we may need to know how many devices or systems were harmed, what kind of data was stolen, and how the attacker got into a network. By identifying the scope of the attack, investigators can determine the appropriate course of action to mitigate the damage and prevent future incidents.

2.2.2 GATHERING DIGITAL EVIDENCE

Gathering digital evidence involves collecting and preserving data that can be used to support or refute claims in a legal or investigative context. This includes identifying potential sources of digital evidence, such as computers, mobile devices, or network logs, and collecting relevant data forensically soundly.

To ensure that the original information is not corrupted during the evidence collection process, computer forensic analysts employ specialized tools and procedures to build a bit-by-bit replica of the storage medium, like hard discs or flash drives. To guarantee the accuracy of the evidence, they also establish a chain of custody and document their processes.

The collected digital evidence is then analyzed to identify potential links between individuals, devices, and events. This process can include recovering deleted files, analyzing network traffic logs, and examining system logs and metadata to piece together a timeline of events leading up to the incident.

2.2.3 ANALYZING DIGITAL EVIDENCE

Analyzing digital evidence in digital forensics involves examining the digital data collected during the evidence-gathering process to draw meaningful conclusions about the incident being investigated. This process requires specialized tools and techniques to extract valuable information from the digital evidence forensically soundly.

Digital forensic analysts use a variety of techniques to analyze digital evidence, which include:

- File carving—This involves recovering deleted or fragmented files from storage media.
- Metadata analysis—A file's creation date, creator, and file type are just a few of the details that can be found in metadata. Analysts can utilize this information to determine the date and method of a file's creation or modification.
- Keyword searching—Analysts search for specific keywords or phrases in files or system logs to identify potentially relevant information.
- Timeline analysis—This involves piecing together a timeline of events leading up to and during the incident being investigated.
- Malware analysis—Analysts examine the malware code to determine how it functions and the potential impact on the compromised system.

Once the analysis is complete, the findings are documented and presented in a format that can be easily understood by nontechnical stakeholders. This report may include a summary of the findings, including potential causes of the incident, a timeline of events, and recommendations for remediation of future incidents.

2.2.4 IDENTIFYING THE SOURCE AND EXTENT OF THE ATTACK

This is the process of identifying the source of an incident, the scope of the damage, and the potential impact on the organization. It involves analyzing the digital evidence collected during an investigation, which includes logs, system images, network traffic, and other data. By analyzing this data, investigators determine the cause of the incident, the systems or networks affected, and the extent of the damage.

2.2.5 DEVELOPING REMEDIATION STRATEGIES

Once investigators have identified the source and extent of an incident, they develop a strategy for remediation. This strategy involves restoring systems and data from backups, applying patches or updates, or implementing additional security controls. Digital forensics can assist in the development of these strategies by providing insight into the specific vulnerabilities exploited by the attacker and the techniques used to gain access to the network or system.

2.3 ALGORITHMS IN MACHINE LEARNING

Machine learning (ML) techniques are used by digital forensics investigators to examine big datasets that are housed across multiple networks. The resulting information set is then utilized to forecast user behaviour. ML techniques are utilized by investigators to uncover potential criminal activities by employing a set of rules and approaches that can be used to locate anomalous data patterns. This section discusses numerous algorithms that have been proposed to find digital proof and enhance the investigation phase.

2.3.1 Support Vector Machine Algorithm in Digital Forensics

Using the local binary pattern (LBP) and discrete cosine transformation (DCT) operators, this model identifies copy-move and splice assaults in colour images. The support vector machine (SVM) kernel is used to evaluate this system. The DCT and LBP operators detect micropatterns by capturing changes in the local frequency distribution pattern. The LBP blocks intercell values are also taken into account, and they are organized as feature vectors. The SVM and radial basis function (RBF) are then used to classify the generated images into authentic and altered ones. This approach is ideal for detecting image fraud and calculating accuracy metrics (Roy et al. 2019). We can also tell the difference between authentic and false digital photographs and movies. It extracts features from data acquired by a discrete Fourier transform computation using an SVM-based technique. The SVM processing produces a classification system for the provided data using the Scikit-Learn package in Python 3.9. The pictures in the experimental dataset are then predicted by this model. A collection of Python programs is aimed at processing photo features and extracting video frames. They are additionally utilized to build a model using SVM, which can classify photos.

The pictures and videos are analyzed using the density functional theory (DFT)-SVM algorithm, and the results demonstrate that it takes less time when compared to convolutional neural networks (CNN). The low processing time and great performance of the DFT-SVM approach constitute a great tool for detecting fraudulent multimedia content during the digital forensics phase.

2.3.2 Decision Tree Algorithm in Digital Forensics

The decision tree algorithm employs a framework that integrates the MapReduce structure, Hadoop's distributed storage system, with the decision tree algorithms. This framework is capable of handling massive amounts of private information, which can be gathered and stored. It is divided into five steps: gathering network traffic, transforming it into a readable format, filtering packets, analyzing the information for malicious activity, and analysis and representation. A decision tree distinguishes between undesirable and nonmalicious traffic, increasing accuracy and efficiency in each phase. According to the results, this algorithm detects 99% of both fraudulent and not fraudulent traffic (Sharma et al. 2019).

2.3.3 K-Means Algorithm in Digital Forensics

The K-means technique is used to analyze data obtained from diverse sources. It is simple to sift the data gathered from various cybercrime analysis methods to extract the features. The K-means method finds interactions between characteristics during the clustering stage. This strategy provides concrete ways to prevent similar crimes from occurring in subsequent years.

2.3.4 Principal Component Analysis Algorithm in Digital Forensics

Principal component analysis uses a random forest-based digital forensics system to investigate where and how the image originated. The primary benefit of this feature is that it enables investigators to locate several camera sources responsible for the various JPEG (Joint Photographic Experts Group) compression defects. The principal component analysis technique is used to increase the framework's precision for classification. The degree of dimensionality of the characteristics is significantly reduced using this technique (Chhabra et al. 2018).

2.3.5 Logistic Regression Algorithm in Digital Forensics

This covers every type of malware that frequently targets the registry in Windows operating systems. This method offers insightful information about how these varieties of malware communicate with the registry. Different classifiers, including the logistic regression, the decision tree, and the neural network, are tested. The findings demonstrate that forensic analysis using improved timestamps and ML methods is feasible. Over 70% of the malware is classified correctly using the boosted tree. This technique enables investigators to quickly distinguish between malware types that are present and those that aren't (Huan et al. 2018).

2.3.6 A Priori Algorithm in Digital Forensics

The a priori method increases the mining efficiency by finding the rules that satisfy the minimum confidence threshold and creating frequent item sets. Additionally, representing the data using a vertical format improves the database's intimacy. The relationships between the different people are used to categorize the outcomes of the clustering. The data is examined using association rules, and it is indicated that the high confidence rules demonstrate that the user's daily routines are compatible with the data's properties (Islam et al. 2018).

2.4 MACHINE LEARNING FORENSICS

ML forensics, which can identify and anticipate criminal trends, is a new area that has developed as a result of the application of machine learning. For the purpose of link association, visualization, segmentation, and clustering criminal activity, a framework must be able to acquire and analyze data from servers, the internet, and many other sources. A list of the many techniques applied in ML forensics is provided in the explanations that follow.

- Link Analysis
  By transforming the data into a group of interconnected entities that are linked together, link analysis can be used by ML techniques to ascertain both the content and format of a body of information. In most cases, it enables us to find association patterns. The investigators can learn about the connections between the entities through visualization, as well as the

frequency of contacts and undiscovered correlations. Due to the significance of link analysis, intelligence professionals frequently use it to examine the networks of hacking groups. The link analysis method starts with data and attempts to gather pertinent information and knowledge coming from the links and nodes of a given network to ensure investigation teams and analysts can identify relationships. Several categories can be used to group the network associations. As an illustration, associations between a company and a person can include manager, head, members, and employee.

- Clustering Incidents and Crimes
  On the basis of people's verbal and behavioural patterns, clustering software can establish groupings. With this approach to the use of ML forensics analysis, the program creates the cluster from the data on its own. This type of exploratory research is done to find anomalies that might be present in the data, such as behaviour that might indicate illegal activity, like the use of a suspect port in a computer network for an attack. The main clustering approach is a self-organizing map neural network. One can use emails, call site records, instant messaging, website forms, conversations, phone calls, and texts as examples of word clustering data.

- Crime Forecasting
  Due to the enormous number of incidents that occur each year, investigators find it challenging, tedious, and costly to identify whether an attack was committed by the same offender. Analytical approaches are applied to estimate the potential locations of crime scenes in future years through statistical projections. An online attack often involves a person who chooses a target, a time, and a place for their own illegal act. It may not be able to forecast crimes using simple approaches, but it is possible to do so by using more sophisticated methodologies such as ML. SVMs, naive Bayes, random forests, and other ML methods can all be useful in this process.

- Fraud Identification
  Any action carried out with the intention of misleading is considered to be fraud. The data obtained from organizations that maintain data pertaining to customers, clients, users of websites, and many other entities is the most important source that can be useful to the investigators. Because of these information repositories, we can now create models and rules based on how users interact with websites and organizations on a daily basis. These generated patterns can help to spot fraud and also help businesses fix any potentially flawed procedures. For the identification of fraud procedures to work, a plan must be created. The tactics and procedures must be consistently checked, and the models employed by artificial intelligence forensics must react in real time to find suspicious conduct, as fraud schemes and the methodologies used for identifying fraud constantly evolve.

2.4.1   A Case Study: Recognizing Phishing URLs Using Machine Learning

Phishing is a cybercrime in which targets are contacted by email, telephone, or text message by someone posing as a legitimate agent to attract individuals into providing sensitive or personal data, banking, and credit card details, and passwords, which are then used to access accounts, resulting in identity theft and financial loss. Phishing Uniform Resource Locator (URL) may link some of the websites that are similar in name and appearance to an official website.

In this case study, our team proposed an ensemble approach for binary classification of phishing URLs, which involves combining multiple ML algorithms. This method involves the extraction of various features from the URLs, including length, presence of specific keywords, and similarity to known phishing URLs. These features are used to train multiple classifiers, including random forests, SVMs, and gradient boosting machines. The predictions of these classifiers are then combined using ensemble techniques, such as majority voting or stacking, to obtain the final classification result. Experimental results show that the ensemble approach achieves high accuracy in detecting phishing URLs, outperforming individual classifiers. This method has been evaluated on several benchmark datasets, including the Phishing Websites Data Set, and achieved an accuracy of up to 99.9%. Overall, our proposed ensemble approach provides an effective solution for binary classification of phishing URLs, which can be used to improve the security of online services and protect users from cyberattacks.

A random forest is a collection of decision trees. Based on various features, we created four decision trees.

- Features based on the address bar
- Abnormal features
- JavaScript and HTML-based features
- Domain-based features

A randomly chosen portion of the training data is used by the random forest classifier to generate a collection of decision trees. It simply consists of a collection of decision trees from a randomly chosen subset of the training set, which are subsequently used to decide the final prediction. The random forest classifier performs better as there are more trees present but at the expense of more computations. As all the main operations take place in the backend, anyone may easily operate it. The user must enter a URL into the GUI (Graphical User Interface) and then click the submit button. "It is a Phishing Website" is displayed for the phishing URL, and "It is a Legitimate Website; continue using it" is displayed for the legitimate URL in the output. Figure 2.2 shows the result. The ML approach significantly expedited the phishing process. It also allowed for proactive detection and prevention of future phishing and malware attacks by continuously updating the model with new samples.

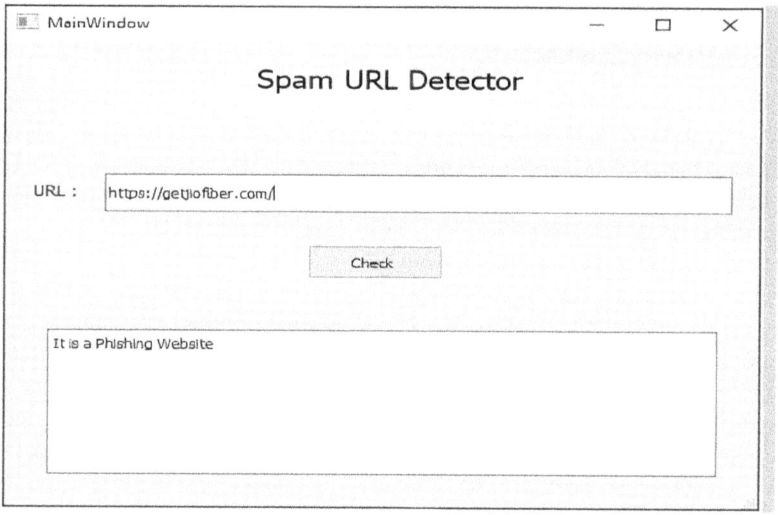

FIGURE 2.2   Result showing whether a given URL is legitimate or phished.

2.5   COMPLIANCE ASSURANCE THROUGH DIGITAL FORENSICS

Organizations now have to comply with new rules and compliance requirements brought about by Industry 4.0 technologies. These laws include the Payment Card Industry Data Security Standard (PCI DSS) and the General Data Protection Regulation (GDPR). Digital forensics assures adherence to these rules by offering proof of the safety precautions implemented with regard to digital assets.

2.5.1

2.5.1.1   General Data Protection Regulation

GDPR is a rule that establishes guidelines for how organizations in the European Union (EU) should gather, keep, and utilize personal data about individuals. Companies that gather or process the personal data of people in the EU are required to abide by the GDPR, which calls for the implementation of suitable security measures to safeguard the data.

2.5.1.2   Cybersecurity Information Sharing Act

The Cybersecurity Information Sharing Act is a US law that encourages public and private entities to share cybersecurity information and intelligence to prevent and respond to cyber threats.

2.5.1.3   National Institute of Standards and Technology Cybersecurity Framework

The National Institute of Standards and Technology Cybersecurity Framework offers suggestions for how organizations can control and lower cybersecurity risk.

Identifying, safeguarding, reacting to, and recovering from cyber incidents comprises a collection of best practices and guidelines.

2.5.1.4 Payment Card Industry Data Security Standard

The PCI DSS is a list of specifications for organizations that deal with credit card data. It contains recommendations for safe payment processing, such as maintaining a safe network, safeguarding cardholder information, and putting in place robust access restrictions.

Compliance with these regulations is essential for organizations operating in Industry 4.0 to protect their assets. Achieving compliance often requires a combination of technical controls, policies and procedures, and staff training to ensure that all aspects of the organization are aligned with the relevant regulations.

2.5.2 HOW DIGITAL FORENSICS ASSISTS IN COMPLIANCE EFFORTS

Digital forensics helps in assisting organizations with compliance efforts, as described in the following.

2.5.2.1 Incident Response

Digital forensics helps organizations respond to cybersecurity incidents by identifying the scope of the attack, gathering digital evidence, and analyzing the evidence to determine the cause and extent of the incident. This information can help organizations take appropriate action to mitigate the impact of the incident and prevent similar incidents from occurring in the future.

2.5.2.2 Forensic Readiness Planning

Digital forensics assists organizations in preparing for potential cyber incidents by developing a forensic readiness plan. This involves identifying critical systems and data, establishing procedures for collecting and preserving evidence, and training staff to ensure they know what to do in the event of a cyber incident.

2.5.2.3 Compliance Monitoring

Digital forensics helps organizations monitor compliance with relevant regulations by performing regular audits of digital systems and data. This helps in identifying potential compliance issues before they become a problem and ensures that the organization is prepared to respond to any incidents that do occur.

2.5.2.4 Investigating Compliance Violations

If an organization is found to be in violation of a compliance regulation, digital forensics assists in the investigation by collecting and analyzing digital evidence to determine the cause and extent of the violation. This information helps organizations take appropriate corrective action and demonstrate to regulatory authorities that they are taking the necessary steps to address the issue.

2.5.2.5 Data Recovery

In the event of data loss or corruption, digital forensics assists in recovering lost or damaged data. This is particularly important in healthcare, where the loss of patient data can have serious consequences.

2.6 PROTECTING INTELLECTUAL PROPERTY: LEVERAGING DIGITAL FORENSICS FOR EFFECTIVE ASSET PROTECTION

Intellectual property (IP) is a valuable asset for many manufacturing organizations, and protecting it from theft or misuse is crucial. Digital forensics helps in identifying any attempts to steal or misuse IP by monitoring access to sensitive files and identifying any unauthorized access attempts.

2.6.1 Importance of Protecting Intellectual Property in Industry 4.0

Creations of the mind, including inventions, literary and creative works, designs, and symbols, are all considered IP. In today's digital age, protecting IP is more important than ever.

Protecting IP is important for organizations to maintain a competitive advantage and secure their economic value. It encourages innovation by providing legal protection for new and innovative ideas. This protection incentivizes organizations to invest in research and development, leading to new and improved products and services. It also helps to protect a company's brand reputation. Counterfeit products and piracy can damage the reputation of a company and result in lost revenue. IP theft can have serious security implications; it can be used by competitors or hackers to gain a competitive advantage or to develop new products.

2.6.2 How Digital Forensics Assists in Identifying and Preventing Intellectual Property Theft

Digital forensics plays an important prole in identifying and stopping intellectual property theft. Techniques include those described in the following sections.

2.6.2.1 Identifying Potential Intellectual Property Theft

Digital forensics can be used to monitor digital systems for any suspicious activity that may indicate an attempted theft of IP. By monitoring networks and digital devices, digital forensics can detect unauthorized access, data exfiltration, or other signs of suspicious activity.

2.6.2.2 Preserving Digital Evidence

If a suspected IP theft occurs, digital forensics can help preserve digital evidence. This involves collecting and analyzing data from affected systems to determine what data has been taken and who is responsible.

2.6.2.3 Analyzing Stolen Intellectual Property

If a theft of IP has occurred, digital forensics can help analyze the stolen IP to determine its extent and how it may be used. This allows organizations to take appropriate legal action and develop strategies to prevent future IP theft.

2.6.2.4 Implementing Security Measures

Digital forensics allows organizations to identify weaknesses in their security systems that may make them vulnerable to IP theft. This information can be used to implement new security measures or to improve existing ones, such as access controls, data encryption, or data loss prevention systems.

2.6.2.5 Training Employees

Employees must be trained in how to identify and prevent IP theft. This involves educating them on the types of data, how to report suspicious activity, and the importance of maintaining strong security practices.

2.7 AI AS A CATALYST TO DIGITAL FORENSICS

As Industry 4.0 continues to evolve, new technologies and trends are emerging in digital forensics to address the changing security landscape. Some of the emerging trends in digital forensics for Industry 4.0 cybersecurity are described in the following.

2.7.1 SOCIAL NETWORK FORENSICS

Social media platforms are now widely used and have become a main source of social interaction because of the development of Industry 4.0. Through these websites, users actively exchange their information and participate in social activities. As a result, attackers are given numerous chances to abuse user accounts. Additionally, several social media platforms, including Facebook and Twitter, are vulnerable to numerous cyber dangers. Social media platforms may be attacked from the outside or from inside the network. Attacks against external systems usually involve DDoS (Distributed Denial-of-Service) or DoS (Denial-of-Service), but attacks against internal networks generally involve cookie data retrieval (Ali et al. 2019). Thus, it can be concluded that social network forensics is a rising trend in this domain due to its capability to provide appropriate digital evidence.

Additionally, three functional dimensions are offered by social media forensics: reverse search integration, temporal localization analysis, and metadata visualization and extraction (Zampoglou et al. 2016). The first benefit of the Google Image Search feature is that it displays the results in a new tab in your computer browser. Second, it includes six different tampering localization maps that were created using forensic algorithms and are intended to find various traces of social media manipulation. Third, it shows any possible attached thumbnails and completely enables metadata listing. These features enable forensic professionals to go deeper into the data and retrieve relevant evidence. As a result, social networking forensics is becoming increasingly popular in the field of digital forensics.

2.7.2 FORENSICS IN THE INTERNET OF THINGS

IoT systems are vulnerable to a variety of security threats and assaults, including ransomware, DoS, and mass surveillance. The emergence of IoT forensics is a result of the numerous, complicated, and distinctive problems that IoT systems pose to the discipline of digital forensics. Additionally, IoT-based applications include a significant number of resources and unique devices that produce a huge amount of data, known as big data. Investigators have the chance to track down cybercrimes using this data along with digital forensics tools and procedures, which further aid them in preventing cyberattacks.

IoT forensics responds to user needs without demanding their active participation. Thus, the IoT forensics environment offers contextual data that aids digital forensic investigators in analyzing events that happened in the real world. IoT forensics has, therefore, become one of the dominant developments in the digital forensics field, both for its capability to produce digital and context-based proof and for the multiple challenges this field encounters.

2.8 CONCLUSION

The rise in cybercrimes has drawn significant attention to digital forensics. Although the development of digital technology has been advantageous to many industries, it has also given rise to new strategies for cybercrime. Additionally, every day, malicious software and tools are created and put into use in an effort to threaten private as well as public networks and exploit data storage in order to extract and use valuable data. With the objective of retrieving digital evidence from digital devices, these security flaws and breaches have sparked innovations in the field of digital forensics.

ACKNOWLEDGMENTS

We would like to express our heartfelt thanks to Director Shri B.V. Papa Rao of the Centre for Advanced Systems, DRDO, who has always encouraged us to pursue our passions and provided us with the necessary support. We would also like to extend our appreciation to the editor for his unwavering guidance and expertise in helping us refine and shape our ideas. We are indebted to our colleagues for their thoughtful feedback and unwavering support throughout the writing process. Thank you all for your support and encouragement.

REFERENCES

Ali, M., Shiaeles, S., Clarke, N. and Kontogeorgis, D. 2019. A Proactive Malicious Software Identification Approach for digital Forensic Examiners. Journal of Information Security and Applications, 47(C): 139–155. https://doi.org/10.1016/j.jisa.2019.04.013.

Chhabra, G. S., Singh, V. and Singh, M. 2018. Hadoop-based Analytic Framework for Cyber Forensics. International Journal of Communication Systems, 31(15): e3772. https://doi.org/10.1002/dac.3772.

Islam, M., Karmakar, G., Kamruzzaman, J., Murshed, M., Kahandawa, G. and Parvin, N. 2018. Detecting Splicing and Copy-move Attacks in Color Images. In International Conference on Digital Image Computing: Techniques and Applications, DICTA 2018; Canberra, Australia, pp. 1–3. https://doi.org/10.1109/DICTA.2018.8615874.

Li, H., Xi, B., Wu, S., Jiang, J. and Rao, Y. 2018. The Application of Association Analysis in Mobile Phone Forensics System. In 2nd International Conference on Intelligence Science (ICIS), Beijing, China, pp. 126–133, ff10.1007/978-3-030-01313-4_13ff. Ffhal-02118811. https://inria.hal.science/hal-02118811/document.

Roy, A., Dixit, R., Naskar, R. and Chakraborty, R. S. 2019. Digital Image Forensics. Springer—Technology & Engineering. ISBN: 9811076448, 9789811076442. https://books.google.co.in/books/about/Digital_Image_Forensics.html?id=KruXDwAAQBAJ&redir_esc=y.

Sharma, B. K., Joseph, M. A., Jacob, B. and Miranda, L. C. B. 2019. Emerging Trends in Digital Forensic and Cyber Security-An Overview. In Sixth HCT Information Technology Trends (ITT), pp. 309–313, November. http://dx.doi.org/10.21533/pen.v8i3.1463.
Zampoglou, M., Papadopoulos, S., Kompatsiaris, Y., Bouwmeester, R. and Spangenberg, J. 2016. Web and Social Media Image Forensics for News Professionals. In Tenth International AAAI Conference on Web and Social Sedia, April. http://dx.doi.org/10.21533/pen.v8i3.1463

3 State-of-the-Art Analysis in Reversible Data Hiding Techniques

R. Geetha and D. Kavitha

3.1 INTRODUCTION

Since the 1990s, the investigation of technology that can serve as an alternative to cryptography has attracted extensive attention from both academicians and industries. Information hiding has been widely considered as a fairly promising technology to fulfill this purpose (Cox et al. 1997). Information hiding works by covertly embedding a message within a host digital signal (Lee et al. 2007). The message to be hidden can be anything, such as a personal identification code, electronic patient information, a company logo or any secret information in the form of bit stream based on the application (Wu and Memon 1997). Owing to its future applications, information hiding has become an emerging area of research in the past two decades.

Steganography is known as the art of covert communication between the sender and the receiver, in which the covert message is confidentially embedded in an imperceptible manner into the cover signal (Weinberger et al. 2000). This ensures that no one, other than the sender and receiver, is aware of the existence of the hidden message (Fridrich and Goljan 2002). Concealing the presence of hidden data is the ultimate goal of steganography and it is least bothered about recovering the cover medium at the receiver side. A similar concept to steganography is watermarking. Watermarking refers to the process of embedding messages, which is often called a watermark, into a host signal without causing noticeable degradation to the human eye. The focus of a watermarking system is achieving a high level of robustness against attacks (Dittmann et al. 2002). In other words, a watermarked system makes it highly impossible to remove the injected watermark without causing visual damage to the stego image.

The art of hiding the data and recovering the original data back losslessly after retrieving the hidden content is known as reversible data hiding (RDH) (Honsinger et al. 2001). RDH has been one of the active fields in research for more than 20 years. This is evident as more and more research papers are being published these days. This RDH technique embeds secret message into the cover medium without causing distortion in cover medium that remains unnoticeable to the human eye (Celik et al. 2005). It also restores the cover medium exactly after retrieving the secret data. However, the other data hidden techniques like steganography and digital watermarking (Vleeschouwer et al. 2003) rarely care about the recovery of the cover medium. In a few sensitive domains, such as medical, military, forensic, law

DOI: 10.1201/9781003405436-4

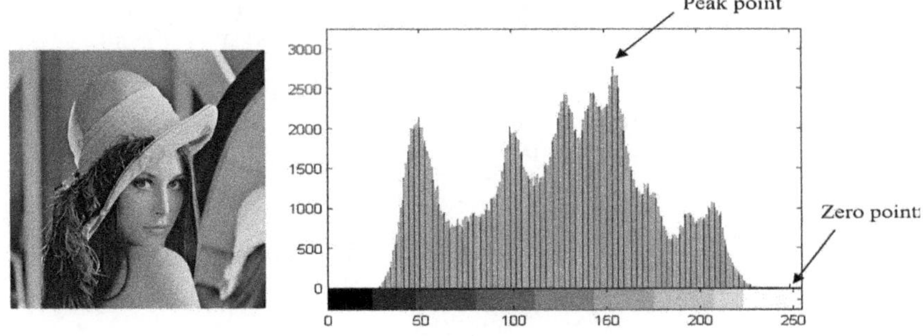

FIGURE 3.1 Lena image and a histogram showing the distribution of intensity values.

enforcement fields, etc., permanent distortion to the cover media is strictly forbidden. To overcome these issues, RDH, which is otherwise called invertible data hiding, is proposed. RDH algorithms can be broadly classified based on the operation domain namely (i) frequency domain, (ii) spatial domain and (iii) encrypted domain.

3.2 HISTOGRAM MODIFICATION APPROACH

In this scheme the histogram of the cover image or the residual image or the difference image is altered. The peak points are identified to embed the secret data without altering the other points (Ni et al. 2006). Many algorithms have been developed in this approach to improve the hiding capacity and to keep the distortion within the allowable limit. Figure 3.1 shows the black and white image of a model (Lena) and a histogram of the image.

3.3 DIFFERENCE EXPANSION APPROACH

In this approach the redundancy of the pixels is looked at and each pair of neighboring pixels is used to embed at least one-bit secret data. Many schemes have been developed by computing the difference and average value of the two neighboring pixels for embedding the secret message (Li et al. 2010). A common feature of this approach is to use the decorrelating operator to produce features with fewer magnitudes. The expansion embeddable technique makes it possible to embed larger amounts of data when compared to the histogram modification approach by maintaining the image quality (Chung et al. 2011).

3.4 PREDICTION AND INTERPOLATION APPROACH

Pixel values are predicted using various techniques available. The difference between the original pixel and the predicted pixel is calculated as the error. The plus or minus error is expanded to embed a single message bit 1 or 0 on the predicted pixel values. The selection of expandable difference values is fixed based on the nonoverlapping regions in the histogram of expandable locations. To avoid overlapping of expanded bins, the histogram of other location bins is shifted in such a way that it remains nonoverlapped

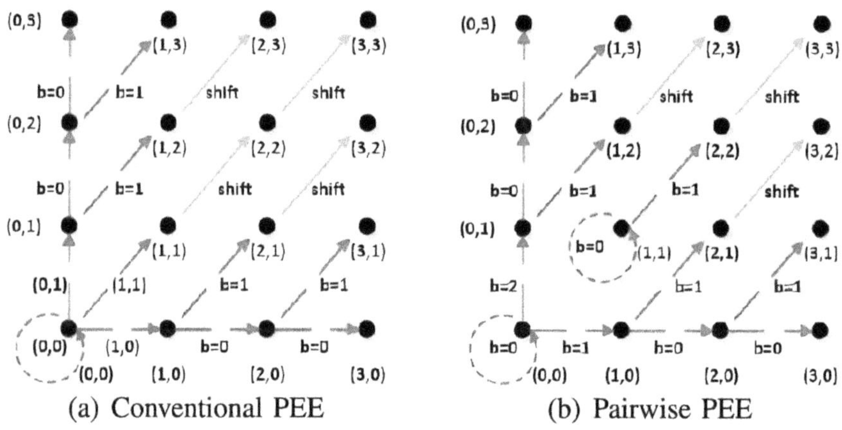

(a) Conventional PEE   (b) Pairwise PEE

FIGURE 3.2   Prediction error expansion (PEE) approach.

in any circumstances. Location map is required to indicate the location of pixels closer to the boundary so that it remains unaltered during the shifting and expansion of the image histogram. Figure 3.2 depicts the Prediction error expansion (PEE) approach.

Interpolation is a technique that is often used to incorporate information insufficiency in digital images (Lee and Huang 2012). It plays a major role in three-dimensional medical imaging for image reconstruction. This technique involves scaling of the image based on the application. Then the interpolated points between the rows and columns are approximated using different means of transformation like bilinear, bicubic, nearest neighbor mean, etc. The better the approximation of the interpolated points, the higher is the quality of the output image that can be used for various applications. Most of the schemes are considered for discussion. The mathematical expressions involved in each scheme are not included for simplicity. Readers are requested to look into the original work for further details.

In the technique proposed by Ni et al. (2006), a histogram of the image is considered for embedding the data. By finding the peak point and zero point of the image from its histogram, the pixel values are shifted by 1 point to vacate the peak point pixels to embed the secret data. The computational complexity of this algorithm is very low since there are no transformations involved in performing the embedding or extraction process. The total computational complexity is O (3kM N), if there are k pairs of minimum and maximum points for an M × N image.

Advantages: The image quality achieved by this method after embedding the data is very high. It involves a simple logic to embed and extract the secret data.

Disadvantage: Since the algorithm is simple to execute, this method is prone to attacks easily. The embedding capacity is very low since this method uses only the peak points in embedding the secret message.

Wang et al. (2013) proposed a scheme for image authentication by histogram modification of the difference image. In this method the secret message is embedded into the pixels having difference values of +1 or −1. The embedding capacity is equal to the number of pixels with difference values of +1 or −1. A cryptographic hash function MD5 is used to verify the integrity of the image. The peak signal-to-noise ratio (PSNR) is assured to be higher than 51.14 dB in this scheme.

Advantage: Distortion is reduced significantly in this scheme. The algorithm involves less computational complexity. PSNR is 51.14 dB.

Disadvantage: Payload is restricted to the number of pixels having differences of ±1. The algorithm is assumed for no overflow and underflow conditions.

Lin and Hsueh (2008) proposed a reversible data hiding scheme using multilevel histogram modification of difference images. In this approach, first the difference image is calculated and then the histogram of the difference is generated to find the peak point of the difference image. Secret data is embedded using the peak point values of the difference image. The extraction procedure is the reverse of the above-mentioned steps. A modulo operation is performed to take care of underflow and overflow problems. This method is well suited for images having a histogram with uniform distribution so that the difference between the neighboring pixels is close to zero since the embedding of data depends upon the peak point of the difference image.

Advantage: This scheme has high embedding capacity with a nominal image quality.

Disadvantage: Solution for handling the underflow and overflow problems results in salt and pepper noise. This scheme will not apply well for all types of images.

Tian (2003) proposed the first method using the difference expansion of pixel pairs to hide the information. This algorithm allows a single bit to be embedded in every pixel pair difference. So a maximum of 0.5 bpp capacity can be embedded using this algorithm with low distortion in the cover image. The location map is created to indicate the modified pixel pairs. The pixel pairs that will create overflow or underflow when modified are left untouched. Then the location map is compressed and included in the payload.

Advantage: Distortion and complexity are low.

Disadvantage: Capacity cannot be exceeded by more than 0.5 bpp. The location map is too large, which hinders the payload even lower. If the location map is more than the expandable pairs then this algorithm doesn't work for embedding the data.

Li et al. (2013) proposed a general framework to construct any histogram-based approach using shifting and embedding functions. In this framework the cover image is divided into nonoverlapping blocks. There are two functions, shifting and embedding functions, g and (f_0, f_1), respectively. Function g is applied on the pixel values to create vacant positions for embedding the secret data bits. f_0, f_1 are embedding functions for embedding 1 and 0. The nonoverlapping blocks are further divided into I1, I2 and I3. Secret data is embedded into I1 and I3, the location map is embedded into LSbs of I1 and original LSbs of I1 are embedded into I2. The reverse procedure is applied for extracting the data.

Zhao et al. (2011) proposed another multilevel histogram modification scheme in which an integer parameter called the embedding level (EL) greater than or equal to zero is introduced to control the capacity of the bits to be hidden. As EL increases the hiding capacity also increases, and the embedding process also gets complicated. Initially the difference is computed, then the right bins of the histogram are shifted by ELs. The EL is decreased by 1 and this continues till the EL becomes zero.

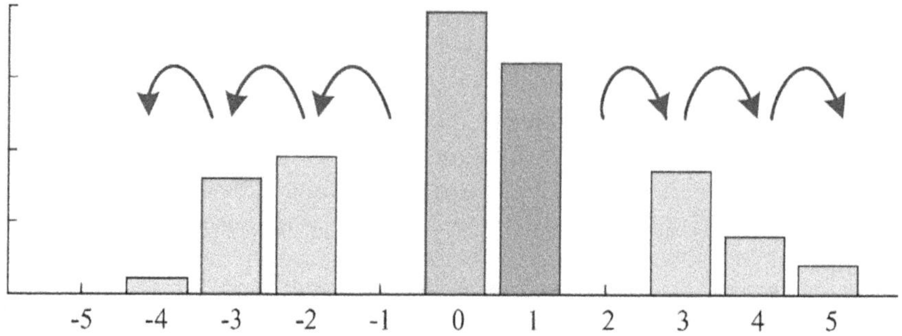

FIGURE 3.3 Traditional histogram shifting operation.

A sequential recovery line of approach is used for extracting the hidden data. During extraction EL is considered as auxiliary information and the pixel value difference is computed. The value of EL is most favorably selected so that underflow and overflow problems can be taken care of. A location map is used to identify the underflow and overflow pixels. An arithmetic coding technique is used to compress the location map and is embedded along with the payload.

Advantage: Embedding capacity is controlled by introducing a parameter called the embedding level.

Disadvantage: As the embedding level increases, distortion in the image increases more aggressively.

Thodi and Rodriguez (2007) proposed an algorithm using a histogram shifting technique in which the need to send the location map is not required as in Tian's algorithm. The locations for embedding are selected by having nonoverlapping blocks in the histogram where there are embeddable locations. To compensate for the overlap of bins with the expanded locations, a histogram shifting method is introduced to eliminate this issue. The bins having smaller differences are preferred during the selection of the embeddable location because if the difference magnitude is smaller, the smaller will be the distortion introduced. Figure 3.3 explains the traditional histogram shifting operation.

Comments: This scheme had better performance in payload and image distortion level when compared to previous works by Tian and Alattar.

Alattar (2004) proposed a method that is an improvement to Tian's (2003) difference expansion method. In this scheme the algorithm is developed for sets of vectors instead of pixel pairs (Lee et al. 2008). Integer transform is performed on these vectors. The location map is compressed and is embedded along with the payload. A maximum payload of 1 bpp can be achieved using this algorithm. This algorithm outperforms most of the drawbacks in Tian's (2003) algorithm. For a given vector size and image size the embedding capacity is determined by the value of the threshold used. The approach for recursive embedding is also discussed.

Advantage: Very high-capacity algorithm with low image distortion. Performance of the spatial quad-based method is much higher when compared to the triplet-based method.

Disadvantage: Applying the algorithm for color components does not show better performance.

Tai et al. (2009) proposed an RDH scheme that is an extension to the difference-based scheme. The main issue of communicating the multiple peak points is addressed by introducing a binary tree structure. The difference value histogram is concentrated around zero. After finding out the peak point in the histogram, it is shifted and modified. To increase the hiding capacity this process is repeated a number of times. A tree level is calculated from the number of peak points such that it is equal to 2^L. The only side information that is transmitted is the binary level. The problem of underflow and overflow is taken care of by narrowing the histogram. The location map is compressed and embedded along with the payload.

3.5 ANALYSIS AND RECOMMENDATIONS

More reversible data hiding techniques have been discussed in the literature. From the survey, most of the reversible data hiding schemes use raw image as their source. The payload of the compressed domain schemes is less when compared to spatial or frequency domain schemes. Generally, in histogram modification techniques the hiding capacity is low when compared to difference expansion–based techniques and prediction-based techniques. But the side information required to recover the data and the original cover image is more in other techniques when compared to histogram techniques. Table 3.1 shows the performance comparison of various existing schemes. Lot of work has been done to improve the quality of the stego image and to increase the capacity of the secret data. Location mapping is a major issue that thwarts the payload. A combined approach of histogram with pixel difference and predictive coding gives a better performance.

3.6 REVERSIBLE DATA HIDING USING AI METHODS

The development of artificial intelligence and other cutting-edge techniques has led to significant advancements in informed data hiding methods using deep learning models. These methods have demonstrated breakthroughs in terms of techniques and performance, making them the prevailing trend in the field. Among these novel approaches, some allow for the effective hiding and extraction of secret data without altering the original cover image, making them undetectable by steganalysis algorithms. In contrast, RDH with deep learning has received less attention and lacks methods capable of hiding data without modification. Currently, RDH methods can be categorized into two types: RDH in unencrypted images and RDH in encrypted images. However, data hiding based on cover image modification is becoming increasingly susceptible to detection by advanced machine learning tools. Figure 3.4 shows the architecture of a convolutional neural network (CNN; Yang and Huang 2022).

3.6.1 Reversible Data Hiding Using Convolutional Neural Networks

The input image is the feature extracted in the beginning followed by the image prediction part. These are the two main components of the proposed predictor shown

TABLE 3.1

Performance Analysis of Existing Reversible Data Hiding Methods

Sl No.	Hiding Scheme	Image Type	No. of Levels	Rate (bpp)	Average Peak Signal-to-Noise Ratio (dB)	Availability of Location Map	Complexity
1	Multilevel histogram modification (Zhao et al. 2011)	Raw image	Multilevel	0.35	37	Yes	High
2	Histogram modification (Ni et al. 2006)	Raw image	Single	0.02	48.3	Yes	Low
3	Prediction and interpolation (Hong 2012)	Interpolated raw image	Single	1.5	20	No	Low
4	Residual histogram shifting (Hong et al. 2009)	Compressed image	Single	1.25	32	No	High
5	Histogram shifting (Li et al. 2011)	Raw image	Single	0.5	40	Yes	Low
6	Histogram modification (Tai et al. 2009)	Raw image	Multilevel	0.2	40	Yes	Low
7	Histogram modification on prediction (Zhou and Oscar 2012)	Raw image	Single	0.5	40	Yes	Low
8	Integer transform (Peng et al. 2012)	Raw image	Single	1.2	28	Yes	Low
9	Prediction error expansion (Zhou and Oscar 2012)	Raw image	Single	0.3	40	Yes	Low
10	Integer to integer wavelet transform (Lee et al. 2007)	Wavelet transformed image	Single	0.3	45	Yes	High
11	Histogram shifting (Xuan et al. 2007)	Raw image	Single	0.2	40	Yes	Low
12	Histogram shifting (Tsai et al. 2013)	Raw image	Single	0.2	48	Yes	Low
13	Difference expansion (Alattar 2004)	Raw image	Single	0.2	32	Yes	Low
14	Difference expansion (Tian 2003)	Raw image	Single	0.5	36	Yes	Low
15	Difference expansion (Hu et al. 2009)	Raw color image	Single	0.5	38	Yes	Low
16	Prediction error (Gao et al. 2011)	Raw image	Single	1	35	Yes	Low
17	Sorting and prediction (Sachnev et al. 2009)	Raw image	Single	0.7	34	Yes	Low
18	Prediction error (Fallahpour 2008)	Raw image	Single	0.5	40	Yes	Low
19	Histogram shifting (Chang et al. 2015)	Raw image	Single	0.2	48.3	Yes	High
20	Predictive coding (Yu et al. 2005)	Compressed image	Single	1	50	Yes	High

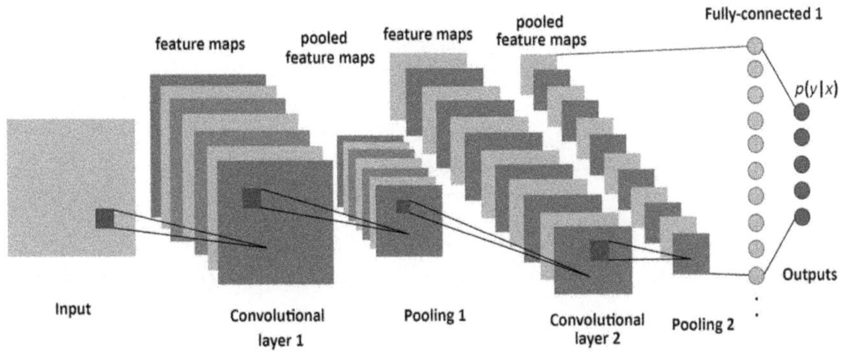

FIGURE 3.4 Convolutional neural network architecture.

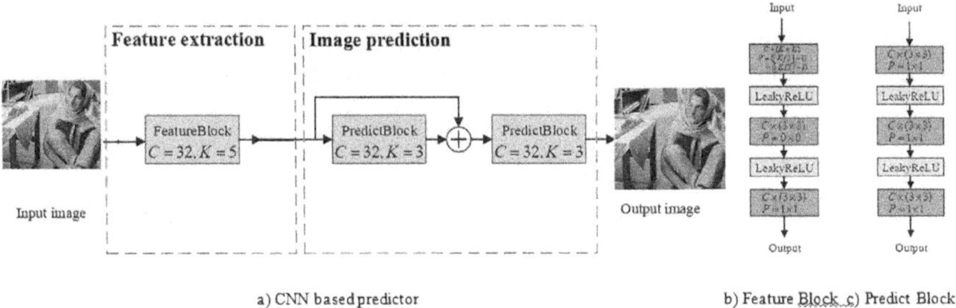

FIGURE 3.5 (a) Convolutional neural network (CNN)-based predictor; (b) feature block; (c) predict block.

in Figure 3.5(a). The feature extraction comprises many parallel convolution blocks, titled feature blocks. The designed arrangement of the feature block is as shown in Figure 3.5(b), in which three convolution layers are connected, with a Leaky ReLu activation in between each two convolution operations. The first, second and third convolution layer kernel sizes are set to $n \times n, 3 \times 3$ and 3×3, respectively. The factor n is any odd number that is greater than 1 and less than the size of the image. To achieve accurate pixel by pixel prediction, the convolution operation step size is set

to 1×1. $\left(\left\lfloor \dfrac{n}{2} \right\rfloor - 1\right) \times \left(\left\lfloor \dfrac{n}{2} \right\rfloor - 1\right), (0 \times 0)$ and (1×1) are the padding sizes of the feature

block convolution operations, respectively. The symbol $\lfloor \ \rfloor$ represents the flooring function. C is set as the output channel count. Dissimilar high-dimensional feature maps are generated once the input images are fed into the feature extraction part. These high-dimensional feature maps are fed as input to the prediction part. In the prediction part, named the predict block, two convolutions are involved, as shown in Figure 3.5(c).

Another approach in image steganography method utilizing deep convolutional generative adversarial networks (DCGAN) is introduced (shown in Figure 3.6). This

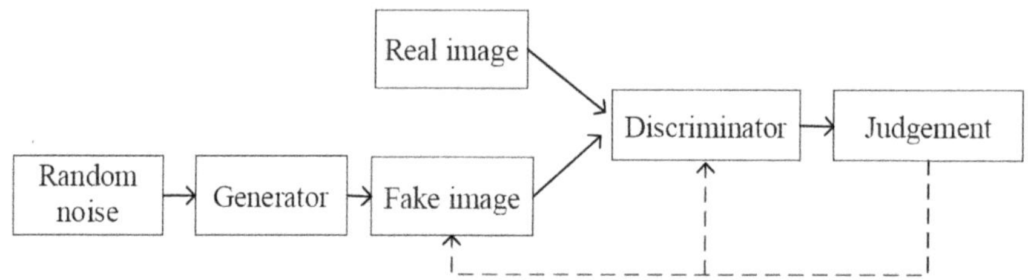

FIGURE 3.6 Block diagram representation of a generative adversarial network (GAN) model.

method establishes a mapping between secret data and random noise, enabling the generation of a corresponding stego image using the DCGAN model (Zhuo et al. 2019; Tsai et al. 2021). Additionally, an extractor is trained to retrieve the hidden data. Notably, this method exhibits a robust ability to evade state-of-the-art detection tools, providing valuable inspiration for RDH techniques without image modification.

3.7 THEORETICAL INVESTIGATIONS IN REVERSIBLE DATA HIDING METHODOLOGY

One basic issue in RDH schemes is the maximum allowable payload given a distortion limit, given by Equation 3.1:

$$C_{rev}\left(\Delta\right) = maximum\left\{H\left(CI'\right)\right\} - H\left(CI\right) \tag{3.1}$$

where C_{rev} is the capacity of the RDH scheme. CI, CI' denotes cover and stego image points, respectively. H represents an entropy function. The equation specifies the amount of message data embedded into stego sequence reversibility. In practical embedding schemes, to derive an independent and identically distributed integer sequence, the cover image is usually projected at a lower-dimensional space and then the generated sequence is applied to the equation. However, practically, the cover medium sequence is usually correlated and not independent and identically distributed. Bridging the gap between the theoretical upper bound capacity and practical performance methods has been achieved by various embedding schemes that have been discussed in the literature.

Most of the RDH works discussed in the literature are based on PEE, histogram modification or difference expansion in the spatial domain. It is easier to work with the algorithm and also easier to achieve reversibility in the spatial domain than in the frequency domain. Even though the literature survey showed that the issue of location mapping and histogram shrinking has been improved, the solution to eliminate such issues remains open. The payload has been increased quite well over the years. But still there are more chances to improve the embedding capacity maintaining the imperceptibility of the marked image.

Although a few investigations have been carried out in JPEG images, still RDH in JPEG is challenging when compared to uncompressed images. Information redundancy in JPEG images is less when compared to uncompressed images. So, the process becomes more complicated for JPEG images. When a stego image undergoes incidental alterations, it may be possible to extract the hidden information, but the recovery of the cover image becomes impossible. Hence developing a robust reversible data hiding method in the future would be more useful.

3.8　CONCLUSION

Maintaining the stego image indiscernible is still a worry in reversible data hiding. Most of the data hiding schemes are vulnerable to attacks. Much work has been done in the basic approaches like histogram shifting, difference expansion, prediction and interpolation. Almost all the three basic approaches have achieved their peak performance. It's time to hit upon a new method considering the nature of the image and improving the quality of it by hiding the secret image which also preserves security of the covert information. Interleaving the secret message before embedding and deinterleaving after recovering the cover would be more secure for the covert message. RDH using machine algorithms is the preferred choice currently since the accuracy achieved is much better than for convolutional methods.

REFERENCES

A.M. Alattar. 2004. Reversible watermark using the difference expansion of a generalized integer transform, IEEE Transactions on Image Processing, vol. 13, no. 8, pp. 1147–1156.

C. De Vleeschouwer, J.F. Delaigle, and B. Macq. 2003. Circular interpretation of bijective transformations in lossless watermarking for media asset management, IEEE Transactions on Multimedia, vol. 5, no. 1, pp. 97–105.

C.C. Lin and N.L. Hsueh. 2008. A lossless data hiding scheme based on three-pixel block differences, Pattern Recognition, vol. 41, no. 4, pp. 1415–1425.

C.-F. Lee and Y.-L. Huang. 2012. An efficient image interpolation increasing payload in reversible data hiding, Expert Systems with Applications, vol. 39, pp. 6712–6719.

C.-S. Tsai, H.-C. Wu, Y.-W. Li, and J.J.-C. Ying. 2021. Applying GMEI-GAN to generate meaningful encrypted images in reversible data hiding techniques, Symmetry, vol. 13, p. 2438.

C.W. Honsinger, P.W. Jones, M. Rabbani, and J.C. Stoffel. 2001. Lossless recovery of an original image containing embedded data, U.S. Patent 6,278,791.

D.M. Thodi and J.J. Rodriguez. 2007. Expansion embedding techniques for reversible watermarking, IEEE Transactions on Image Processing, vol. 16, no. 3, pp. 721–730.

F. Peng, X. Li, and B. Yang. 2012. Adaptive reversible data hiding scheme based on integer transform, Signal Processing, vol. 92, pp. 54–62.

G. Xuan, Y.Q. Shi, P. Chai, X. Cui, Z. Ni, and X. Tong. 2007. Optimum histogram pair based image lossless data embedding, Proceedings of the International Workshop Digital Watermarking Series in Springer Lecture Notes Computing Science, pp. 264–278.

I.-C. Chang, Y.-C. Hu, W.-L. Chen, and C.-C. Lo. 2015. High capacity reversible data hiding scheme based on residual histogram shifting for block truncation coding, Signal Processing, vol. 108, pp. 376–388.

J. Cox, J. Kilian, T. Leighton, and T. Shamoon. 1997. Secure spread spectrum watermarking for multimedia, IEEE Transactions on Image Processing, vol. 6, no. 12, pp. 1673–1687.

J. Dittmann, M. Steinebach, and L.C. Ferri. 2002. Watermarking protocols for authentication and ownership protection based on timestamps and holograms, Proceedings of SPIE, no. 1, pp. 240–251,

J. Fridrich and M. Goljan. 2002. Lossless data embedding for all image formats, SPIE Proceedings of Photonics West, Electronic Imaging, Security and Watermarking of Multimedia Contents, vol. 4675, pp. 572–583.

J. Tian. 2003. Reversible data embedding using a difference expansion, IEEE Transactions Circuits and System Video Technology, vol. 3, no. 8, pp. 890–896.

J. Zhou and C. Oscar. 2012. Determining the capacity parameters in PEE-Based reversible Image watermarking, IEEE Signal Processing Letters, vol. 19, no. 5.

K.L. Chung, Y.H. Huang, W.M. Yen, and W.C. Teng. 2011. Distortion reduction for histogram modification based reversible data hiding, Applied Mathematics and Computation, vol. 218, no. 22, pp. 5819–5826.

M. Fallahpour. 2008. Reversible image data hiding based on gradient adjusted prediction, Institute of Electronics, Information and Communication Engineers Electronic Express, vol. 5, no. 20, pp. 870–876.

M.J. Weinberger, G. Seroussi, and G. Sapiro. 2000. The LOCO-I lossless image compression algorithm: Principles and standardization into JPEGLS, IEEE Transactions on Image Processing, vol. 9, no. 8, pp. 1309–1324.

M.U. Celik, G. Sharma, A.M. Tekalp, and E. Saber. 2005. Lossless generalized-LSB data embedding, IEEE Transactions on Image Processing, vol. 14, no. 2, pp. 253–266.

S. Lee, C.D. Yoo, and T. Kalker. 2007. Reversible image watermarking based on integer-to-integer wavelet transform, IEEE Transactions on Information Forensics and Security, vol. 2, no. 3, pp. 321–330.

V. Sachnev, H.J. Kim, J. Nam, S. Suresh, and Y.Q. Shi. 2009. Reversible watermarking algorithm using sorting and prediction, IEEE Transactions on Circuits and Systems for Video Technology, vol. 19, no. 7, pp. 989–999, July.

W. Hong. 2012. Adaptive reversible data hiding method based on error energy control and histogram shifting, Optics Communications, vol. 285, no. 2, pp. 101–108.

W. Hong, T.S. Chen, and C.W. Chiu. 2009. Reversible data hiding for high quality images using modification of prediction errors, Journal for Systems and Software, vol. 82, pp. 1833–1842.

W.L. Tai, C.M. Yeh, and C.C. Chang. 2009. Reversible data hiding based on histogram modification of pixel differences, IEEE Transactions on Circuits and Systems for Video Technology, vol. 19, no. 6.

X. Gao, L. An, Y. Yuan, D. Tao, and X. Li. 2011. Lossless data embedding using generalized statistical quantity histogram, IEEE Transactions on Circuits and Systems for Video Technology, vol. 21, no. 8, pp. 1061–1070.

X. Li, B. Yang, and T. Zeng. 2011. Efficient reversible watermarking based on adaptive prediction error expansion and pixel selection, IEEE Transactions on Image Processing, vol. 20, no. 12, December.

X. Li, W. Zhang, X. Gui, and B. Yang. 2013. General framework to histogram shifting based reversible data hiding, IEEE Transactions on Image Processing, vol. 22, no. 6.

X. Wu and N. Memon. 1997. Context-based, adaptive, lossless image codec, IEEE Transactions on Communications, vol. 45, no. 4, pp. 437–444

X. Yang and F. Huang. 2022. New CNN-based predictor for reversible data hiding, IEEE Signal Processing Letters, vol. 29, pp. 2627–2631, https://doi.org/10.1109/LSP.2022.3231193.

Y.C. Li, C.M. Yeh, and C.C. Chang. 2010. Data hiding based on the similarity between neighboring pixels with reversibility, Digital Signal Process, vol. 20, no. 4, pp. 1116–1128.

Y. Hu, H.-K. Lee, and J. Li. 2009. DE-based reversible data hiding with improved overflow location map, IEEE Transactions on Circuits and Systems for Video Technology, vol. 19, no. 2.

Y.-H. Yu, C.-C. Chang, and Y.-C. Hu. 2005. Hiding secret data in images via predictive coding, Pattern Recognition, vol. 38.

Y.Y. Tsai, D.S. Tsai, and C.-L. Liu. 2013. Reversible data hiding scheme based on neighboring pixel differences, Digital Signal Processing, vol. 23, pp. 919–927.

Z. Ni, Y.Q. Shi, N. Ansari, and W. Su. 2006. Reversible data hiding, IEEE Transactions on Circuits and Systems for Video Technology, vol. 16, no. 3, pp. 354–362.

Z. Zhang, G. Fu, F. Di, C. Li, and J. Liu. 2019. Generative reversible data hiding by image-to-image translation via GANs, Security and Communication Networks, vol. 2019, Article ID: 4932782

Z. Zhao, H. Luo, Z.M. Lu, and J.S. Pan. 2011. Reversible data hiding based on multilevel histogram modification and sequential recovery, International Journal of Electronics and Communication, vol. 65, pp. 814–826.

Z.H. Wang, C.F. Lee, and C.Y. Chang. 2013. Histogram shifting-imitated reversible data hiding, Journal of Systems and Software, vol. 86, no. 2, pp. 315–232.

4 The Intelligent Research Laboratory

Artificial Intelligence/ Machine Learning Methods for Chemists

Bhawani Narayan and Vinod Seshadri

4.1 INTRODUCTION

A scientific research laboratory is the hub of ideas and birthplace of many technological inventions. Underpinning many branches of science, chemistry as a subject has revolutionized the modern world, giving livelihood and employment to a considerable fraction of the human population. A deep knowledge of chemistry is required in multiple industrial sectors, for example, pharmaceuticals, drug discovery, cosmetics, food, breweries, clothing, polymers, metallurgy, cement, oil, semiconductors, to name a few (Weissermel and Arpe 2003). Moreover, many emerging branches of modern science, such as nanotechnology, chemical biology, laser technology, food technology, space technology and materials science require quintessential knowledge of chemistry (Yue et al. 2023). It is therefore required that research in chemical science gets a boost. Being the study of elements and compounds and the reactions that transform them, chemistry has a vast scope and is complicated in nature. It encompasses the properties and reactions of both nonliving matter and living matter. Chemistry can, thus, be defined as 'nonlinear' in nature, wherein defined sets of rules coexist with conditional exceptional behavior of elements and compounds. Two compounds may react to form a whole new unpredictable compound by subtle tuning of reaction conditions. Chemistry thus embraces rigid universal rules, such as physics and mathematics, and is yet open to the behavioral anomalies of matter. It therefore requires years of expertise in human knowledge and skills to master this subject. A perfect balance of deep knowledge of the subject, extensive hands-on experience and a subtle 'chemical intuition' is essential to be capable of performing research in chemistry.

About half a century ago, E.J. Corey, a well-renowned chemist, realized that "humanity and creativity bring with them certain unavoidable short-sightedness and prejudices" (Corey and Wipke 1969). It was then that the idea of 'formalization' and 'automation' of laboratory chemistry was conceived. In the late 1970s, there were enthusiastic attempts to employ AI methods to facilitate the synthesis of organic

DOI: 10.1201/9781003405436-5

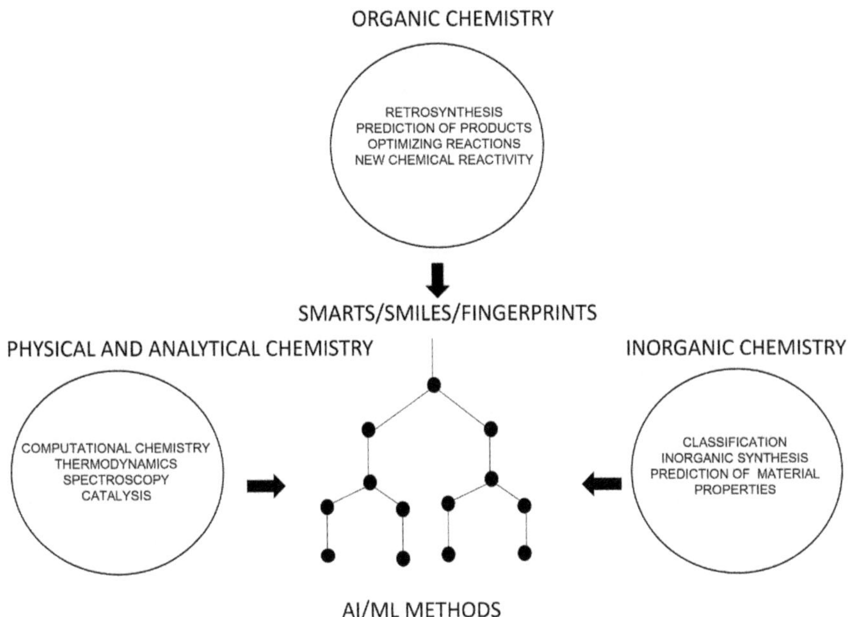

FIGURE 4.1 Schematic showing various problems in chemistry that can be solved by artificial intelligence/machine learning (AI/ML) methods.

molecules (Eakin and Warr 1977; Sridharan 1977). 'Organic synthesis' (Nantermet 2016; Nicolaou and Chen 2009; de Almeida et al. 2019; Baskin et al. 2017), which lies at the heart of chemistry, is one the most cumbersome and time-consuming processes. The synthesis of new molecules, though tedious, creates breakthroughs in the world. Earlier attempts to automate chemical synthesis did not meet with success due to lack of data, appropriate algorithms, and computing power. This dormant seedling sprouted 50 years later and was embraced by pharmaceutical companies, accelerating drug discovery (Struble et al. 2020). With the availability of 'big data', advanced AI/ML (artificial intelligence/machine learning) algorithms, availability of open-source tools and databases, faster speed of computation and reduced hardware costs, it is now much easier to apply AI in chemistry. It is, therefore, not surprising that within a span of ten years, AI methods have become indispensable in the chemical research sector.

Various types of problems in Chemistry can be solved by AI in an efficient and unbiased manner (Figure 4.1.). Depending on the data at hand, various algorithms powered by fast computing speed can serve as a time-saving guide for an active researcher. AI has demonstrated huge success in assisting synthetic chemists with the automation of retrosynthetic analyses (Jiang et al. 2023; Ucak et al. 2022), prediction of optimum reaction conditions (Gao et al. 2018) and prediction of the products of a given chemical reaction (Fitzner et al. 2023) in an unbiased manner. Furthermore, AI is able to assist in the analysis of complicated spectral patterns and assign them to possible molecules which sometimes might be unperceivable by the human mind

(Kuhn 2022). With the current demand of quick research outputs, the human mind gets stressed with the vast expanse of available data and makes erroneous conclusions in mundane pattern recognitions. Although the ingenuity of the human mind is required for generating new ideas for impactful research, repetitive tasks with huge datasets can be assigned to computers. AI/ML-enabled tools not only enable computers to assist with repetitive jobs, but also can lead to useful predictions correlated with the experimental data many times in the recent past. This chapter highlights the utility of AI for bench and theoretical chemists. Researchers often fear the replacement of human intelligence with artificial intelligence. It should be understood that AI, as it stands today, is driven by data provided and generated by humans. AI cannot be run without the assistance of humans. Researchers should, therefore, embrace this innovation to seek assistance for their tasks instead of being intimidated by this technology. This chapter aims to demystify AI for bench chemists and aims at building a bridge between research and emerging technology for faster and better progress. We demonstrate the detailed applications of AI algorithms in various research areas of chemistry, such as organic, inorganic and physical chemistry. Besides, we also highlight the gaps existing in the field, hindering the rapid progress of chemistry, and suggest some remedial steps that could be incorporated for a stronger AI-chemistry collaboration.

4.2 AI/ML METHODS FOR SYNTHETIC ORGANIC CHEMISTRY

'Synthetic organic chemistry' is one of the most challenging branches of chemistry. Being one of the most useful areas for industrial applications, such as drug development, materials science, polymers and the textile industry, organic synthesis has taken an indispensable place in the field of chemical research (Fitzpatrick et al. 2016). Organic chemists often require years of practice to gain expertise in this field. The work requires a deep understanding of the language of organic chemistry, insights into structure-property relationships, retrosynthetic analysis, optimization of reaction conditions, prediction of products of a chemical reaction and finding 'greener' synthetic routes that can be achieved in the best economically possible manner (Anastas and Eghbali 2010). An organic chemist's mind, therefore, has to process the optimum correlation of databases of small molecules, reactants, reaction conditions, target molecule, commercially available chemicals, cost-effective, environmentally friendly synthetic routes to come up with a solution. It is here that AI has started assisting humans and made synthetic organic chemistry much easier than before. This section discusses the application of specific AI/ML methods in various stages of organic synthesis in detail.

4.2.1 EMBEDDING CHEMICAL MOLECULAR STRUCTURES
AND CHEMICAL REACTIONS

4.2.1.1 Embedding Chemical Molecular Structures

Natural language processing (NLP; Chowdhury 2003) is a branch of AI that deals with understanding and processing natural language data. NLP is widely used in internet search engines, parsing, automated summarization and translation, and extraction of information (Bangalore and Joshi 1999). However, as chemistry has a language of its own, it was a challenging task to be interpreted by traditional NLP methods directly.

Gryzbowski and coworkers formally established the similarity between English (a natural language) and organic chemistry in terms of 'text fragments' and 'molecular fragments' (Cadeddu et al. 2014). They showed how the 'words' to create a sentence in organic chemistry are the 'substructures of a variety of molecules' and not the functional groups (like hydroxy, methyl, ethyl), which determine the chemical reactivity of the molecule. To establish this, a variety of chemical structures were examined and the most commonly occurring substructures were analyzed.

Further, to identify information-rich bonds of organic molecules, a common NLP technique, tf-idf (term frequency–inverse document frequency) scoring of the chemical substructure was done. This approach is commonly used in NLP to determine the relevance of a word based on numerical statistics. The algorithm developed by Gryzbowski and coworkers based on td-idf scoring was applied to 68 molecules from the natural product database. The substructures with the highest td-idf score were also tested for retrosynthetic relevance. AI/ML methods indicated the top retrosynthetic strategies, which were then verified with human chemists. In 97% of the cases, there was coherence of results, which establishes the success of the tf-idf algorithm. Furthermore, the tf-idf scores of all the individual bonds/substructures in the overall structure were summed up and averaged to determine the scores of the most relevant bonds. This algorithm is applied in retrosynthetic software, for example, Chematica. However, due to the molecular structural features to be considered in each task, it often made the embedding complicated.

NLP-based approaches to applying AI/ML in organic chemistry have been very useful in advancing the domains of drug discovery and biomedical chemistry, but a discussion of these topics is beyond the scope of this chapter. A more intuitive expert rule-based means of encoding a chemical structure is through the SMILES (Simplified-Molecular-Line-Entry-System), which provides a notation for the identification of substructural features and atom typing. Similar to natural language, SMILES uses a linear string of symbols. The atoms are represented by their symbols, normal carbon by capitalized 'C' and aromatic carbon by lowercase 'c'. Attached hydrogens are bracketed. For longer structures, the hydrogens are omitted and only the other atoms (carbon and heteroatoms) are shown. This notation, designed by Weininger in the 1980s (Weininger et al. 1988), provides a human-computer interface to understand chemistry. Humans find it simple as it has very few rules of notation. In terms of computers, SMILES provides a tree that can be understood in a single pass. The language syntax is easy to follow. For example, the structure of morphine can be stored with just 41 characters instead of 1,000 to 2,000 characters, thus making computation very efficient (Figure 4.2a). There is a 100-fold reduction in the processing times over conventional table formats used by CAS or MOL software. Over the years, specialized SMILES packages have been developed through AI/ML algorithms. For example, BigSMILES pertaining to macromolecules (Lin et al. 2019), CurlySMILES to describe supramolecular structures and nanodevices (Drefahl 2011), CXSMILES, especially as a storage for special features of molecules (ChemAxon.com), OpenSMILES, popular with pharmaceutical

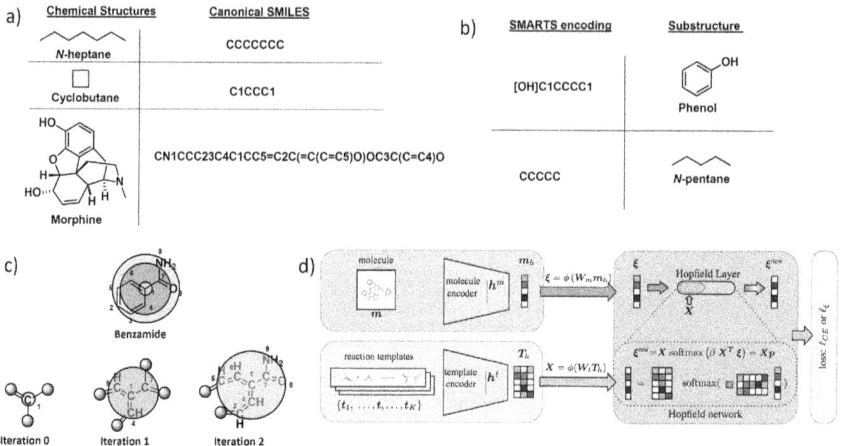

FIGURE 4.2   AI/ML methods for embedding molecular structures. (a) Canonical SMILES of some chemical structures; (b) corresponding substructure results for SMARTS encoding; (c) ECFP of benzamide as a simple case study; (d) detailed schematic diagram of reaction template developed by Seidl and coworkers. See text for explanation of acronyms. (Reproduced with permission from Seidl et al. [2022]. Copyright 2022 American Chemical Society.)

industry catering to the stereochemistry and chirality of molecules (opensmiles.org), DeepSMILES (O'Boyle et al. 2018) and SELFIES (Krenn et al. 2020) designed for ML applications, and canonicalization algorithms (O'Boyle 2012; Schneider et al. 2015). 'Tokenization', an important step taken to train NLP models, has been used for partitioning molecular structures based on SMILES strings. Cheminformatics primarily encompasses atom-wise tokenization to train models based on chemical language. A general drawback of these tokens is their easy degeneration. Recently, atom-in-SMILES (AIS) tokenization was proposed wherein token spaces of AIS and SMILES are created and both 'one-to-one' and 'onto' mapping is done. A function $f(x)$ pulls the atoms either as a central atom or as an identity operator in the token space. $f(x)$ is an invertible function, enabling the recovery of SMILES strings from SMILES projection. The algorithm (Ucak et al. 2023) involves iteration over all the atoms of a SMILES string producing atom-environment relevant variants.

While SMILES focuses on embedding a molecular structure efficiently, SMARTS (SMILES Arbitary Target Specification) generates a precise textual representation for substructures of molecules. SMARTS patterns are similar to drawing subgraphs from a molecular graph (Sayle et al. 1997). Many a times, a substructure identification in the synthetic course of a molecule opens up many different ways of synthesis (Figure 4.2b). This is the utility of SMARTS over SMILES. SMARTS enables substructure identification, which is then matched with various databases to find the most effective route (Coley et al. 2017b).

In the initial phases of molecular embedding, isomorphism (molecules with the same formula but different atom numberings) posed a challenge. The Morgan fingerprint was developed to solve this problem (Morgan 1965). The Morgan algorithm uses an iterative process to assign unique numbers for each atom, which is encoded into an initial atom identifier (Capecchi et al. 2020). The process is repeated over multiple iterations, generating identifiers independent of the numbers originally assigned to the atoms. The iterations are performed until unique numbers are assigned to every atom (limited by the symmetry of the molecule). The results of intermediate iterations are discarded. A variant of this algorithm, Extended Connectivity Fingerprints (ECFPs; Rogers and Hahn 2010) runs for a predetermined number of iterations, storing the results of the intermediate iterations into a set (Figure 4.2c). This set is the basis for determining the ECFP of molecules. The ECFP algorithm, unlike the Morgan algorithm, gives scope for algorithmic optimizations. ECFP reduces computational effort by avoiding rigorous canonicalization. The rapidly generated topological fingerprints are useful for determining structure-activity relationship. A major advantage of ECFPs is their ability to represent intricate molecular features, such as stereochemistry.

The use of 'reaction templates' has been very useful in embedding molecules and entire chemical reactions. In this approach, a template is created. This is either hand-coded by synthetic chemists or extracted from chemical reaction databases. A template-relevance prediction is then carried out by training the predictor with ML methods based on a database of previously recorded reactions. Template-based approaches visualize different templates as categories and the overall task as a classification task. The template-based approach is, however, unsuccessful in predicting rare reactions, due to the unavailability of large training datasets to develop ML algorithms.

A recent model developed by Seidl and coworkers (Seidl et al. 2022) extracts information about the templates instead of considering them as categories. Using the Modern Hopfield Network (MHN), this algorithmic model helps in the prediction of less experimented reactions and in the generalization of reaction templates. In the traditional template-based methods, templates are treated as separate categories, and a neural network is used to predict templates based on the molecule representation (Figure 4.2d). The neural network has a matrix W that maps the molecule encoder output to predictive scores for each template. However, this approach doesn't consider similarities between templates, which limits its ability to generalize across different templates. Additionally, if there are many unique templates or rare templates in the dataset, they don't contribute effectively to the model's performance. To address these limitations, (Seidl et al. 2022) propose an alternative approach that involves using a template encoder, represented by h^t, to map each template to a vector of size d^t. These encoded template vectors are then concatenated row-wise to form a new matrix T_h, representing the relationship between templates and their encoded representations. If the molecule encoder output size d^m is the same as the template encoder output size dt, then the matrix W can be replaced with Th, and the predictive scores for templates can be obtained using the equation:

$$\hat{y} = \mathrm{softmax}(T_h * h^m(m)) \quad \ldots \ldots \ldots \ldots \ldots \qquad (4.1)$$

This new equation associates each molecule m with each template using the dot product between their encoded representations. By doing so, the model can leverage the structural information of the templates to better represent the classes and learn from similarities between different templates. This approach allows for better generalization across templates and can effectively utilize information from unique or rare templates. To make this association between molecules and templates, the authors adapt modern Hopfield networks. Hopfield networks are a type of recurrent neural network that can learn and retrieve patterns based on their associations. In this context, Hopfield networks help establish the relationship between the molecular representations and the encoded templates, allowing the model to generalize effectively and improve template prediction.

4.2.1.2   Embedding Chemical Reactions

Embedding a representation for a chemical reaction, though challenging, is extremely important for planning a chemical synthesis. Three main approaches have been developed to represent a chemical reaction. In the first approach, called the 'Dugunji-Ugi matrix formalism', the chemical reactions are represented by 'R-matrices'. The reactants and products are represented as bond-electron (BE) matrices, wherein B matrices represent the reactants and E matrices represent the products. A primary requirement for this method is that the number of atoms in the reactant and product must be equal and each atom in the reactant must be mapped to each atom in the product. Based on an atom-to-atom mapping, this formalism yields the R-matrix (reaction matrix) by subtracting the E-matrix (product matrix) from the B-matrix (reactant matrix) (Dugunji and Ugi 1973). This method was used to plan synthetic pathways, classify reaction types, search for the shortest route (distance) to reach the product and design new reactions. However, this method has its limitations. The Dugunji-Ugi formalism lacks a compact representation for storing chemical reactions in databases. As chemical reactions can involve multiple reactants, reagents, catalysts and products, the encoding may result in inefficient data storage and retrieval. Due to the absence of well-defined indexing and search mechanisms, querying chemical reactions encoded using the Dugunji-Ugi formalism may lead to suboptimal search capabilities, resulting in slower and less accurate retrieval of specific reactions or reaction patterns. The Dugunji-Ugi formalism primarily focuses on the representation of chemical reactions and lacks explicit provisions for modeling reactivity accurately. This limitation restricts its ability to provide insights into the underlying mechanisms and predict reaction outcomes effectively. Unlike other reaction formalisms, the Dugunji-Ugi model does not inherently support the definition of reaction rules or patterns that describe commonly observed transformations. The absence of such rules may impede pattern matching, classification and predictive capabilities. Moreover, this method does not provide a direct method to calculate reaction descriptors.

With the second approach, a chemical reaction is encoded as a combination of numerical values representing the structural changes during a chemical reaction. These are called molecular descriptors (Hu et al. 2012). This approach overcomes the shortcomings of the first model by providing opportunities for efficient storage of reactions in databases. The major drawback of the 'molecular descriptor' approach is its inability to provide information about the reaction center. The third and the most widely used approach is the 'condensed graph of reaction' method (Bort et al. 2021).

The focus of this method is to represent a reaction as a graph of the reaction center and the key atoms (atoms that undergo transformation during a chemical reaction). It involves superimposition of the key atoms of the reactant onto the product and marking the changes in the bonds between them during the reaction.

4.2.2 AI/ML Methods in Automation of Retrosynthetic Analysis

'Retrosynthesis' or the 'disconnection approach' is utilized by synthetic chemists to break down a target molecule into 'synthons' from which commercially available reactants can be determined (McCowen et al. 2020) This approach relies on recognizing the sites of 'potential disconnection' of the molecule that could be easily formed by a chemical reaction. The increasing complexity of the target molecules makes this task more challenging.

The first application of computer-assisted retrosynthetic planning was in 1960 by E.J. Corey and coworkers, who introduced LHASA (Corey et al. 1972). However, this was limited by a lack of computational resources. Since the last decade there have been outstanding improvements in this area.

Computer-assisted retrosynthesis involves an expert system comprising a knowledge base and an interface engine. As retrosynthesis is a breakdown approach, the representation of knowledge bases involves transformations inverse to chemical reactions, that is, conversion of products to reactants. The input for these methods is the chemical structure of the target molecule and the expected output is a retrosynthetic analysis tree. The target molecule forms the root of the tree and the plausible reactants forms the leaves. If more than one synthetic pathway is available, the alternative pathways form the branches. Retrosynthesis does not consider the mechanism of organic reactions. Rules governing chemical transformation occupy the central position in this method. These rules may be empirical (rules defining specific chemical transformations) or nonempirical (rules defining general reaction principles). While the LHASA program implements the empirical approach, with five general reaction descriptors, the EROS (Gasteiger et al. 1987) program implements the nonempirical approach. Gryzbowski and coworkers developed a commercial retrosynthetic software called Chematica, published in 2017 and purchased by MERCK in 2017. It is now called SYNTHIA (Klucznik et al. 2018). This software is capable of performing retrosynthetic analysis, suggesting the commercial reactants for a particular chemical reaction, predicting the products of a chemical reaction and suggesting economical methods for industrial scaling up of organic target molecules. The software also performs batch processing of queries. It works on tree search algorithms and undisclosed AI heuristics.

Modern AI-based techniques used in chemical retrosynthesis are broadly classified into template-based methods and template-free methods.

Template-based methods use a library of known reaction templates to break down a target molecule into its component parts. The AI model then learns to match the target molecule to a set of templates and to predict the reactants that would be required to produce the target molecule from those templates.

One example of this is the work done by Segler and coworkers, who implemented a Monte Carlo tree search (MCTS)-based algorithm to achieve this. It was trained

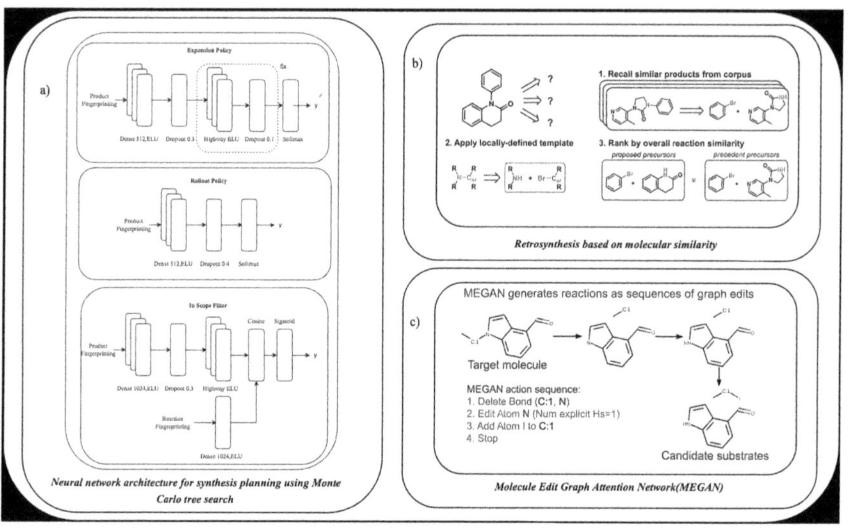

FIGURE 4.3 Various AI-based retrosynthetic approaches. ([b] reproduced with permission from Coley et al. [2017], Copyright 2017 American Chemical Society; [c] reproduced with permission from Sacha et al. [2021], Copyright 2021 American Chemical Society.)

on a dataset of 12.5 million single-step reactions (Segler et al. 2017). It is capable of predicting any step of a multistep synthesis. This program dissects the target molecule sequentially until it reaches the commercially available reagents. It is built on three coordinated neural networks that implement an MCTS (Figure 4.3a). The first neural network (expansion policy neural network), which is a deep highway network, creates the expansion policy determining which are the best transformations for a given product or rather best reactions for a given product. The second neural network, which is the rollout policy network, estimates the next best position to explore among the various paths determined by the expansion policy. The third neural network (in policy filter neural network), which is a binary classifier, effectively acts as a filter that filters out nonprobable and nonsensical reactions. Using these three networks, the MCTS was performed by iterating over four phases: (1) the selection phase wherein the promising node is picked up for analysis; (2) the expansion phase, where the potential best transformations are determined by the expansion policy neural network and the in policy neural network; (3) the rollout phase, where using the rollout policy neural network, a next best position is determined; and finally (4) an update phase, where the search tree is updated before moving back to the selection phase. This is repeated up to a maximum number of iterations to determine the synthesis plan.

Another innovative approach was suggested by Jensen and colleagues (Coley et al. 2017a). They proposed molecular similarity as an effective metric to use to select and rank the reaction precedents. Their approach involved three main steps: (1) identify similar products to the molecule under test; (2) using the template of those products, identify possible reactants and precedent reactions; and (3) the order of the most matching template can then be ranked based on both product similarity and the similarity with precedent reactions using which the final disconnection is generated (Figure 4.3b). The similarity was calculated using Morgan fingerprinting and the Tanimoto index (Coley et al. 2018). Morgan fingerprinting is a technique to encode molecular structures into binary vector representation by the enumeration of submolecular neighbourhoods. The Tanimoto index calculates similarity by normalizing the prevalence of overlapping substructures by the total number of unique substructures.

A drawback with template-based methods is that, considering there are a limited number of templates, the coverage of chemical reactions is limited by such methods.

Template-free methods, in contrast, do not use any preexisting reaction templates. Instead, they use a deep learning model to directly predict the reactants that would be required to produce the target molecule. This is accomplished by training the model on a large dataset of reactions.

One of the most popular approaches is to formulate retrosynthesis prediction as a neural network–based text sequence-to-sequence (seq2seq) learning that is used in complex NLP, such as language translation and summarization. This can be repurposed for tackling the retrosynthesis problem by representing the molecules in SMILES representation as input and the reactants in SMILES representation as the output. This has enabled researchers to take advantage of the latest advancements in the text seq2seq learning and apply them in the area of chemical retrosynthesis.

One of the significant milestones in the text seq2seq problem has been the introduction of transformers (Vaswani et al. 2017). Transformers enabled faster training and inference by eliminating the need for recurrent or convolutional operations, compared to earlier recurrence-based seq2seq approaches. Karpov et al. (2019) presented an approach to use transformers in the retrosynthesis, resulting in the model outperforming previous seq2seq model–based approaches.

A drawback of modelling retrosynthesis as sequence-to-sequence models is that they tend to generate a lot of nonplausible chemical reactions for a given input as there is no sequential or guided disconnection of the molecules to a plausible set of reactions.

A study by Tetko and colleagues (Tetko et al. 2020) showed that data augmentation helps to reduce the search space and invalid reaction rate and improve the accuracy of a transformer model for retrosynthesis, making it one of the best-performing template-free retrosynthesis approaches. Data augmentation has long been used in neural networks to improve performance of the models by eliminating the effect of data memorization during the prediction of new sequences. Data augmentation in the context of retrosynthesis involves creating synthetic data similar to known reactions, products and targets and feeding those as inputs to the transformers.

In a very different approach, Sacha et al. proposed a new template-free approach that uses molecular graph editing to predict reactants for retrosynthesis—MEGAN (Molecule Edit Graph Attention Network; Sacha et al. 2021). MEGAN is an encoder-decoder architecture that internally uses a graph convolutional network for predicting a sequence of graph edit actions on atoms and bonds of a chemical compound. Graph edit actions are a set of operations that can be performed on reactants to convert them into products. For instance, one edit action could be to add a bond between two atoms, or to remove an atom from the molecule.

The input to MEGAN is a molecular graph that is represented as bonds and atoms matrices. The basic block of the encoder and the decoder is an attention-based GCN (Graph Convolutional Network) layer called GCN-att. An attention-based GCN is a type of neural network that uses attention mechanisms to focus on the most relevant information in a graph. The encoder here creates a hidden representation of the input structure and is called just once at the start of the operation. The decoder sequentially generates a series of graph edit actions that creates the intermediate and final substratesuntil a Stop action is output. The final output is a series of graph edit actions, such as EditAtom, EditBond, AddAtom, AddBenzene and Stop (Figure 4.3c). The graph editing operations prevent the model from generating invalid predictions and allow it to explore chemical space more efficiently by avoiding the need to generate reactants from scratch.

4.2.3 AI/ML Methods in Prediction of Products of a Chemical Reaction

A successful planning of an organic reaction involves prediction of products. The prediction of reaction products is a critical task in organic chemistry, as it facilitates the design of new molecules and drug discovery. Chemical reactions generally follow certain principles of chemical reactivity, for example, regioselectivity and stereoselectivity to yield a molecule as the 'major product' of the reaction (Sanderson 1964). Chemists often plan synthesis out of expert knowledge and intuition, which may result in poor yields.

To improve the chances of success, the assistance of computational methods proves very helpful. Classically established computational methods have assisted scientists in understanding the molecular level picture of a chemical reaction for decades. These methods utilize empirical optimization and correlation between molecular structure and reactivity (Glossman-Mitnik 2013). Using energy minimization tools and concepts of molecular orbital compatibilities, theoretical chemistry has become an integral part of synthetic organic chemistry. In this regard, density functional theory (DFT) deserves special mention (Burke 2012). This approach involves investigating the electronic structure of molecules yielding physicochemical descriptors correlating with chemical reactivity. DFT involves various levels of computations, depending upon the complexity of the problem at hand. Many times, this method becomes unscalable and expensive for practical applications. Newer and more efficient methods, therefore, need to be explored.

AI/ML methods offer promising alternatives, leveraging large datasets and complex algorithms to predict reaction outcomes more efficiently. The following subsection discusses the development of AI/ML methods for the prediction of products of a chemical reaction.

4.2.3.1 Development of AI/ML Methods for Prediction of Products of a Chemical Reaction

AI/ML methods for predicting products utilize the molecular structures of the reactants and the conditions of the specific reaction at hand as the input data and yield the possible product(s) of the chemical reaction as the output. Though there is no

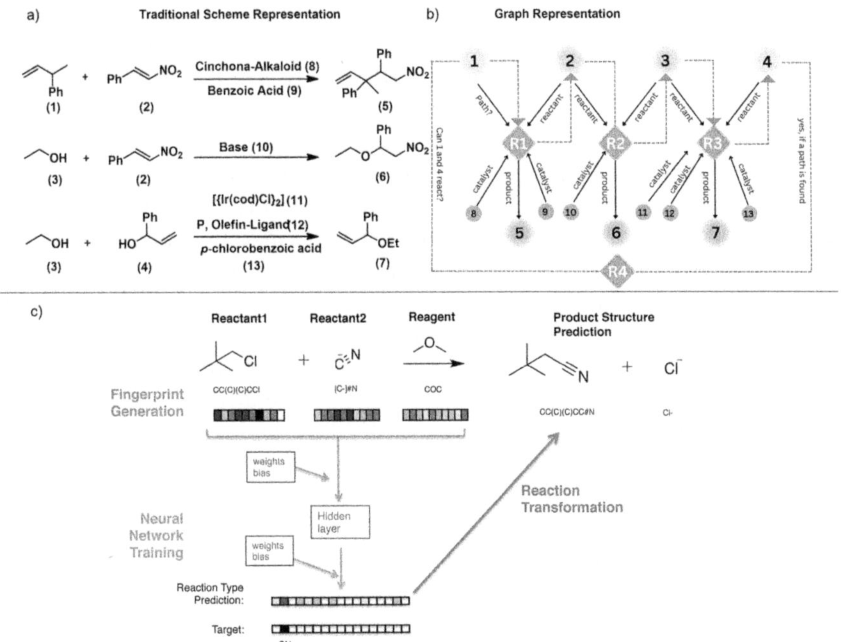

FIGURE 4.4 AI/ML methods in the prediction of products: (a) selected chemical reactions; (b) a knowledge graph for the prediction of products and reaction pathway; (c) use of concatenated molecular fingerprints to predict the products and feasibility of a sample reaction. ([c] reproduced with permission from Wei et al. [2016].)

'perfect method' available, the development of various AI/ML models for predicting products of a chemical reaction is commendable.

Segler and Waller (Segler et al. 2017) proposed a model mimicking chemical reasoning. Knowledge graphs were created based on 8.2 million binary reactions. The products were identified as missing links in the knowledge graph. Similar to MEGAN discussed in Section 4.2.2, the model constructs graphs where molecules (represented by nodes) are connected by reactions (edges). This model is capable of predicting a chemical reaction, the product of a chemical reaction and reaction conditions. To achieve a more detailed depiction of reaction roles, employing a directed bipartite graph, denoted as $G = (M, R, E)$, emerges as a highly advantageous approach (Figure 4.4a, b). Within this representation, M corresponds to the set of molecule nodes, $m_i \in M$, while R denotes the set of reaction nodes, $r_j \in R$. Notably, this traditional framework is expanded to encompass edges, $e_k = (mi, rj, t) \in E$, which bear an additional type 't', representing the role (reactant/reagent/catalyst/solvent/product) played by a molecule, m_i, within a reaction, r_j. Crucially, edges with the first four designated roles are directed from the respective molecule towards the associated reaction, enabling clear delineation of molecular involvement. Conversely, when a molecule assumes the role of a reaction product, the edge originates from the reaction node, pointing towards the molecule. To predict the molecular structure of products resulting from a predicted reaction involving two molecules, m_1 and m_L, the concept of half-reactions is employed. In this approach, a binary reaction is disassembled into two separate half-reactions, each corresponding to one of the reactant molecules. These half-reactions encompass vital information concerning the transformations and merging of atoms and bonds in the reactant, culminating in the formation of the product. Subsequently, the combination of two matching half-reactions leads to the reconstruction of the 'full' reaction. Subsequent to the establishment of a valid path, $iL = (m_1, r_A, \ldots, r_Z, m_L)$, linking the two molecules, m_1 and m_L, wherein the path length $L = 4n + 2$ (with n denoting the complementary reactivity), the determination of the potential product structure follows a meticulous protocol. Initially, the pertinent half-reaction, h_1, concerning the molecular entity m_1 in reaction r_A, is extracted, alongside the corresponding half-reaction, h_L, attributed to molecule m_L in reaction r_Z. The structure of the possible product is generated by merging h_1 and h_L. This 'pathfinding model' was successful on 180,000 chemical reactions. Apart from known reactions, this model can also predict the products of novel/less performed reactions. There are two major limitations of this model. First, this model has only been applied to binary reactions (containing two reactants). Second, the results of the model are heavily biased by the input database. With a knowledge base of eight million reactions, the model performs with an accuracy of 67.5%. Limiting the database to 'only' one million reactions, brings down the performance to 15%.

In the same year, Aspuru-Guzik and coworkers reported an AI/ML algorithm using a neural network to predict reaction types (Wei et al. 2016). They devised a neural fingerprinting method in which the reactants (input molecules) were converted to 'concatenated reaction fingerprints' using a convolutional neural network (Figure 4.4c). The ECFP and neural fingerprint method were utilized to generate the molecular fingerprints. These fingerprints were then fed to a neural network for classifying the reaction type. After the reaction type was predicted, the product was predicted using a SMARTS transformation associated with the reaction type. If the reactants (input) met the requirements of the SMARTS transformation, the predicted products are those generated as a result of these transformations. If the reactants do not meet the requirements of the SMARTS transformation, the predicted products by the algorithm are the same as the reactants, which implies that no reaction took place. This method was applied on 16 basic reactions involving alkyl halides and alkenes, yielding 80% accurate prediction of major product(s).

A major drawback of most of the studies (except that of Segler and Waller) is the consideration of only 'positive reactions' for the creation of the knowledge base for the algorithms. The available literature has a huge bias of only 'high yielding' reactions. Unsuccessful reactions are neither reported nor published. Models based on these 'available reactions' fail to classify a reaction as 'successful' or 'unsuccessful. In 2018, Jensen and colleagues used a data augmentation strategy supplementing the reaction databases with failed plausible reactions from patents (Coley et al. 2017b). This enabled ML to generate the plausible virtual products which were passed through neural-network architecture for determining the likelihood of the proposed reactions wherein atom and bond level changes were represented as features. Overgeneralized forward reaction templates are applied to a pool of reactants. A set of chemically probable products are generated. Depending on the rates at which the products are produced, the ML model then solves a 'classification problem', wherein it classifies the reaction according to the scores generated. The product generated from the reaction of highest likelihood as classified by the ML model is considered to be the major product. The comparison of the scores is done in a softmax network layer. A major advantage of this model is the prediction of alternate products ('false products') besides the expected reported products in the literature ('true products'). The method was applied to multiple chemical reactions, such as the conversion of alcohol to acid chloride, carbamate to carbamide, Suzuki coupling and the Grignard reaction. The accuracy of the models ranged from 68.5% to 71.8%. Including the reaction conditions and using a convolutional neural network improved the model's accuracy to 85% and reduced the calculation time to 100 ms per molecule (Coley et al. 2019). The major limitation of this model is that it only considers the reactants (those molecules that contribute atoms to the product) and does not consider reaction conditions (temperature, catalyst, reactant concentrations, pressure, solvent[s] or time of reaction). Another limitation of the model is that it can identify the most popular reagents for a transformation but cannot exactly predict the pathways to make the reaction realistic and practical in a scalable manner.

Linear regression is a tool traditionally used to predict the product of a chemical reaction in industries and academia. This approach assumes a linear relationship between reactants (reagent descriptors) and products (output) and predicts the products based on the goodness of fit in a linear curve. The classical Hammett equation is based on this model. Selecting relevant reagent descriptors from a large multidimensional dataset is challenging for this model. To address the problem of predicting the products of chemical reactions involving a large, multidimensional dataset, Doyle and coworkers have reported a remarkably improved performance of a random forest model over the classical linear regression analysis model. The model was trained over 4,608 Pd-catalyzed Buchwald-Hartwig cross-coupling reactions of aryl halides with 4-methylaniline in the presence of potentially inhibitory additives (Ahneman et al. 2018). The descriptors required for the model (atomic, molecular and vibrational) were extracted from the text data files of Spartan calculations. The random forest model outperformed the regression model with an average R^2 of 0.83.

4.2.4 Applications of AI/ML Methods in Optimizing Reaction Conditions

The 'optimum conditions of a chemical reaction' refers to the best method of performing the reaction. Optimum conditions ensure the efficient conversion of reactants to products in a time-saving manner. The variables involved in a chemical reaction are solvent, temperature, reaction time and catalyst. In an experimental laboratory, reaction optimization is performed by addressing one variable at a time and is often driven by acquired experience and intuition. This approach is unsuitable for complex reactions, which require simultaneous tuning of multiple reaction parameters. This section deals with AI/ML methods for the optimization of reaction conditions.

Before the rapid progress of AI/ML methods for optimizing reaction conditions, Struebing et al. (2013) proposed a systematic methodology to rapidly identify suitable solvents from a pool of multiple solvents by combining quantum mechanical calculation of reaction rate constants in a select number of solvents and a computer-aided molecular design procedure. By employing this approach, solvents that enhance reaction rates were identified from an extensive pool of potential molecules, ultimately leading to the identification of a high-performance solvent. The method combines rate constant calculations based on DFT and a linear regression model. Application of this model to the Menschutkin reaction, a well-known nucleophilic substitution reaction used to study solvent effects, enabled the selection of 9 out of 1,341 solvents that increased the rate constant of the reaction by 40%.

Data-driven approaches were soon employed to predict the optimum conditions for a chemical reaction. Marcou et al. (2015) developed a QSRR (quantitative structure reactivity relationships) model to predict the occurrence of Michael reactions with a set of given conditions (solvents and catalysts). The models were built using different ML algorithms, including SVM, naive Bayes and random forest. Additionally, various types of descriptors were employed in the model building process, such as ISIDA fragments derived from Condensed Graphs of Reactions, MOLMAP, Electronic Effect Descriptors, and Chemistry Development Kit. A threefold validation determined the performance of the model. The problem was conceived as a binary classification problem. The outcome is a reaction condition profile presented as a set of binary values: 1 indicates that the associated reaction condition is predicted to promote the reaction, while 0 suggests that it may be unfavorable for the reaction. The model was trained over 198 known reactions. The results obtained by employing the model were found to be highly accurate for the reactions (accuracy scale between 0.7 and 1). The drawback of the model was that only 8 out of 52 results were successful on an external test set (both solvent and catalyst predictions simultaneously matched with the experimental ones).

Apart from predicting retrosynthetic analysis and product of a chemical reaction, the model built by Segler and Waller (2017) was also successful in predicting the optimum conditions of a chemical reaction. The knowledge graph model (discussed in detail in Section 4.2.3.1) successfully predicted the feasibility of chemical reactions. A small knowledge graph comprising 30,000 reactions was built. Upon testing the method over 11 known reactions from the literature that were not in the knowledge graph dataset (validation dataset), the model was able to predict the same or similar reagent as reported in the literature for most of the reactions.

All of the AI/ML models discussed so far pertain to prediction of only reaction catalysts and solvents. The models were also focused on only a few reaction types. Jensen and coworkers (Gao et al. 2018) developed an algorithmic model based on a neural network that was trained on roughly 10 million reactions. The dataset involved a broad selection of different classes of reactions. The predictive outputs of the model included temperature of the reaction along with catalysts, reagents and appropriate solvents. The results were evaluated both qualitatively and quantitatively, generating better accuracy metrics for use. The model structure comprises a neural network that takes two inputs: the reaction fingerprint and the product fingerprint (Figure 4.5). These fingerprints are concatenated and passed through connected layers. This generates a dense fingerprint. This dense fingerprint predicts the catalyst upon passing through two connected layers. A one-hot vector generated from the catalyst prediction is again merged with the dense fingerprint and passed through fully connected

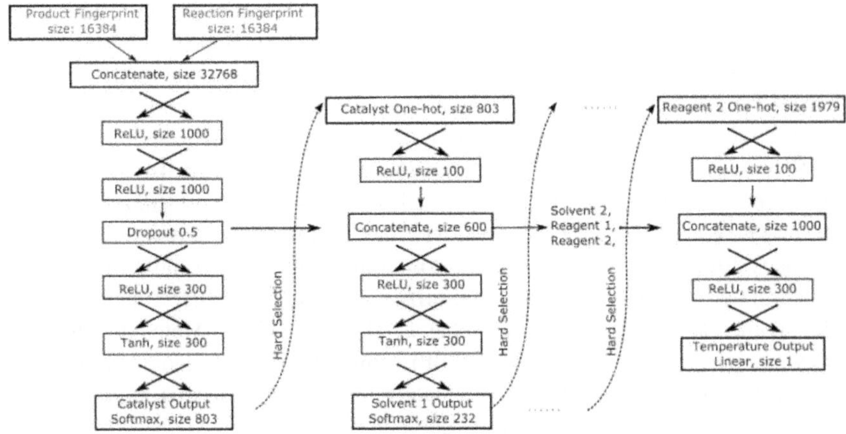

FIGURE 4.5 AI/ML algorithm structure for optimizing reaction conditions. (Reproduced with permission from Gao et al. [2018], Copyright 2018, American Chemical Society.)

layers to generate the first predicted solvent. This step is repeated to generate all the reagents and solvents for the reaction. The one-hot vectors of the generated catalyst, solvents and reagents are then concatenated and passed through fully connected layers to predict the temperature of the reaction. Though the validation accuracies of predictions vary, this model was able to yield all the suitable conditions of a given reaction. The model was tested on one million reactions outside the training set and predicted closely at least one solvent and reagent matching with the experimental data in 69.6% of the cases.

Zhou and colleagues utilized reinforcement learning via a continuous feedback method to optimize reaction conditions (Zhou et al. 2017) using a model called deep reaction optimizer (DRO). Reinforcement learning is a type of machine learning where an agent learns an optimum policy function that guides the optimum actions to be taken in order to maximize 'rewards' that are awarded upon taking each action. In the context of reaction optimization, the 'environment' is the reaction, 'state' refers to the reaction condition, 'actions' are the changes to reaction conditions and 'rewards' is the yield. The idea is to repeatedly run an experiment under a specific condition (state) and then record the outcome (reward) based on which an action is selected. As part of the training, the policy function will then use all of the historical data on which conditions led to which outcomes to define a policy that will tell what experimental condition to use next. The DRO uses a recurrent neural network to train the policy function. The recurrent neural network is a type of neural network that is able to capture the temporal events of the inputs. The DRO results were compared with several other state-of-the-art optimization algorithms, such as Covariance Matrix Adaptation Evolution Strategy (CMA-ES) and it outperformed them with fewer steps to find the optimum conditions. Zhou and colleagues demonstrated the success of their model by application in four different synthetic reactions.

Techniques such as Bayesian optimization have also been applied extensively to optimizing reaction conditions. Bayesian optimization works by building a probabilistic model of the objective function (often a Gaussian process). This model is updated as more data is collected, and it is used to select the next point to evaluate. This process is repeated iteratively unitl the stopping criterion is met. 'Phoenics' (Probabilistic Harvard Optimizer Exploring Non-Intuitive Complex Surfaces), a Bayesian optimization method, was proposed by Aspuru-Guzik and coworkers in 2018 that uses a Bayesian neural network (BNN) to yield a probabilistic model (Häse et al. 2018). The uniqueness of the approach is to use BNNs to estimate the kernel distributions associated with the objective function instead of predicting the objective values directly. This reduces the cost of computing that is usually the drawback of a neural network-based Bayesian optimizer for data with higher dimensions.

In another advanced approach, Reker et al. designed LabMate.AI, which is a self-evolving algorithm (Reker et al. 2018) using an adaptive random forest model to select experiments at each step. This algorithm requires approximately ten data points to start with (200 to 100 times lesser than previous models), that consume only 0.04% of the total search space. The algorithm does not require any pretraining or expert-based rules. The parameters of the random forest model dynamically change with each reaction to give the optimum output. As a case study, the Ugi reaction (three-component reaction) was chosen. This reaction, though extremely useful, is known to have less than 50% yield. For an initial training set, ten random conditions were provided to Labmate.AI. The method could then predict which conditions favour good yield and which do not. Moreover, it could predict ideal conditions for high yield of Ugi reactions. When this model was applied to C-N cross coupling reactions in drug development, it resulted in optimizing conditions that led to average yield improvements by 1.4 times the current yield. By exhaustively optimizing hyperparameters, including the number of trees, tree depth and number of features, over 600,000 decision trees were screened to construct a robust prediction model. Employing ten-fold cross-validation ensures the model's reliability and generalization. The model then predicts conversion values for unexplored reaction conditions, enabling the selection of the most promising experiments. An innovative exploration approach, emphasizing high prediction variance, drives model refinement during the initial iterations. Subsequently, an exploitative strategy takes charge, optimizing the target reaction value to achieve chemical reaction perfection. LabMate.AI's pioneering methodology exemplifies the potent synergy between AI and chemistry, unlocking unprecedented opportunities in the realm of chemical experimentation and discovery.

4.2.5 AI/ML Methods in Exploring Novel Reactivity

Serendipity, or the accidental occurrence of events, has led to repeated breakthroughs in all fields of science. Apart from expert-rule-based tasks, AI/ML can be utilized for exploring reactivity spaces, leading to surprising results. An under-researched aspect of AI/ML is the 'formalization of serendipity'.

A recent work by Granda et al. (2018) utilized coupling of a robot capable of performing organic synthesis with spectroscopic analytical techniques and feedback mechanisms. This was configured to perform 6 reactions in parallel and limited to perform 36 reactions per day. After performing the reaction, the products pass through spectroscopic tools to obtain the recorded spectra of the products. Then the robot uses an ML algorithm to solve a classification problem of whether the reaction was successful or not. The model was trained over 72 organic reactions, and could predict the occurrence of a reaction with an accuracy of 86%. By leveraging machine learning and binary encoding of chemical inputs, this intelligent system enables rapid and efficient execution of chemical reactions and analysis, outpacing manual processes. Moreover, the robot exhibits remarkable capability in predicting the reactivity of potential reagent combinations after performing a limited number of experiments, effectively navigating the vast chemical reaction space. The incorporation of nuclear magnetic resonance (NMR) and infrared spectroscopy allows real-time assessment of reactions, facilitating swift decision-making. Impressively, the ML system demonstrates exceptional predictive accuracy, surpassing 80% for approximately 1,000 reaction combinations after examining only 10% of the dataset. This methodology also extends to calculating the reactivity of published datasets. The successful validation of predictions by manual experimentation led by a chemist culminated in the discovery of four novel reactions. The algorithm employs linear discriminant analysis (LDA), which classifies which combinations would yield a product. The ML algorithm involved the creation of molecular descriptors, similar to one-hot encoding in vector form. ML simulations were performed to predict the feasibility of a chemical reaction. The sampling was random, with 50% of successful and unsuccessful reactions.

Another model using a neural network consisting of two layers, 50 neurons in the first layer and 7 in the second layer, was employed to investigate Suzuki-Miyayura reactions (Granda et al. 2018). To ensure the effectiveness and generalization of the neural network model, the dataset was carefully partitioned into three distinct subsets. The training set comprised 3,456 reactions, providing a substantial foundation for the neural network to learn from. The validation set, consisting of 576 reactions, served as a means of fine-tuning the model and tuning hyperparameters to prevent overfitting. Finally, the test set contained 1,728 reactions and was used to assess the neural network's performance on previously unseen data, offering an unbiased evaluation of its predictive capabilities. This strategic partitioning enabled robust training, validation and testing of the neural network, ultimately contributing to its reliable and accurate predictions in the domain of organic synthesis. Undoubtedly, this transformative fusion of AI and organic synthesis holds immense promise for accelerating chemical discovery and advancing the frontiers of science and technology.

4.3 AI/ML METHODS IN PHYSICAL, ANALYTICAL AND INORGANIC CHEMISTRY

AI/ML methods have proved very useful in all branches of chemistry. The major improvement being in the field of synthetic organic chemistry (discussed in detail in the previous section), other branches of chemistry too have immensely benefitted from the applications of AI/ML algorithms. This section focuses on a brief overview of AI/ML applications in physical, analytical and inorganic chemistry.

4.3.1 AI/ML Methods in Physical Chemistry

4.3.1.1 Incorporation of AI/ML Algorithms in Physical Computational Chemistry

Immense developments in physical chemistry have occurred in the past decade due to the rapid development of AI/ML algorithms. Incorporation of ML in quantum chemical calculations has improved the accuracy and speed of calculations.

AI/ML methods have immensely contributed to the charge delocalization problems using DFT in boosting the speed of calculations and calibrating the results obtained. Though there are multiple examples of applications of AI/ML assisting DFT calculations (Prezhdo 2020), a recent work in this regard is worth mentioning. Zhou and coworkers (Zhou et al. 2019) created a deep neural network to determine the exact exchange correlation potentials of molecules. Though the input needed only the data for small molecules, the algorithm was able to process and generate the parameters for large molecules. An important parameter to be kept in mind while designing the model was the easy transferability of the resulting neural network from small molecules to large molecules. As an exemplary case study, the model was trained and validated on the data for H_2 and HeH molecules (diatomic molecules). The model was capable of determining the ground state electron density of tri- and tetra-atomic species containing the atoms H and He (stretched HeH^+, linear H_3^+ and $H–He–He–H_2^+$.

4.3.1.2 Application of AI/ML Algorithms in Understanding Thermodynamic Processes

ML techniques have been used to generate reliable and effective potential energy surfaces (PES) for large-scale molecular dynamics. A very few case studies have focused on calculating the PES of nonadiabatic (processes that allow the exchange of heat) processes in the excited state. Two independent studies published in 2018 (Hu et al. 2018; Dral et al. 2018) have focused on different variants of surface-hopping dynamics (dynamics followed by molecules to hop between two potential energy surfaces) with an ML algorithm based on the Zhu-Nakamura method and kernel ridge regression (KRR). The Zhu-Nakamura method was used to avoid adiabatic couplings in the calculations. The latter method combined kernel trick and ridge regression, two concepts used in ML. The model was successful in predicting the PES of one-dimensional systems. However, it had severe limitations when applied on multidimensional systems. In this context, Cui and coworkers (Chen et al. 2018) have used deep-learning methods based on deep neural networks combined with the Zhu-Nakamura method to understand the nonadiabatic dynamics of methylene amine (CH_2NH). Their approach is based purely on ML methods. The summation of potential energies of a polyatomic molecule is used as the basis of the neural network. The atomic coordinates are transformed to vectors and used as input data for the deep neural network. Several hidden layers, with 240, 960, 480, 240, 60, 30 and 10 nodes are used to generate very accurate results. The model has been trained over 90,000 data points.

4.3.1.3 Application of AI/ML Algorithms in Predicting Catalytic Properties

An ML model was employed as an alternative to DFT in studying the gas adsorption properties of catalysts. Li and coworkers (Li et al. 2021) used a graph neural network (GNN) and Bayesian optimization to develop the algorithm. The dataset was divided into 90% training set and 10% validation set. The predictions are very close to the values generated by DFT calculations (highly accurate).

4.3.2 AI/ML Methods in Analytical Chemistry

4.3.2.1 Application of AI/ML Algorithms in Interpreting Nuclear Magnetic Resonance Data

Spectroscopy is a very important tool for the characterization of chemical compounds. Spectroscopy works on the principle of interaction of molecules with electromagnetic radiation. NMR, a spectroscopic technique, works on the resonance of spin active nuclei with radiofrequency. This interaction then generates signals, which are fingerprint identifications of a molecule. This technique is extremely popular for use in the characterization of organic molecules. NMR gives information about the purity of the sample. However, expert knowledge is required to read and interpret the data. In 2022, Rzepa and Kuhn (2022) proposed a methodical approach to collect NMR data from repositories, collecting the metadata by running queries so that this data can be read by a computer easily. Though there is no AI/ML model used in this work, it offers a pathway to researchers to obtain readable data so that advanced AI/ML methods can be applied to it. Recent advancements include the identification of molecules by key features of NMR data (Li et al. 2022) by using two different methods, first equidistant sampling and second peak sampling. SVMs and a K-nearest-neighbor model (KNN) ensured the correctness of feature classifications. Finally, the authors built a recurrent neural network (RNN) and proved its improved accuracy over conventional methods.

4.3.2.2 Application of AI/ML Algorithms in Interpreting Electron Paramagnetic Resonance Data

Taguchi and coworkers have designed a convolutional neural network (CNN) to extract information from the electron paramagnetic resonance (EPR) spectrum (Taguchi et al. 2019). The CNN was trained on 95,000 simulated data points from a software called HYSCORE. The neural network comprises layers to recognize peaks and their patterns; subsequently fully connected layers extract the desired parameters from the spectra. Complicated spectra containing multiple nitrogen atoms were resolved and presented as one nitrogen EPR signal at a time. Upon concatenating the HYSCORE spectrum with the experimental conditions, a tensor was generated that was passed through three convolutional networks. The CNN rendered unnormalized probability distributions as output.

4.3.2.3 Application of AI/ML Algorithms in Predicting Surface Enhanced Raman Spectral Data

In the same year, Hu and coworkers (Hu et al. 2019) came up with an ML protocol to predict surface enhanced Raman spectra (SERS) of *trans*-1,2-bis (4-pyridyl) ethylene (BPE) molecules adsorbed on a gold substrate as a case study. The ML model developed was based on a random forest algorithm, consisting of 300 decision trees with a maximum depth of 3, to predict frequencies and Raman intensities of a given system. The dataset used for training the model comprised 3,600 randomly selected data points, while 400 data points were reserved for testing using the scikit-learn framework. A Pearson correlation coefficient (r) of 0.95 between the ML predictions and the results obtained from DFT calculations validated the accuracy of the model. The high correlation indicated that the final ML model is capable of reproducing good results.

4.3.3 AI/ML Methods in Inorganic Chemistry

4.3.3.1 Pattern Recognition in Classification of Inorganic Compounds

AI/ML methods have been applied in inorganic chemistry since the 1970s. As inorganic chemistry deals primarily with classification of molecules into categories such as elements and compounds, acids and bases, pattern recognition, which was a developing method in AI/ML, was rightly applicable (Kowalski and Bender 1972). The nonparametric pattern recognition can be further subdivided into preprocessing and learning and classification. Preprocessing involves the numerical transformation of data, which can then be understood by computers for learning and classification. Using this approach, oxides of elements were classified into acidic, basic and amphoteric. The major features were valence, ionic radius, covalent radius, melting point, electronegativity and enthalpy of fusion. Since then, AI/ML methods have shown rapid progression in development.

4.3.3.2 AI/ML Algorithms to Optimize the Conditions for Inorganic Material Synthesis

In the last few years, many AI/ML algorithms have been developed for boosting research in the field of inorganic chemistry. Single-walled carbon nanotubes (SWCNTs) are highly promising materials in electronic devices owing to their high conductivity, heat resistance and stability. A transparent conductive film (TCF) based on SWCNTs is being developed to replace metal oxides in transparent electronics. The experimental methods have not been able to match the theoretically predicted values so far. In a detailed investigation, Krasnikov and coworkers (Khabushev et al. 2019) have employed ML methods to optimize the two key conditions (temperature and mole fraction of carbon dioxide) to obtain high-performing TCFs. The algorithm was developed using a support vector regression (SVR) model developed in scikit learn, which is an open-source Python library. The SVR model was chosen as it required low amounts of data and works well with nonlinear data, giving results close to the accuracy of artificial neural networks. The Gaussian radial basis function served as a kernel for the model. Various fittings were attempted using linear kernel fittings, leading to poor output, suggesting the nonlinear nature of the data. The Gaussian radial basis function $(\exp(-\gamma\|x-x_0\|)$ was found to be most appropriate as the regression fitted the input data as a linear composition of the radial basis function in every support vector. This was followed by hyperparameter optimization yielding cross-validation R^2 scores of 0.81. The conclusion derived from this work was that lower growth temperatures resulted in the growth of SWCNTs with higher conductivity, the optimum being 850 °C. The optimum carbon dioxide was found to be 0.5%.

4.3.3.3 AI/ML Algorithms for Predicting Inorganic Material Properties

The prediction of properties of materials with limited amounts of data through AI/ML methods has been reported by Yoshida and coworkers (Yamada et al. 2019). 'Transfer learning', an ML framework, can be used to overcome the issue of working with low amounts of data. The authors developed neural networks based on transfer learning. Usually, two methods are applied for neural transfer learning; the 'frozen featurizer' and 'fine-tuning'. Both these methods were employed. For each task, approximately 1,000 neural networks were generated, which involved different bootstrap datasets. With the applied algorithms, predictions of thermal conductivity of inorganic crystals were made with high degrees of accuracy.

4.3.3.4 AI/ML Algorithms to Answer Classification Problems in Inorganic Chemistry

In a recent work, Mar and coworkers (Selvaratnam et al. 2023) picked up questions in inorganic chemistry that could be addressed as classification problems, including whether the structure type of a given compound is rutile or fluorite, and whether a compound has superconducting properties or whether it would have ferromagnetic property or not. A decision tree (DT) approach was used to answer these questions. The classification and regression tree (CART) algorithm was used to implement the DT model. The Gini impurity was employed to evaluate the best features. The results were quite promising, with an accuracy of ~0.95.

4.4 CURRENT CHALLENGES IN AI/ML APPLICATIONS IN CHEMISTRY

4.4.1 Restricted Access to Data and Results

One of the biggest enablers in modern software development has been the proliferation of open-source-based software and tools. Open-source foundations like Linux Foundation (linuxfoundation.org) and Apache Software Foundation (Apache.org) have helped usher in rapid progress in software development. As computing and the software industry were the vanguards of the latest AI/ML revolution, this paradigm of free and open software, along with readily available code for new algorithms, helped bring in rapid progress in the field of AI/ML as well. Almost all the software frameworks used in the fields of machine learning and deep learning are open source, and the availability of large curated open datasets made it easier to find and replicate research papers readily with open-source code bases (Langenkamp and Yue 2022).

In contrast to this, the chemical sciences industry in general has long safeguarded its progress behind patents and licensed software. This has also affected the progress and implementations done with AI/ML algorithms in the chemical sciences (de Almeida et al. 2019; Struble et al. 2020). That said, there have been serious efforts in remediating this through open-source initiatives like ASKCOS (Automated System for Knowledge-based Continuous Organic Synthesis; askcos.mit.edu). But algorithms developed through only these initiatives still lack in performance and accuracy due to the nonavailability of training data, common benchmarks and an industry-wide standard for a programmatic interface.

4.4.1.1   Proprietary Training Data

AI applied in chemistry today relies mostly on curated commercial databases and the extraction of schemes from the literature with manual custom-made code.

Though there are publicly available data repositories such as Drug Bank, PubChem and ChEMBL, they are limited in their utility (de Almeida et al. 2019). The past literature has shown that retraining the models developed in open source with proprietary data has resulted consistently in giving better results on both accuracy and efficiency (Struble et al. 2020).

4.4.1.2   Common Benchmarks to Evaluate Efficacy

Though there are commonplace metrics to evaluate the efficacy of the models, a common benchmark across the industry, such as the GLUE (General Language Understanding Evaluation; Wang et al. 2018) for Natural Language Understanding (NLU) models, is lacking in the implementation of ML in chemistry. This restricts comparison of models and their improvements over previously developed models (Struble et al. 2020).

4.4.1.3   Industry-Wide Standard for a Programmatic Interface

AI tools used in chemistry more often than not provide a graphical user interface (GUI) that is not readily integrable with other components and systems. A lack of industry-wide standard for a programmatic interface restricts exploration and seamless integration with many of the in-house tools (Struble et al. 2020). This would also enable the development of automated synthesis platforms that can one day perform end-to-end planning and synthesis of the molecules.

4.4.2   Commercial Availability of Predicted Compounds by AI Algorithms

AI-driven chemistry does not stop just at the prediction of potential molecules and synthesis paths. The availability of the starting and the intermediate materials is very critical to the final synthesis of the molecule (Struble et al. 2020). Moreover, these influence the way models are created as well. The commercial availability of the starting materials can influence the stopping criteria in a search-based algorithm (Struble et al. 2020). This is further complicated by the fact that the database of buyable materials is often proprietary and close sourced, with the data augmented by the internal vendors and purchasers.

4.4.3   Data Issues

Due to the limited availability of high quality data, much of the AI development in chemistry depends on data that is extracted from research papers and patents, and curated. This brings bias into the data that in turn affects the quality of the predictions generated by the models trained on the data.

In most cases, only successful data is published; this results in an imbalanced dataset containing data from reactions yielding successful or high-yield outputs (Struble et al. 2020; Rodrigues 2019). Since most of the reactions are influenced by external factors such as the availability of specific solvents and chemicals in an organization's laboratory this results in literature that is skewed towards a specific set of solvents and reagents. Also, conditions for a particular reaction are often selected based on how often those are reported in previous experiments by human specialists, further propagating similar data (Beker et al. 2022). For these reasons, the literature shows selection bias towards common reactions such as alkylation, arylation and acylation. Also, many common errors such as invalid valences, concentration and incorrectly annotated tautomers are widespread in the data sources and papers (Rodrigues 2019). This restricts how generalized an algorithm developed on this data can get. Varying machine learning model techniques have also failed to prove useful in improving the results due to the above constraints (Beker et al. 2022). A way out of this would be to augment the data with standardized experiments under multiple conditions and multiple yields (Beker et al. 2022).

4.5 CONCLUSION

This chapter presents the application of AI/ML methods in chemistry. Though the methods discussed here are relevant to the problems in chemistry, these are of general applicability as well. The pharmaceutical industry has shown rapid progress by adopting AI/ML strategies. Other research and development sectors too, are gradually adopting these modern tools. However, the true revolution with AI/ML cannot be brought about until bench chemists in experimental laboratories embrace this technique. It must be understood that AI/ML can never replace human intelligence. When this understanding is internalized, the fear of AI would be replaced by rapid progress. Human-AI-robot interfaces must be welcomed with more openness to address the grand challenges of chemistry. An important step required in this regard is the availability of open-source data. Though chemists have created wonderful databases, such as chemical abstracts services (CAS) and Reaxys, commercialization is hampering growth in this field. The bright future of chemistry lies in the effective mutual collaboration of AI/ML and human intelligence.

REFERENCES

About ASKCOS. https://askcos.mit.edu/
About the Apache Foundation. www.apache.org/foundation/
About the Linux Foundation. www.linuxfoundation.org/about/
Ahneman, D. T., et al. 2018. Predicting reaction performance in C–N cross-coupling using machine learning, Science **360**(6385): 186–190.
Anastas, P. and N. Eghbali. 2010. Green chemistry: Principles and practice, Chemical Society Reviews **39**(1): 301–312.
Bangalore, S. and A. K. Joshi. 1999. Supertagging: An approach to almost parsing, Computational Linguistics **25**: 237–265
Baskin, I. I., et al. 2017. Artificial intelligence in synthetic chemistry: Achievements and prospects, Russian Chemical Reviews **86**(11): 1127.

Beker, W., et al. 2022. Machine learning may sometimes simply capture literature popularity trends: A case study of heterocyclic Suzuki–Miyaura coupling, Journal of the American Chemical Society **144**(11): 4819–4827.

Bort, W., et al. 2021. Discovery of novel chemical reactions by deep generative recurrent neural network, Scientific Reports **11**(1): 3178.

Burke, K. 2012. Perspective on density functional theory, The Journal of Chemical Physics **136**(15).

Cadeddu, A., et al. 2014. Organic chemistry as a language and the implications of chemical linguistics for structural and retrosynthetic analyses, Angewandte Chemie International Edition **53**(31): 8108–8112.

Capecchi, A., et al. 2020. One molecular fingerprint to rule them all: Drugs, biomolecules, and the metabolome, Journal of Cheminformatics **12**(1): 43.

ChemAxon Extended SMILES and SMARTS—CXSMILES and CXSMARTS—Documentation. https://docs.chemaxon.com/display/docs/chemaxonsmiles-extensions.md.

Chen, W.-K., et al. 2018. Deep learning for nonadiabatic excited-state dynamics, The Journal of Physical Chemistry Letters **9**(23): 6702–6708.

Chowdhury, G. 2003. Natural language processing, Annual Review of Information Science and Technology **37**: 51–89.

Coley, C. W., et al. 2017a. Computer-assisted retrosynthesis based on molecular similarity, ACS Central Science **3**(12): 1237–1245. https://pubs.acs.org/doi/10.1021/acscentsci.7b00355.

Coley, C. W., et al. 2017b. Prediction of organic reaction outcomes using machine learning, ACS Central Science **3**(5): 434–443.

Coley, C. W., et al. 2018. Machine learning in computer-aided synthesis planning, Accounts of Chemical Research **51**(5): 1281–1289.

Coley, C. W., et al. 2019. A graph-convolutional neural network model for the prediction of chemical reactivity, Chemical Science **10**(2): 370–377.

Corey, E. J. and W. T. Wipke. 1969. Computer-assisted design of complex organic syntheses, Science **166**(3902): 178–192.

Corey, E. J., et al. 1972. Computer-assisted synthetic analysis. Facile man-machine communication of chemical structure by interactive computer graphics., Journal of the American Chemical Society **94**(2): 421–430.

de Almeida, A. F., et al. 2019. Synthetic organic chemistry driven by artificial intelligence, Nature Reviews Chemistry **3**(10): 589–604.

Dral, P. O., et al. 2018. Nonadiabatic excited-state dynamics with machine learning, The Journal of Physical Chemistry Letters **9**(19): 5660–5663.

Drefahl, A. 2011. CurlySMILES: A chemical language to customize and annotate encodings of molecular and nanodevice structures, Journal of Cheminformatics **3**(1): 1.

Dugunji, J. and I. Ugi. 1973. An algebraic model of constitutional chemistry as a basis for chemical computer programs, Berlin, Heidelberg, Springer Berlin Heidelberg.

Eakin, D. R. and W. A. Warr. 1977. Computerized aids to organic synthesis in a pharmaceutical research company, American Chemical Society **61**: 217–226.

Fitzner, M., et al. 2023. Machine learning C–N couplings: Obstacles for a general-purpose reaction yield prediction, ACS Omega **8**(3): 3017–3025.

Fitzpatrick, D. E., et al. 2016. Enabling technologies for the future of chemical synthesis, ACS Central Science **2**(3): 131–138.

Gao, H., et al. 2018. Using machine learning to predict suitable conditions for organic reactions, ACS Central Science **4**(11): 1465–1476. https://pubs.acs.org/doi/10.1021/acscentsci.8b00357.

Gasteiger, J., et al. 1987. A new treatment of chemical reactivity: Development of EROS, an expert system for reaction prediction and synthesis design, Berlin, Heidelberg, Springer Berlin Heidelberg.

Glossman-Mitnik, D. 2013. Computational study of the chemical reactivity properties of the rhodamine B molecule, Procedia Computer Science 18: 816–825.

Granda, J. M., et al. 2018. Controlling an organic synthesis robot with machine learning to search for new reactivity, Nature **559**(7714): 377–381.

Häse, F., et al. 2018. Phoenics: A Bayesian optimizer for chemistry, ACS Central Science **4**(9): 1134–1145.

Hu, D., et al. 2018. Inclusion of machine learning Kernel Ridge regression potential energy surfaces in on-the-fly nonadiabatic molecular dynamics simulation, The Journal of Physical Chemistry Letters **9**(11): 2725–2732.

Hu, Q.-N., et al. 2012. Assignment of EC numbers to enzymatic reactions with reaction difference fingerprints, PLOS ONE **7**(12): e52901.

Hu, W., et al. 2019. Machine learning protocol for surface-enhanced Raman spectroscopy, The Journal of Physical Chemistry Letters **10**(20): 6026–6031.

Jiang, Y., et al. 2023. Artificial intelligence for retrosynthesis prediction, Engineering **25**: 32–50, ISSN 2095–8099. https://doi.org/10.1016/j.eng.2022.04.021.

Karpov, P., et al. 2019. A transformer model for retrosynthesis, Cham, Springer International Publishing.

Khabushev, E. M., et al. 2019. Machine learning for tailoring optoelectronic properties of single-walled carbon nanotube films, The Journal of Physical Chemistry Letters **10**(21): 6962–6966.

Klucznik, T., et al. 2018. Efficient syntheses of diverse, medicinally relevant targets planned by computer and executed in the laboratory, Chem **4**(3): 522–532.

Kowalski, B. R. and C. F. Bender. 1972. Pattern recognition. Powerful approach to interpreting chemical data, Journal of the American Chemical Society **94**(16): 5632–5639.

Krenn, M., et al. 2020. Self-referencing embedded strings (SELFIES): A 100% robust molecular string representation, Machine Learning: Science and Technology **1**(4): 045024.

Kuhn, S. 2022. Applications of machine learning and artificial intelligence in NMR, Magnetic Resonance in Chemistry **60**(11): 1019–1020

Langenkamp, M. and D. Yue. 2022. How open source machine learning software shapes AI. Proceedings of the 2022 AAAI/ACM Conference on AI, Ethics, and Society. Oxford, UK, Association for Computing Machinery: 385–395.

Li, C., et al. 2022. Identifying molecular functional groups of organic compounds by deep learning of NMR data, Magnetic Resonance in Chemistry **60**(11): 1061–1069.

Li, X., et al. 2021. Group and period-based representations for improved machine learning prediction of heterogeneous alloy catalysts, The Journal of Physical Chemistry Letters **12**(21): 5156–5162.

Lin, T.-S., et al. 2019. BigSMILES: A structurally-based line notation for describing macromolecules, ACS Central Science **5**(9): 1523–1531.

Marcou, G., et al. 2015. Expert system for predicting reaction conditions: The Michael reaction case, Journal of Chemical Information and Modeling **55**(2): 239–250.

McCowen, S. V., et al. 2020. Retrosynthetic strategies and their impact on synthesis of arcutane natural products, Chemical Science **11**(29): 7538–7552.

Morgan, H. L. 1965. The generation of a unique machine description for chemical structures-a technique developed at chemical abstracts service, Journal of Chemical Documentation **5**(2): 107–113.

Nantermet, P. G. 2016. Reaction: The art of synthetic chemistry, Chem **1**(3): 335–336

Nicolaou, K. C. and J. S. Chen. 2009. The art of total synthesis through cascade reactions, Chemical Society Reviews **38**(11): 2993–3009

O'Boyle, N. M. 2012. Towards a Universal SMILES representation—A standard method to generate canonical SMILES based on the InChI, Journal of Cheminformatics **4**(1): 22.

O'Boyle, N. M. and A. Dalke. 2018. DeepSMILES: An adaptation of SMILES for use in machine-learning of chemical structures. ChemRxiv. https://doi.org/10.26434/chemrxiv.7097960.v1

OpenSMILES. Home page. http://opensmiles.Org

Prezhdo, O. V. 2020. Advancing physical chemistry with machine learning, The Journal of Physical Chemistry Letters **11**(22): 9656–9658;

Reker, D., G. Bernardes, and T. Rodrigues. 2018. Evolving and nano data enabled machine intelligence for chemical reaction optimization, ChemRxiv, Cambridge, Cambridge Open Engage; This content is a preprint and has not been peer-reviewed.

Rodrigues, T. 2020. The good, the bad, and the ugly in chemical and biological data for machine learning. Drug Discover Today Technology **32–33**(December 2019): 3–8. https://doi.org/10.1016/j.ddtec.2020.07.001. Epub PMID: 33386092; PMCID: PMC7382642.

Rogers, D. and M. Hahn. 2010. Extended-connectivity fingerprints, Journal of Chemical Information and Modeling **50**(5): 742–754.

Rzepa, H. S. and S. Kuhn. 2022. A data-oriented approach to making new molecules as a student experiment: Artificial intelligence-enabling FAIR publication of NMR data for organic esters, Magnetic Resonance in Chemistry **60**(11): 93–103.

Sacha, M., et al. 2021. Molecule edit graph attention network: Modeling chemical reactions as sequences of graph edits, Journal of Chemical Information and Modeling **61**(7): 3273–3284. https://pubs.acs.org/doi/abs/10.1021/acs.jcim.1c00537.

Sanderson, R. T. 1964. Principles of chemical reaction, Journal of Chemical Education **41**(1): 13.

Sayle, R., et al. 1997. www.daylight.com/meetings/emug97/Sayle/slides.html

Schneider, N., et al. 2015. Get your atoms in order—An open-source implementation of a novel and robust molecular canonicalization algorithm, Journal of Chemical Information and Modeling **55**(10): 2111–2120.

Segler, M. H. S. and M. P. Waller. 2017. Modelling chemical reasoning to predict and invent reactions, Chemistry—A European Journal **23**(25): 6118–6128.

Segler, M. H. S., et al. 2017. Learning to plan chemical syntheses, arXiv preprint, arXiv:1708.04202.

Seidl, P., et al. 2022. Improving few- and zero-shot reaction template prediction using modern hopfield networks, journal of Chemical Information and Modeling **62**(9): 2111–2120. https://pubs.acs.org/doi/10.1021/acs.jcim.1c01065

Selvaratnam, B., et al. 2023. Interpretable machine learning in solid-state chemistry, with applications to perovskites, spinels, and rare-earth intermetallics: Finding descriptors using decision trees, Inorganic Chemistry **62**(28): 10865–10875.

Sridharan, N. S. 1977. An artificial intelligence system to model and guide chemical synthesis planning by computer: A proposal, American Chemical Society. **61:** 148–178.

Struble, T. J., et al. 2020. Current and future roles of artificial intelligence in medicinal chemistry synthesis, Journal of Medicinal Chemistry **63**(16): 8667–8682.

Struebing, H., et al. 2013. Computer-aided molecular design of solvents for accelerated reaction kinetics, Nature Chemistry **5**(11): 952–957.

Taguchi, A. T., et al. 2019. Convolutional neural network analysis of two-dimensional hyperfine sublevel correlation electron paramagnetic resonance spectra, The Journal of Physical Chemistry Letters **10**(5): 1115–1119.

Tetko, I. V., et al. 2020. State-of-the-art augmented NLP transformer models for direct and single-step retrosynthesis, Nature Communications **11**(1): 5575.

Ucak, U. V., et al. 2022. Retrosynthetic reaction pathway prediction through neural machine translation of atomic environments, Nature Communications **13**(1): 1186.

Ucak, U. V., et al. 2023. Improving the quality of chemical language model outcomes with atom-in-SMILES tokenization, Journal of Cheminformatics **15**(1): 55.

Vaswani, A., et al. 2017. Attention is all you need, Advances in Neural Information Processing Systems 30.

Wang, A., et al. 2018. GLUE: A multi-task benchmark and analysis platform for natural language understanding, arXiv preprint, arXiv:1804.07461.

Wei, J. N., et al. 2016. Neural networks for the prediction of organic chemistry reactions, ACS Central Science 2(10): 725–732. https://pubs.acs.org/doi/10.1021/acscentsci.6b00219.

Weininger, D. 1988. SMILES, a chemical language and information system. 1. Introduction to methodology and encoding rules, Journal of Chemical Information and Computer Sciences 28(1): 31–36.

Weissermel, K. and H. Arpe. 2003. Basic products of industrial syntheses, Industrial Organic Chemistry: 15–57.

Yamada, H., et al. 2019. Predicting materials properties with little data using shotgun transfer learning, ACS Central Science 5(10): 1717–1730.

Yue, K., et al. 2023. Trends and opportunities in organic synthesis: Global state of research metrics and advances in precision, efficiency, and green chemistry, The Journal of Organic Chemistry 88(7): 4031–4035.

Zhou, Y., et al. 2019. Toward the exact exchange–correlation potential: A three-dimensional convolutional neural network construct, The Journal of Physical Chemistry Letters 10(22): 7264–7269.

Zhou, Z., et al. 2017. Optimizing chemical reactions with deep reinforcement learning, ACS Central Science 3(12): 1337–1344.

Section II

AI and Its Applications in Public Health

5 MLACP 2.0
Utilizing Machine Learning to Predict Anticancer Peptide Activity from Protein Peptide Patterns

Sivakannan Subramani and A. Sharwin

5.1 INTRODUCTION

Millions of people are affected by cancer, a sneaky enemy, and those who suffer from it need our sympathy. Beyond the figures and technical language, cancer poses a very human challenge—a conflict between life's fragility and the resiliency of the human spirit. Our collaborative resolve to improve healthcare and lengthen life becomes even more crucial as we negotiate this complex environment (Phan et al. 2022). The World Health Organization and the International Agency for Research on Cancer (IARC) disclosed gloomy figures in 2018: 18.1 million new cases of cancer and 9.6 million deaths from cancer. These numbers serve as a sobering reminder of the critical need for new and efficient cancer treatment methods. Traditional cancer treatments have always depended on surgery, radiation, and chemotherapy. While essential to human advancement, these techniques have certain inherent drawbacks (Bray et al. 2018). Despite their importance, surgical treatments may not always guarantee a perfect outcome, leaving room for ambiguity and the possibility that malignant tissue will continue to exist. Although radiotherapy is efficient in reducing tumour size, it can be difficult to discern between cancer and healthy cells, which occasionally causes unintended damage to the tissues around the tumour. Chemotherapy's all-encompassing strategy causes severe adverse effects and the development of drug resistance. Despite these difficulties, there is optimism because of recent advancements in science and medicine. Researchers and healthcare professionals have focused on the fascinating realm of anticancer peptides (ACPs) to conquer the difficulties of cancer. These extraordinary compounds, which are a subclass of antimicrobial peptides (AMPs), have a great deal of potential as cancer-targeted therapy (Wang et al. 2015).

ACPs stand out due to their extraordinary capacity to target cancer cells only while leaving healthy cells unharmed. Their special makeup, which includes hydrophobic and positive residues, enables them to interact with the anionic membranes of cancer cells, destroying their integrity and leading to cell death (Wei et al. 2018). By lessening the adverse impacts on healthy tissues and easing the difficulties caused by traditional medicines, this selectivity offers a ray of hope for precision therapy.

DOI: 10.1201/9781003405436-7

ACPs also provide several benefits. They are excellent candidates for treatments since they occur naturally as biological inhibitors. Since their chemical production and modification are very simple, they may be tailored and optimized to increase their effectiveness (Feng and Wang 2019). ACPs also have low toxicity, high selectivity, membrane-penetrating properties, and are simple to modify chemically—a potent combination for the creation of targeted cancer treatments (Hu et al. 2019). To maximize the effectiveness of ACPs, researchers have created computer models that can forecast their existence. These complex models include a variety of feature encodings and sophisticated techniques, including support vector machines (SVMs), random forests (RF), and convolutional neural networks (CNN; Basith et al. 2020a). These prediction models seek to improve accuracy, robustness, and performance via ongoing improvement, ultimately opening the door for the creation of ground-breaking ACP-based medicines. The limits of conventional cancer treatments are explored in this chapter, and the difficulties that chemotherapy, radiation, and surgical methods must overcome are also highlighted. Phan et al. (2022) dive deeply into ACPs, revealing their enormous promise as game-changing therapeutics in the fight against cancer (Schweizer 2009). Additionally, Phan et al. (2022) looked at the benefits that peptide-based therapies have over more traditional methods, offering insight into their potential to revolutionize cancer care. The authors also review the state-of-the-art computational models of prediction for ACPs (Feng and Wang 2019), offering crucial insights into the creation of MLACP 2.0 (Machine Learning-based Anticancer Peptide Predictor), a cutting-edge predictor that outperforms current methods in precision and stability. MLACP 2.0 displays its ability to change the field of ACP prediction research and bring us closer to personalized and effective cancer therapies by utilizing a hybrid ensemble model and well-curated training datasets. The hope is that this investigation will spark a paradigm change in cancer treatment—one that embraces the special qualities of ACPs (Manavalan et al. 2018), recognizes the value of computational forecasts, and ushers in a new era of precise medicine.

5.2 ANTIMICROBIAL PEPTIDES: OVERVIEW

Living things have a great defence mechanism against microbial invasion thanks to nature. Antimicrobial peptides (AMPs), an exceptional family of bioactive chemicals, are at the forefront of this defence (Chung et al. 2019). AMPs are found in a wide variety of species, including bacteria, fungi, plants, and mammals. They operate as nature's first line of defence against pathogenic threats (Raffatellu 2018).

This section delves further into the intriguing world of AMPs, examining their core traits, biological purposes, and many ways of action. Phan et al. (2022) provided the groundwork for understanding the distinctive subgroup referred to as anticancer peptides (ACPs) by understanding the vast spectrum of AMPs and their function in the complex balance of microbial defence.

5.2.1 ANTICANCER PEPTIDES AS A SUBSET OF ANTIMICROBIAL PEPTIDES

ACPs, a subset of the broad range of AMPs, have enormous potential as cancer treatments. These extraordinary molecules have a variety of physicochemical characteristics that give them strong anticancer action (Veltri et al. 2018). They are generally made up of fewer than 50 amino acid residues. The distinguishing traits that set ACPs apart from their antibacterial counterparts are highlighted in this section. Phan et al. (2002) investigated the structural and functional characteristics that allow ACPs to target cancer cells only while sparing normal cells (Schweizer 2009). The authors uncovered the possibility of a novel strategy for treating cancer by looking at the special qualities of ACPs—focusing on specificity, effectiveness, and minimal toxicity (Fosgerau and Hoffmann 2015).

5.2.2 Mechanisms of Action of Anticancer Peptides

A thorough grasp of their mechanisms of action is necessary to completely harness the potential of ACPs. In this section, we explore the complex operations of ACPs within cancer cells. Phan et al. (2022) dissect the many tactics used by ACPs to interfere with the delicate equilibrium of cancer cell membranes. ACPs specifically target anionic membranes by their amphiphilic character and electrostatic interactions, which results in membrane breakdown, the activation of apoptosis, and the prevention of cancer cell growth (Soon et al. 2020). Understanding these pathways helps better understand the exceptional potential of ACPs as powerful tools in the fight against cancer.

5.2.3 Advantages and Potential Applications of Anticancer Peptides

ACPs are appealing for reasons other than their powerful anticancer abilities. Researchers have explored the numerous benefits and prospective uses that make ACPs an exciting therapeutic option. The inherent biological inhibitors, simplicity in synthesis, and chemical adaptability have been explored, emphasizing the potential for highly specialized and focused cancer therapies (Lau and Dunn 2018). We investigate their higher selectivity for cancer cells, lower toxicity relative to standard medicines, and capacity to cross cell membranes. By comprehending the benefits provided by ACPs, we see a time when cancer therapies may be specifically customized to each patient, increasing efficacy while reducing adverse effects.

5.3 MLACP 2.0 CONSTRUCTION

Phan et al. (2022) provide a thorough review of the creation of MLACP 2.0, a unique anticancer peptide prediction model, shown in Figure 5.1. Several procedures were followed in the creation of MLACP 2.0, including dataset curation, baseline model evaluation, and the use of cutting-edge computational methods (Phan et al. 2022). Here, we summarize the main actions performed by the authors to develop MLACP 2.0, which considerably increased the reliability and accuracy of ACP prediction.

5.3.1 Dataset Curation

Curating a high-quality and non-redundant training dataset was the first and most important stage in building MLACP 2.0. ACPs were gathered from a variety of sources, including CancerPPD (Tyagi et al. 2014), APD3 (Wang et al. 2015), PlantPepDB (Das et al. 2020), DBAASP v3.0 (Pirtskhalava et al. 2020), SATPdb

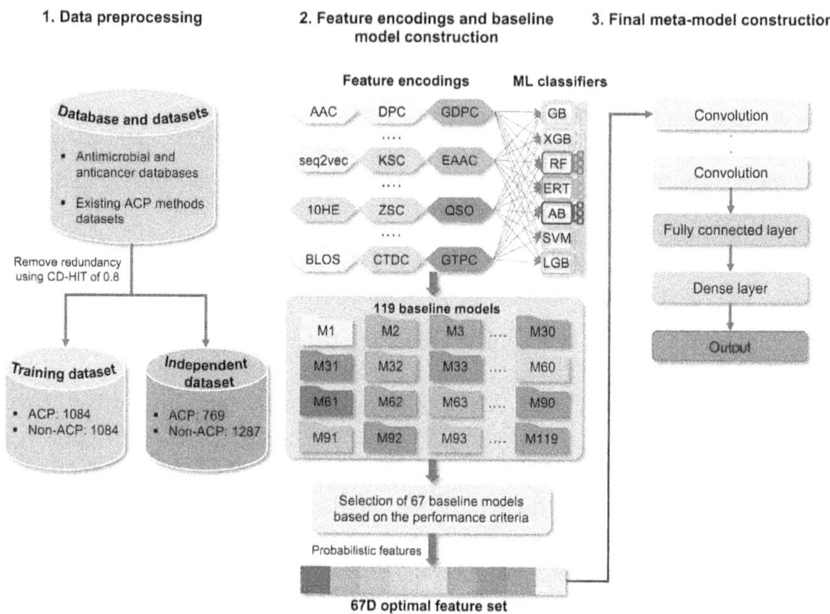

FIGURE 5.1 Overview of MLACP 2.0 (From Phan et al. 2022).

(Singh et al. 2015), ADAM, DRAMP 3.0 (Shi et al. 2021), LAMP (Zhao et al. 2013), Peptipedia (Quiroz et al. 2021), DbAMP (Jhong et al. 2018), and AMPfun (Chung et al. 2019), through extensive literature and database searches. A dataset with 3,725 experimentally validated ACPs was produced after thorough curation.

Phan et al. (2022) also took into account additional functional peptides, a small percentage of random peptides, AMPs, and non-AMPs, as well as experimentally verified non-proinflammatory causing peptides as non-ACPs, to assure the variety and robustness of the training dataset. The authors maintained balance and prevented class bias during model construction by picking 1,084 non-ACPs at random from the curated dataset.

5.3.2 Feature Encodings

An essential step in creating a reliable ACP prediction model is choosing the right feature encodings. To find the best feature encodings for ACP prediction, we investigated a broad range of feature encodings in this study. The composition of k-spaced amino acid group pairs (CKSAAGP), quasi sequence order (QSO), dipeptide deviation from the expected mean (DDE), grouped DPC (GDPC), enhanced grouped AAC (EAAC), grouped tripeptide composition (GTPC), BLOSUM62 (BLOS), enhanced AAC (EAAC), K-spaced conjoint triad (KSC), and Z scale were among the 15 conventional encodings that were tested. Of these, 11 encodings—AAC, CKSAAGP, CTDC, CTDD, CTDT, DDE, DPC, KSC, QSO, 1OHE, and seq2vec—emerged as the most significant contributions to ACP prediction. These encodings were critical in obtaining the sequence data required for precise ACP prediction. It is noteworthy that nine of the conventional encodings, whose encoding details were exhaustively detailed in prior works, were notably significant in the final prediction model (Manavalan and Patra 2022). To represent the peptide sequences, each encoding produced a distinct feature vector representation with variable dimensions. For instance, the feature vectors created by AAC, CKSAAGP, CTDC, CTDD, CTDT, DDE, DPC, KSC, and QSO, respectively, had dimensions of 20, 275, 39, 195, 39, 400, 400, 343, and 100. The machine learning algorithms used in our study used these feature vectors as their input data. To effectively capture various facets of sequence information and improve the prediction performance of ACP models, feature encodings must be explored and chosen. Phan et al. were able to identify pertinent information that considerably aided in the correct prediction of ACPs by utilizing the strengths of various encodings.

5.3.3 Baseline Model Evaluation

The study team concentrated on assessing several baseline models to choose the most promising candidates for the creation of the meta-model. A thorough evaluation of 17 different feature encodings, including both traditional and word embeddings, was done. Seven well-known classifiers—random forest (RF), gradient boosting (GB), light gradient boosting (LGB), extreme gradient boosting (XGB), AdaBoost (AB), support vector machine (SVM), and extremely randomized tree (ERT)—were used to build these models. However, throughout the evaluation process, distinct performance differences were found across the classifiers, raising questions about whether or not all 119 baseline models should have been taken into account (Malik et al. 2022). To solve this issue, the average Matthews correlation coefficient (MCC), based on the 17 baseline models, was determined for each classifier. Thereafter, only models that performed better than the average MCC were kept, yielding a more specialized subset of 67 baseline models. These chosen models showed an accuracy (ACC) range of 0.800 to 0.835, highlighting their potential efficacy in anticipating and categorizing (ACPs. The research team attempted to improve the performance and dependability of the final meta-model by systematically reducing the number of candidate models.

5.3.4 Meta-Model Construction Using Convolutional Neural Network

The prediction probabilities of ACPs were combined with the chosen baseline models and encodings to create a brand-new feature vector (Sharma et al 2022). The final prediction model, MLACP 2.0, was created using a convolutional neural network (CNN) with this vector as its input. Due to its proven reliability and effectiveness in managing complicated and high-dimensional data, CNN was chosen as the recommended meta-model creation method (Sharma et al. 2021).

5.4 COMPARISON OF MLACP 2.0 WITH OTHER DIFFERENT APPROACHES

Phan et al. (2022) also created two CNN-based word embedding models, seq2vec-CNN and 1OHE-CNN, as well as a CNN model based on hybrid features (a linear integration of 11 encodings), to demonstrate the benefits of employing probabilistic features in MLACP 2.0. The effectiveness of MLACP 2.0 is contrasted with that of the top five baseline models, CNN-based word embeddings, and hybrid feature models. The CNN-hybrid model performs similarly to the best five baseline models and much better than the CNN-word embedding model when compared to the best five baseline models, showing that automated word embedding features are less efficient than feature engineering in ACP prediction. MLACP 2.0 beats CNN-based models and the top five baseline models. More specifically, improvement measures for MLACP 2.0 range from 2.3% to 10.3% MCC, 1.1% to 2.2% ACC, and 0.5% to 4.8% AUC, proving that a methodical evaluation of multiple encodings in conjunction with the choice of a set of baseline models used for meta-model construction led to better performance.

5.4.1   MLACP

The abbreviation MLACP means "Machine Learning-Based Anticancer Peptide Predictor." It alludes to a computer approach or model designed to forecast possible ACPs utilizing machine learning techniques. ACPs are attractive possibilities for cancer treatment because of the propensity of their short amino acid sequences to target and destroy cancer cells (Manavalan et al. 2017). The MLACP model is made to accept as input several aspects of peptides determined from their amino acid sequence, such as their atomic makeup, dipeptide composition, amino acid composition, and physicochemical qualities. To create a prediction model for recognizing ACPs, these variables are utilized to train machine learning algorithms like SVMs and random forest.

MLACP tries to get beyond the drawbacks of conventional wet-lab testing, which may be expensive and time-consuming. The model is a useful tool for discovering promising ACP candidates and directing additional in vitro tests, improving cancer research and therapy development as a result of its high accuracy and efficiency. The flowchart for MLACP prediction model is shown in Figure 5.2.

5.4.2   mACPpred

A new prediction model called mACPpred was created in the study to accurately predict ACPs. It is created in two steps, utilizing feature selection methods and machine learning algorithms.

In the first stage, seven distinct feature encodings that capture various facets of sequence information pertaining to peptides are subjected to a two-step feature selection methodology. Composition-based features, physicochemical attributes, and profiles are some of these feature encodings. Finding the best feature-based models for each encoding is the aim. These models then produce projected probabilities of ACPs, which are used as feature vectors. The researchers employ an SVM as the machine learning approach in the second stage (Basith et al. 2020b). They provide the SVM with the expected probability feature vectors that they acquired in the first phase. As a result, they are able to create the ultimate prediction model, known as mACPpred.

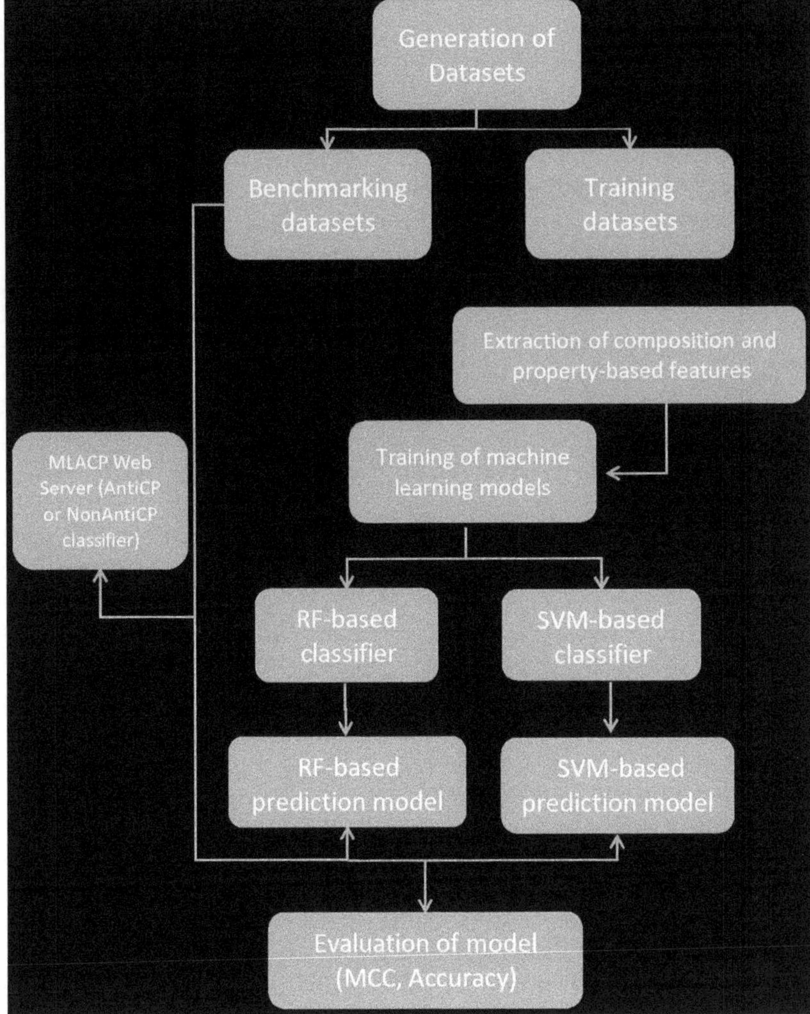

FIGURE 5.2   Flowchart outlining the processes in the MLACP approach used to construct a prediction model. MCC = Matthews correlation coefficient; RF = random forest; SVM = support vector machine.

The goal of mACPpred was to provide a strong tool for precisely predicting ACPs from peptide sequences. Through cross-validation investigation, the researchers have demonstrated that mACPpred works better than employing solitary feature encodings. Additionally, mACPpred performs better than other techniques compared to those examined in the research when assessed on an independent dataset.

In conclusion, mACPpred is a cutting-edge predictive model that advances knowledge about ACPs by enhancing the precision of ACP prediction through the use of machine learning algorithms and feature selection approaches.

5.4.3 ACPredStackL

An algorithm called ACPredStackL (Liang et al. 2021) was created to predict ACPs precisely from peptide sequences. To enhance prediction performance, the method employs a stacking ensemble learning strategy, which integrates many machine learning models.

There are three main phases in the framework:

Step 1: Representing features

AAC (Amino Acid Composition), PAAC (Pseudo Amino Acid Composition), CTD (Composition Transition Distribution), CKSAAP (Conjoint Triad Feature), and QSOrder (Quasi-Sequence Order) are only a few of the feature encoding approaches used to encode peptide sequences in the input datasets.

Each peptide sequence is converted into a high-dimensional feature vector with a total of 2,289 dimensions, where each encoding strategy produces a certain number of features.

Step 2: Choosing a feature

Using a two-step feature selection technique, high-dimensional features and possible noise are dealt with. First, the characteristics are ranked using the F-score method according to how well they can separate ACPs from non-ACPs. Then, a suboptimal selection of characteristics with a high degree of ACP identification is chosen using a sequential forward search technique (FSP). The top 600 features obtained from the FSP make up the final feature subset.

Step 3: Stacking ensemble learning

Multiple machine learning models are combined in this stage using the stacking ensemble learning technique. As the first-level learners, four base classifiers are used: KNN (K-nearest neighbours), NB (naive Bayes), Light-GBM (light gradient boosting Machine), and SVM, with various hyper-parameters. The meta-classifier at the second level is a logistic regression classifier. The input datasets go through a stacking cross-validation procedure before the stacking model is built, ensuring the model's robustness and lowering the danger of overfitting. Cross-validation and jackknife tests are used to assess the stacking ensemble model's prediction ability after it has been trained on a subset of the data. The model's generalizability is evaluated using the final test dataset.

To improve the prediction accuracy of ACPs from peptide sequences, ACPredStackL is a comprehensive and cutting-edge computational framework that combines multiple feature encodings, makes use of feature selection techniques, and stacks ensemble learning with different base classifiers and a meta-classifier in Figure 5.3.

The previous version, as well as the two top ACP predictors (mACPpred and ACPpredStackL), were all evaluated using the independent dataset in addition to MLACP 2.0. Notably, independent assessments of earlier studies found that two of the approaches were the best predictors. Rather than the usual independent datasets utilized in earlier investigations, a difficult dataset was developed.

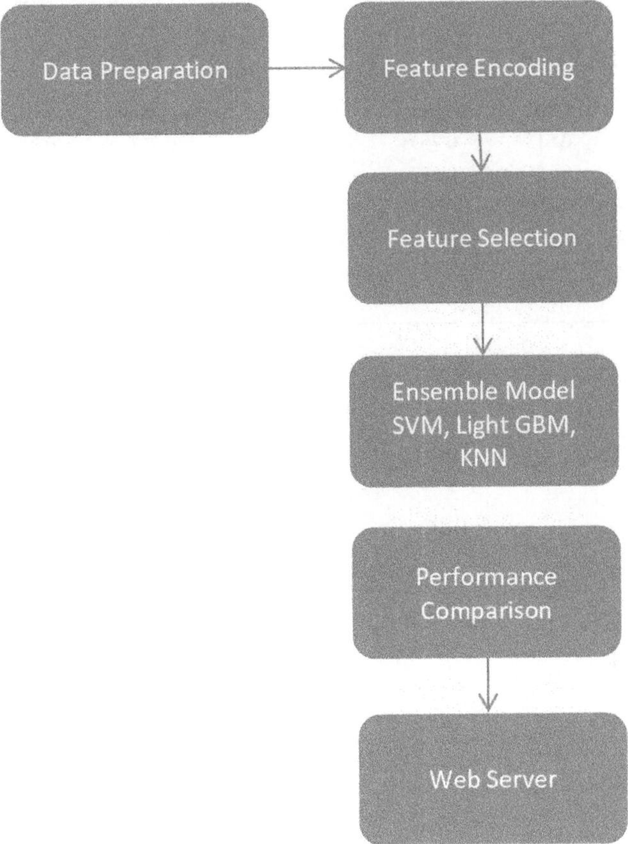

FIGURE 5.3 ACPredStackL's overarching structure. Data preparation, feature encoding, ensemble modelling, performance comparison, and webserver implementation are just a few of the key phases that go into its creation. GBM = gradient boosting machine; KNN = K-nearest neighbours; SVM = support vector machine.

Independent datasets possess the following crucial traits: The non-ACPs were developed taking into account a number of realistic circumstances, including other functional peptides and empirically characterized negative instances. None of the ACPs share >80% sequence identity with the training dataset. The MCC, ACC, sensitivity (Sn), specificity (Sp), and AUC values for MLACP 2.0 are 0.513, 0.765, 0.750, and 0.773, respectively, as shown in Table 5.1.

The MCC of 16.2% to 28.0%, the ACC of 0.8% to 13.7%, and the AUC of 7.3% to 17.7% were specifically enhanced by MLACP 2.0 in comparison to the existing predictors. A further indication that MLACP 2.0 works well on unseen data and is better suited for real applications is that its predictor performance is more evenly distributed (minimal difference between Sn and Sp) when compared to the previous predictors. Inferring statistical estimates from the threshold-based comparison mentioned above is challenging. As a consequence, we used ROC to compare two AUC

TABLE 5.1

Performance of Different Methods on Independent Datasets

Method	MCC	Accuracy	Sensitivity	Specificity	AUC	*P*-value
MLACP 2.0	0.513	0.765	0.750	0.773	0.817	
MLACP	0.256	0.677	0.283	0.911	0.744	0.000003
mACPpred	0.351	0.677	0.700	0.663	0.704	<0.000001
ACPredStackL	0.233	0.628	0.588	0.651	0.640	<0.000001

Source: Phan et al. (2022).

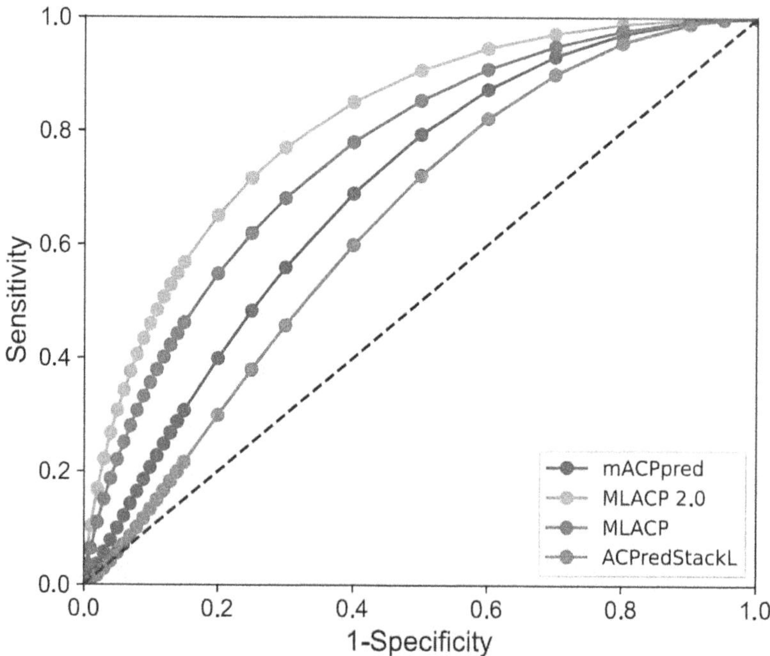

FIGURE 5.4   Comparison of binormal receiver operating characteristics (ROC) curves for ACP prediction using different methods on an independent dataset. (From Phan et al. 2022.)

values from various approaches, and we then used the findings of a two-tailed test to determine the P value for any differences that were found (Hanley and McNeil 1982)

Table 5.1 and Figure 5.4 show that the MLACP 2.0 significantly outperformed the current predictors on the independent dataset. The suggested technique has the drawback of being unable to predict peptides with more than 50 amino acid residues.

5.5   MODEL INTERPRETATION

A model interpretation study was carried out using SHapley Additive exPlanation (SHAP) (Štrumbelj and Kononenko 2013) to acquire insights into the inner workings

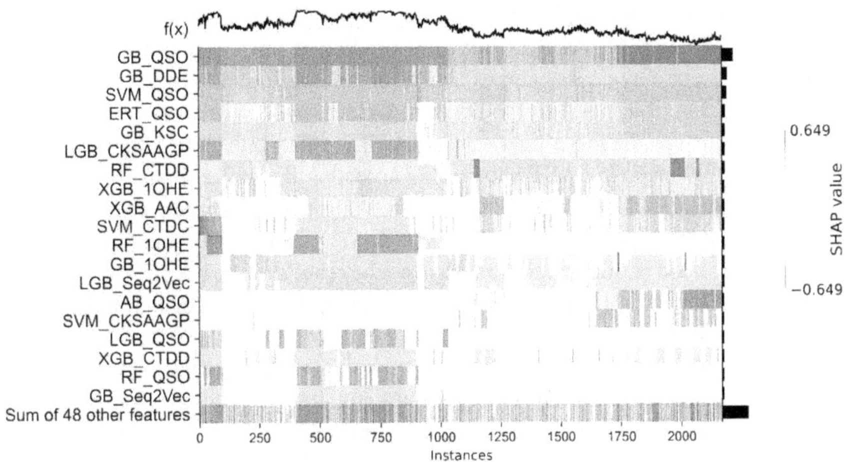

FIGURE 5.5 A heatmap of the top 20 probabilistic features based on the training dataset's SHAP values for locating anticancer peptides. See the text for an explanation of SHAP. (From Phan et al. 2022).

of MLACP 2.0 and comprehend the contribution of various aspects to its predictive performance. The most important characteristics and how they related to the outcomes of MLACP 2.0 were the main objectives. The investigation showed that MLACP 2.0 produced predictions as line charts above a heatmap matrix, where each feature's global value was shown as bar graphs on the right-hand side of the heatmap. The top 20 characteristics were ranked according to their worldwide significance (Figure 5.5). Shades of grey represent the strength of correlation. Darker shades can indicate stronger correlations, while lighter shades can represent weaker correlations. Based on the findings, it was discovered that MLACP 2.0 benefited significantly from baseline models based on particular encodings, including QSO, 1OHE, CTDD, seq2vec, and CKSAAGP. CKSAAGP, one of the encodings, showed promise as a performance improvement for MLACP 2.0. This result was consistent with other research that stressed the significance of physicochemical characteristics in predicting anticancer peptides (Chen et al. 2021a).

The model interpretation and analysis provided valuable insights into the underlying mechanisms of MLACP 2.0 and shed light on the key features contributing to its predictive performance. The findings further supported the effectiveness of the selected encodings and the importance of physicochemical properties in accurately predicting anticancer peptides.

5.6 WEB SERVER IMPLEMENTATION

A user-friendly web server has been created, which can be visited at https://balalab-skku.org/mlacp2, to guarantee the MLACP 2.0 algorithm is widely accessible and usable. The programming languages Django, Python, CSS, HTML, and JavaScript were combined to build the web server. Additionally, for effective work result storage and retrieval, a PostgreSQL database was used. Users can find comprehensive instructions on using MLACP 2.0 for their prediction purposes on the web server's home page. Users may access and examine the underlying data by clicking on links on the home page that point to the curated datasets used in the study. The user has two choices for submitting a prediction task. They can either directly paste one or more query sequences in FASTA format or upload a file containing several FASTA-formatted sequences. Users may simply submit their peptide sequences for prediction using this adaptable input format. The web server displays the results in a specific interface once a job is finished. Users may see the results of the predictions directly in the user interface. Additionally, to conduct additional research or maintain records, users have the option to download the results in CSV format.

The web server also has a function for retrieving the outcomes of previously finished tasks. Users may simply retrieve and evaluate the outcomes of their earlier submissions by entering the job ID into the "find job" field on the submission page. The creation of the MLACP 2.0 web server ensures the creation of an accessible and user-friendly platform for the prediction of anticancer peptides. The practical application of MLACP 2.0 for peptide prediction tasks is made easier by its user-friendly interface, thorough instructions, and result retrieval capabilities.

5.7 CONCLUSION

This chapter describes the work of Phan et al. (2022) to improve the ACP predictor, MLACP 2.0. MLACP 2.0 exhibits enhanced performance and represents a substantial advancement in the field of ACP prediction by using peptide sequence information. To create MLACP 2.0, the authors conducted a number of crucial actions. First, using a variety of databases and literature sources, they created nonredundant training and independent datasets. The use of such a sizable, nonredundant dataset sets our work apart from earlier research projects. Second, to create a broad pool of baseline models, the authors used seven distinct classifiers and 17 different feature encodings. After carefully choosing a subset of baseline models, Phan et al. combined the projected ACP values, trained a convolutional neural network (CNN), and then developed the final MLACP 2.0 model. There are a number of reasons why MLACP 2.0 performs better now. Its success is a result of a meta-model technique and a smaller training sample. Its performance is further improved by the projected probabilistic features' excellent inherent discriminating ability on both the training and independent datasets. This strategy shows the potential for expanding the prediction ability to incorporate other peptide therapeutic activities. Even though MLACP 2.0 shows promising outcomes, there are still certain things that may be done better. The creation and use of a novel sequence-based encoding method that is independent of composition and physicochemical qualities is one of the future research objectives (Chen et al. 2021b). To measure each encoding's contribution to identifying ACPs from non-ACPs, feature selection strategies might be investigated (Chen et al. 2019). When more datasets are made available, integrating ensemble deep learning models or hybrid models that combine deep learning and conventional modelling techniques may also improve ACP prediction ability (Hasan et al. 2022).

In conclusion, the creation of MLACP 2.0 represents a major improvement in ACP prediction. Accuracy and performance have increased as a consequence of the use of a meta-model approach, significant dataset development, and various feature encodings. The future of ACP prediction has enormous promise as investigation continues into innovative encoding methods (Phan et al. 2022), using feature selection strategies, and leveraging the capabilities of ensemble or hybrid models.

REFERENCES

Basith, S., Manavalan, B., Shin, T. J., & Lee, G. 2020a. Machine intelligence in peptide therapeutics: A next-generation tool for rapid disease screening. *Medicinal Research Reviews*, *40*(4), 1276–1314. https://doi.org/10.1002/med.21658

Basith, S., Manavalan, B., Shin, T. J., Lee, D. Y., & Lee, G. 2020b. Evolution of machine learning algorithms in the prediction and design of anticancer peptides. *Current Protein & Peptide Science, 21*(12), 1242–1250. https://doi.org/10.2174/1389203721666200117171403

Bray, F., Ferlay, J., Soerjomataram, I., Siegel, R. L., Torre, L. A., & Jemal, A. 2018. Global cancer statistics: GLOBOCAN estimates of incidence and mortality worldwide for 36 cancers in 185 countries. *CA: A Cancer Journal for Clinicians*, *68*(6), 394–424. https://doi.org/10.3322/caac.21492

Chen, J., Cheong, H., & Siu, S. W. I. 2021a. xDeep-AcPEP: Deep learning method for anticancer peptide activity prediction based on convolutional neural network and multitask learning. *Journal of Chemical Information and Modeling*, *61*(8), 3789–3803. https://doi.org/10.1021/acs.jcim.1c00181

Chen, Z., Zhao, P., Li, C., Li, F., Xiang, D., Chen, Y., Akutsu, T., Daly, R. J., Webb, G. I., Zhao, Q., Kurgan, L., & Song, J. 2021b. *iLearnPlus:* A comprehensive and automated machine-learning platform for nucleic acid and protein sequence analysis, prediction and visualization. *Nucleic Acids Research*, *49*(10), e60. https://doi.org/10.1093/nar/gkab122

Chen, Z., Zhao, P., Li, F., Marquez-Lago, T. T., Leier, A., Revote, J., Zhu, Y., Powell, D. A., Akutsu, T., Webb, G. I., Chou, K., Smith, A. M., Daly, R. J., Li, J., & Song, J. 2019. iLearn: An integrated platform and meta-learner for feature engineering, machine-learning analysis and modelling of DNA, RNA and protein sequence data. *Briefings in Bioinformatics*, *21*(3), 1047–1057. https://doi.org/10.1093/bib/bbz041

Chung, C., Kuo, T. S., Wu, L., Lee, T., & Horng, J. 2019. Characterization and identification of antimicrobial peptides with different functional activities. *Briefings in Bioinformatics*, *21*(3), 1098–1114. https://doi.org/10.1093/bib/bbz043

Das, D., Jaiswal, M., Khan, F., Ahamad, S., & Kumar, S. 2020. PlantPepDB: A manually curated plant peptide database. *Scientific Reports*, *10*(1). https://doi.org/10.1038/s41598-020-59165-2

Feng, P., & Wang, Z. 2019. Recent advances in computational methods for identifying anticancer peptides. *Current Drug Targets, 20*(5), 481–487. https://doi.org/10.2174/1389450119666180801121548

Fosgerau, K., & Hoffmann, T. 2015. Peptide therapeutics: Current status and future directions. *Drug Discovery Today*, *20*(1), 122–128. https://doi.org/10.1016/j.drudis.2014.10.003

Hanley, J. A., & McNeil, B. J. 1982. The meaning and use of the area under a receiver operating characteristic (ROC) curve. *Radiology*, *143*(1), 29–36. https://doi.org/10.1148/radiology.143.1.7063747

Hasan, M., Tsukiyama, S., Cho, J. Y., Kurata, H., Alam, M. A., Liu, X., Manavalan, B., & Deng, H. 2022. Deepm5C: A deep-learning-based hybrid framework for identifying human RNA N5-methylcytosine sites using a stacking strategy. *Molecular Therapy*, *30*(8), 2856–2867. https://doi.org/10.1016/j.ymthe.2022.05.001

Hu, L., Bell, D., Antani, S., Xue, Z., Yu, K., Horning, M. P., Gachuhi, N., Wilson, B., Jaiswal, M. S., Befano, B., Long, L. R., Herrero, R., Einstein, M. H., Burk, R. D., Demarco, M., Gage, J. C., Rodriguez, A. C., Wentzensen, N., & Schiffman, M. 2019. An observational study of deep learning and automated evaluation of cervical images for cancer screening. *Journal of the National Cancer Institute*, *111*(9), 923–932. https://doi.org/10.1093/jnci/djy225. PMID: 30629194; PMCID: PMC6748814.

Jhong, J., Chi, Y., Li, W. J., Lin, T., Huang, K., & Lee, T. 2018. dbAMP: An integrated resource for exploring antimicrobial peptides with functional activities and physicochemical properties on transcriptome and proteome data. *Nucleic Acids Research*, *47*(D1), D285–D297. https://doi.org/10.1093/nar/gky1030

Lau, J. L., & Dunn, M. J. 2018. Therapeutic peptides: Historical perspectives, current development trends, and future directions. *Bioorganic & Medicinal Chemistry*, *26*(10), 2700–2707. https://doi.org/10.1016/j.bmc.2017.06.052

Malik, A., Subramaniyam, S., Kim, C., & Manavalan, B. 2022. SortPred: The first machine learning based predictor to identify bacterial sortases and their classes using sequence-derived information. *Computational and Structural Biotechnology Journal*, *20*, 165–174. https://doi.org/10.1016/j.csbj.2021.12.014

Manavalan, B., Basith, S., Shin, T. J., Choi, S., Kim, D. Y., & Lee, G. 2017. MLACP: Machine-learning-based prediction of anticancer peptides. *Oncotarget*, *8*(44), 77121–77136. https://doi.org/10.18632/oncotarget.20365

Manavalan, B., Basith, S., Shin, T. J., Wei, L., & Lee, G. 2018. mAHTPred: A sequence-based meta-predictor for improving the prediction of anti-hypertensive peptides using effective feature representation. *Bioinformatics*, *35*(16), 2757–2765. https://doi.org/10.1093/bioinformatics/bty1047

Manavalan, B., & Patra, M. C. 2022. MLCPP 2.0: An updated cell-penetrating peptides and their uptake efficiency predictor. *Journal of Molecular Biology*, *434*(11), 167604. https://doi.org/10.1016/j.jmb.2022.167604

Phan, L. M. T., Park, D. H., Pitti, T., Madhavan, T., Jeon, Y., & Manavalan, B. 2022. MLACP 2.0: An updated machine learning tool for anticancer peptide prediction. *Computational and Structural Biotechnology Journal*, *20*, 4473–4480. https://doi.org/10.1016/j.csbj.2022.07.043

Pirtskhalava, M., Amstrong, A. A., Grigolava, M., Chubinidze, M., Alimbarashvili, E., Vishnepolsky, B., Gabrielian, A., Rosenthal, A., Hurt, D. E., & Tartakovsky, M. 2020. DBAASP v3: Database of antimicrobial/cytotoxic activity and structure of peptides as a resource for development of new therapeutics. *Nucleic Acids Research*, *49*(D1), D288–D297. https://doi.org/10.1093/nar/gkaa991

Quiroz, C., Saavedra, Y. B., Armijo-Galdames, B., Amado-Hinojosa, J., Olivera-Nappa, Á., Sanchez-Daza, A., & Medina-Ortiz, D. 2021. Peptipedia: A user-friendly web application and a comprehensive database for peptide research supported by Machine Learning approach. *Database*, *2021*. https://doi.org/10.1093/database/baab055

Raffatellu, M. 2018. Learning from bacterial competition in the host to develop antimicrobials. *Nature Medicine*, *24*(8), 1097–1103. https://doi.org/10.1038/s41591-018-0145-0

Schweizer, F. 2009. Cationic amphiphilic peptides with cancer-selective toxicity. *European Journal of Pharmacology*, *625*(1–3), 190–194. https://doi.org/10.1016/j.ejphar.2009.08.043

Sharma, R. K., Shrivastava, S., Singh, S., Kumar, A., Saxena, S., & Singh, A. 2021. Deep-AFPpred: Identifying novel antifungal peptides using pretrained embeddings from seq2vec with 1DCNN-BiLSTM. *Briefings in Bioinformatics*, *23*(1). https://doi.org/10.1093/bib/bbab422

Sharma, R. K., Shrivastava, S., Singh, S., Kumar, A., Singh, A., & Saxena, S. 2022. Deep-AVPpred: Artificial intelligence driven discovery of peptide drugs for viral infections. *IEEE Journal of Biomedical and Health Informatics*, 26(10), 5067–5074. https://doi.org/10.1109/jbhi.2021.3130825

Shi, G., Kang, X., Dong, F., Liu, Y., Zhu, N., Hu, Y., Xu, H., Lao, X., & Zheng, H. 2021. DRAMP 3.0: An enhanced comprehensive data repository of antimicrobial peptides. *Nucleic Acids Research*, 50(D1), D488–D496. https://doi.org/10.1093/nar/gkab651

Singh, S., Chaudhary, K., Dhanda, S. K., Bhalla, S., Usmani, S. S., Gautam, A., Tuknait, A., Agrawal, P., Mathur, D., & Raghava, G. P. S. 2015. SATPdb: A database of structurally annotated therapeutic peptides. *Nucleic Acids Research*, 44(D1), D1119–D1126. https://doi.org/10.1093/nar/gkv1114

Soon, T. N., Chia, A., Yap, W. H., & Tang, Y. 2020. Anticancer mechanisms of bioactive peptides. *Protein and Peptide Letters*, 27(9), 823–830. https://doi.org/10.2174/0929866527666200409102747

Štrumbelj, E., & Kononenko, I. 2013. Explaining prediction models and individual predictions with feature contributions. *Knowledge and Information Systems*, 41(3), 647–665. https://doi.org/10.1007/s10115-013-0679-x

Tyagi, A., Tuknait, A., Anand, P., Gupta, S., Sharma, M., Mathur, D., Joshi, A., Singh, S., Gautam, A., & Raghava, G. P. S. 2014. CancerPPD: A database of anticancer peptides and proteins. *Nucleic Acids Research*, 43(D1), D837–D843. https://doi.org/10.1093/nar/gku892

Veltri, D., Kamath, U., & Shehu, A. 2018. Deep learning improves antimicrobial peptide recognition. *Bioinformatics*, 34(16), 2740–2747. https://doi.org/10.1093/bioinformatics/bty179

Wang, G., Li, X., & Wang, Z. 2015. APD3: The antimicrobial peptide database as a tool for research and education. *Nucleic Acids Research*, 44(D1), D1087–D1093. https://doi.org/10.1093/nar/gkv1278

Wei, L., Zhou, C., Chen, H., Song, J., & Su, R. 2018. ACPred-FL: A sequence-based predictor using effective feature representation to improve the prediction of anti-cancer peptides. *Bioinformatics*, 34(23), 4007–4016. https://doi.org/10.1093/bioinformatics/bty451. PMID: 29868903; PMCID: PMC6247924. https://doi.org/10.1093/bioinformatics/bty451

Xiao Liang, Fuyi Li, Jinxiang Chen, Junlong Li, Hao Wu, Shuqin Li, Jiangning Song, Quanzhong Liu, 2021. Large-scale comparative review and assessment of computational methods for anti-cancer peptide identification, Briefings in Bioinformatics, Volume 22, Issue 4, bbaa312, https://doi.org/10.1093/bib/bbaa312

Zhao, X., Wu, H., Lu, H., Li, G., & Huang, Q. 2013. LAMP: A database linking antimicrobial peptides. *PLoS ONE*, 8(6), e66557. https://doi.org/10.1371/journal.pone.0066557

6 An AI-Based Diagnostic System to Predict BI-RADS Scores for Detecting Breast Cancer over Mammograms

S. Ruban, Mohammed Jabeer,
and Ram Shenoy Basti

6.1 INTRODUCTION

Insufficient awareness campaigns and other causes make breast cancer one of the most common cancers in India. This cancer is regarded to be the most frequent cancer (World Health Organization 2021) that results in mortality while being simple to diagnose. According to reports, a woman in India receives a breast cancer diagnosis every four minutes. Both rural and urban India are seeing an increase in breast cancer cases. Higher stages of cancer growth make survival more challenging, and stage 3 and stage 4 breast cancer affect more than 50% of Indian women. Women in India have a low survival rate for breast cancer due to a lack of knowledge and inadequate early detection and diagnosis rates.

This cancer ranks second among American women (National Cancer Society 2021). A few variables influencing the surge include changes in lifestyle (Antony et al. 2018) and a lack of awareness among community women (Prusty et al. 2020). Typically, breast cancer cells cause a lump that can be felt or seen on an X-ray. Even though most breast lumps are benign and do not spread outside of the breast, it is advisable to get an opinion from a doctor. Many studies are carried out in different parts of the world to understand the degree of knowledge among women (Abeje et al. 2019). Early detection and progressive cancer treatment are the most important measures for preventing breast cancer mortality (Agide et al. 2018). Mizoram has the highest breast cancer mortality rates, followed by Kerala and Haryana.

Additionally, younger age groups are more prone to it. Ages 25 to 50 account for about half of all cases. Additionally, poor survival and high mortality were seen in more than 70% of patients in the advanced stage. A carcinoma can be treated if discovered early (Sathwara et al. 2017). The most common screening method is a mammogram, often called a low-dose chest X-ray. Years of research have demonstrated that women who get routine mammograms are diagnosed earlier (Vishwakarma et al.

DOI: 10.1201/9781003405436-8

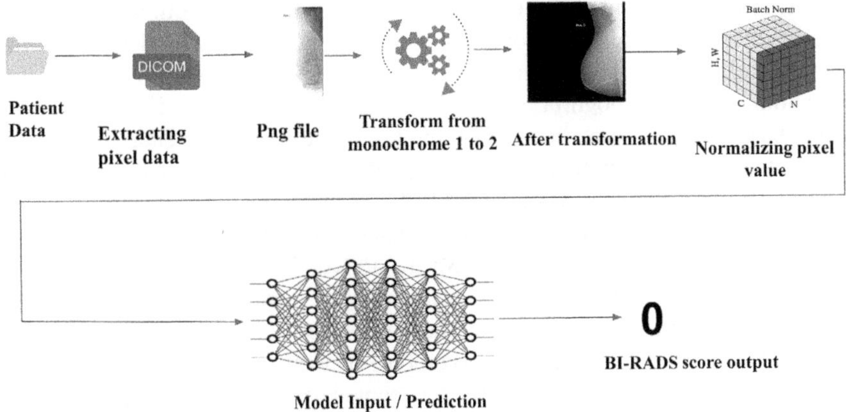

FIGURE 6.1 The framework of the AI-based diagnostic system.

2019), need fewer treatments like mastectomy surgery, and have a higher chance of recovery.

The most popular breast cancer screening test now available is a mammogram. However, additional examinations, such as a breast ultrasound or diagnostic mammography, are needed to confirm.

Even though there is no cancer in the breast, false-positive mammography appears abnormal. Diagnostic mammography, ultrasound, and occasionally magnetic resonance imaging (MRI) or even a breast biopsy are frequently necessary in the case of abnormal mammograms to determine whether the change is cancerous. Additionally, they may necessitate additional testing to rule out malignancy, which would incur extra time, expense, and perhaps even discomfort. However, radiologists have been able to improve diagnostics using AI (Rodríguez-Ruiz et al. 2019a, 2019b). This experimental study examines how radiologists can diagnose breast cancer from mammograms and use that information to inform better judgments. Deep learning algorithms can be quite helpful in improving the precision of cancer detection in its early stages when medical imaging is essential. This experimental study demonstrates how to get around this restriction by applying a deep learning technique called CNN (convolutional neural network). The model developed using this method can assist radiologists with additional machine input and assist them in better diagnosis. The detailed framework that was followed in this study is shown in Figure 6.1.

6.2 EXISTING WORK

Here are a few examples of experimental studies that used deep learning techniques to find breast cancer: The CNN approach was utilized by Yap et al. to locate lesions in ultrasound images (Yap et al. 2018). It was proven that automating the entire procedure improved the accuracy of lesion diagnosis and decreased breast cancer mortality. Similarly to this, Ping and colleagues collected and analyzed breast cancer data using the K-medoid clustering method (Ping et al. 2019). Both research team's experimental investigation employed soft computing techniques for detection.

Vosooghifard and Ebrahimpour created a strategy using decision trees (Ebrahimpour and Vosooghifard 2015). Wang et al. describe a classifier to categorize cancer based on collaborative representation in a distinct experimental investigation (Wang et al. 2015). Machine learning is utilized in the study by Chen et al. to identify and classify the primary sites of cancer (Chen et al. 2015). Similar works on breast cancer have been described by Wahab and Khan (2020) and Gravina et al. (2021).

In the experimental investigation (Murtaza et al. 2020), a few researchers used medical imaging modalities to assess breast images, demonstrating the precision of cancer symptom diagnosis. Similar studies were also used to find lung cancer using deep learning and patch-level categorization (Tripathi et al. 2019; Dang Vu et al. 2019). Imaging techniques are helpful for recognizing questionable spots, figuring out the cancer's stage, approving treatments, and spotting symptoms of recurrence in cancer cells.

The authors of the experimental work classified the breast cancer data using the traditional machine learning classification methods (Yadavendra and Chand 2020). Another comparative study was done using deep learning techniques and is mentioned in Tsochatzidis et al. (2019). A similar study on the Breast Cancer Wisconsin (Diagnostic) Dataset was done by Benbrahim et al. (2020). The authors proposed a deep learning–assisted AdaBoost algorithm (J. Zheng et al. 2020).

A research team developed a strategy for detecting breast cancer from beginning to end in mammography images (Kumar et al. 2020). In a related work (Ramadan 2020), the researchers combined the CNN algorithm with other methods using a hybrid methodology, considerably enhancing its effectiveness. The primary focus of the authors' investigation, which is detailed in Mehmood et al. (2021), is the application of machine learning techniques for early detection. In the experimental research, a further comparable attempt at detection is conducted (Y. Zhang et al. 2020).

6.3 METHODOLOGY

In this experimental study, a total of 1,620 mammography images from 405 participants were used. The necessary reports were also collected from the department before creating the annotation, and the pertinent diagnosis was taken into consideration. This programme has been trained to produce the proper BI-RADS scores. From the above dataset, 256 participants had normal mammograms, 96 participants had benign masses, 20 participants had malignant masses, and 72 patients had extremely dense breasts. Each mammography image is a DICOM format file. The size of each DICOM file is 35 MB, and the size is 2024 × 1024. There were two types of photometric interpretation, that is monochrome1 and monochrome2. Throughout the experimental investigation, there were numerous interactions with the radiologists working in the Father Muller Medical College Department of Radio Diagnosis and Imaging to understand better the process of detecting cancer through mammograms. The radiologist compiles a report by examining all four mammogram images of a patient; they cannot just see one mammogram image and get to a conclusion. Each mammogram image has a link, and by examining all these possibilities, radiologists

TABLE 6.1

Categories of BI-RADS Score

Category	Category 1	Management	Likelihood of Cancer
0 Assessment	Incomplete assessment	Additional imaging required	Not applicable yet
1	Negative	Routine annual screening	No cancer detected
2	Benign	Routine annual screening	0%
3	Probably benign	Follow-up scan after 6 months or earlier, as advised by the doctor	0% to 2%
4	Probably malignant	Breast tissue biopsy recommended by the doctor	4A, 2% to 10%; 4B, 10% to 50%; 4C, 50% to 95%
5	Malignant	Biopsy to be done, essentially	>95 %
6	Biopsy-proven malignancy	Further treatment evaluation is done by the oncologist	Cancer already present

write the report on whether a patient has cancer or not by writing the score of the cancer detection, which is called the BI-RADS score. A BI-RADS score can range from 0 to 6. The various levels are listed in Table 6.1.

6.3.1 DATA GATHERING

The data collected for our study was in the DICOM format, a standard used for data management and workflow in medical image-capturing processes such as X-rays. DICOM also includes a communication protocol and supports TCP/IP. In the world of DICOM, the process of capturing a medical image is known as an acquisition, and the devices used for imaging are called acquisition devices. The size of a DICOM file is typically around 35 MB, with a height and width of 4740 × 2540 pixels.

6.3.2 DATA PREPARATION

You can follow several steps to preprocess DICOM data and extract and keep data based on BI-RADS score by looking at the report. First, you need to read the DICOM files using a DICOM library, such as PyDICOM, which is a popular Python library used for reading and writing DICOM files. DICOM files contain a wealth of metadata, including information about the patient, the imaging study, and the equipment used to capture the images. Therefore, the second step is to extract the relevant metadata from the DICOM files, including the patient ID, study ID, series ID, and image ID. This metadata will be useful later for mapping the BI-RADS score to the corresponding DICOM files.

The third step involves parsing the radiologist's report to extract the BI-RADS score for each patient. The BI-RADS score is a standardized rating system used by radiologists to assess the likelihood of breast cancer based on mammography images. The radiologist report typically contains a section that summarizes the findings of the mammography images and includes the BI-RADS score. Therefore, by

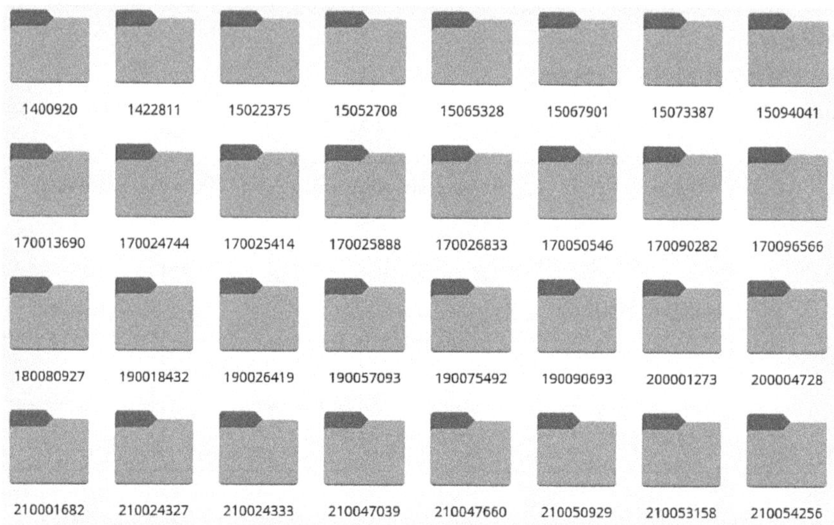

FIGURE 6.2 DICOM repository.

parsing the report, you can extract the BI-RADS score and associate it with the corresponding patient.

Next, you can use the extracted metadata to map the BI-RADS score to the corresponding DICOM files. This mapping will allow you to identify the specific DICOM files corresponding to each patient and their corresponding BI-RADS score. Finally, you can filter the DICOM files based on the BI-RADS score to keep only the relevant files for your analysis. For instance, if you are interested in analyzing cases with a higher likelihood of cancer, you can filter the DICOM files to keep only the ones with a BI-RADS score of 4 or higher. By following these steps, you can preprocess DICOM data and extract and keep data based on the BI-RADS score by looking at the report.

As displayed in Figure 6.2, the patient ID in the DICOM file is matched with the created folder into which the DICOM files and report are moved. So, a folder contains four DICOM images and a description for labelling.

6.3.3 Converting into Photometric Interpretation

In DICOM, monochrome1 and monochrome2 are two possible values for the Photometric Interpretation (0028,0004) attribute, which specifies how pixel data is encoded and displayed. The Photometric Interpretation attribute is critical for image processing and display since it determines whether the image is displayed in a positive or negative format.

The monochrome1 value indicates that the pixel values represent a negative image, whereas the lowest numerical values represent the brightest pixels. In other words, black represents the highest numerical value, and the lowest represents white.

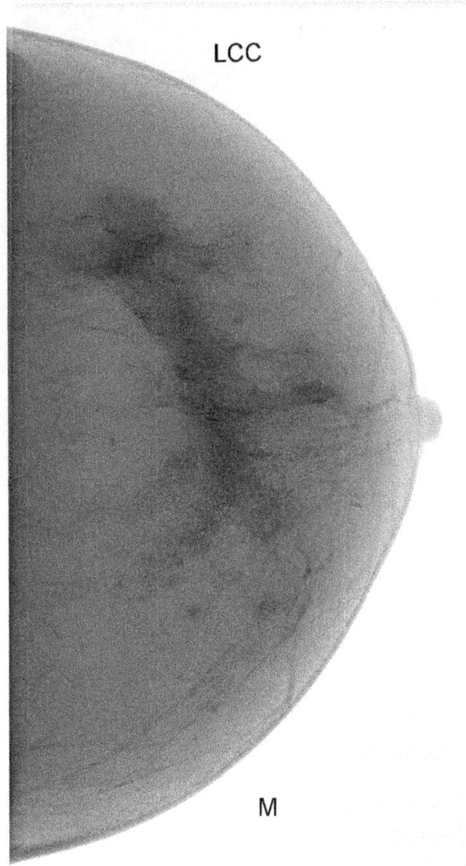

FIGURE 6.3 Sample monochrome1 image.

The monochrome2 value indicates that the pixel values represent a positive image, where the brightest pixels are represented by the highest numerical values, and black is represented by the lowest numerical value.

Monochrome1 is the reversed monochrome image, and higher pixel values are displayed as blacker, such as in Figure 6.3.

Monochrome2 is the normal monochrome image, and higher pixel values are displayed as whiter, such as in Figure 6.4. To proceed further, convert all the monochrome1 to 2.

6.3.4 Labelling the Data

Each DICOM file has the patient ID. Using that ID, match the report and label them according to the BI-RADS score. A bad score can range from 0 to 6, making a total of 7 categories. After labelling, the total number of cases is displayed in Table 6.2.

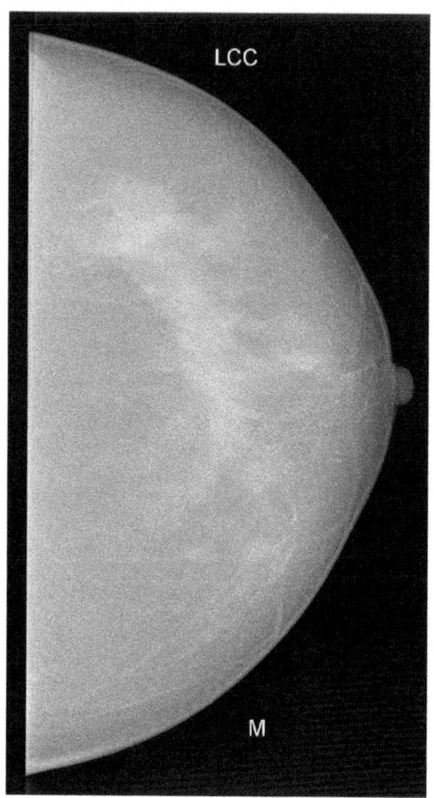

FIGURE 6.4　Sample monochrome2 image.

TABLE 6.2

BI-RADS Score Categories and Total Number of Cases in Each Category

BI-RADS Score	Total No. of Cases
0	72
1	256
2	35
3	22
4	14
5	15
6	1

As the table shows, there were very few abnormal cases. It's not good to train with less data, so instead of making seven categories, there can be four categories: 0, 1, joining BI-RADS 2 and 3 because they're benign and joining BI-RADS 4,5 and 6 because they're malignant, as seen in Table 6.3.

TABLE 6.3

Updated BI-RADS Score Categories and Number of Cases

BI-RADS Score	Total No. of Cases
0	72
1	256
2 and 3	57
4, 5 and 6	22

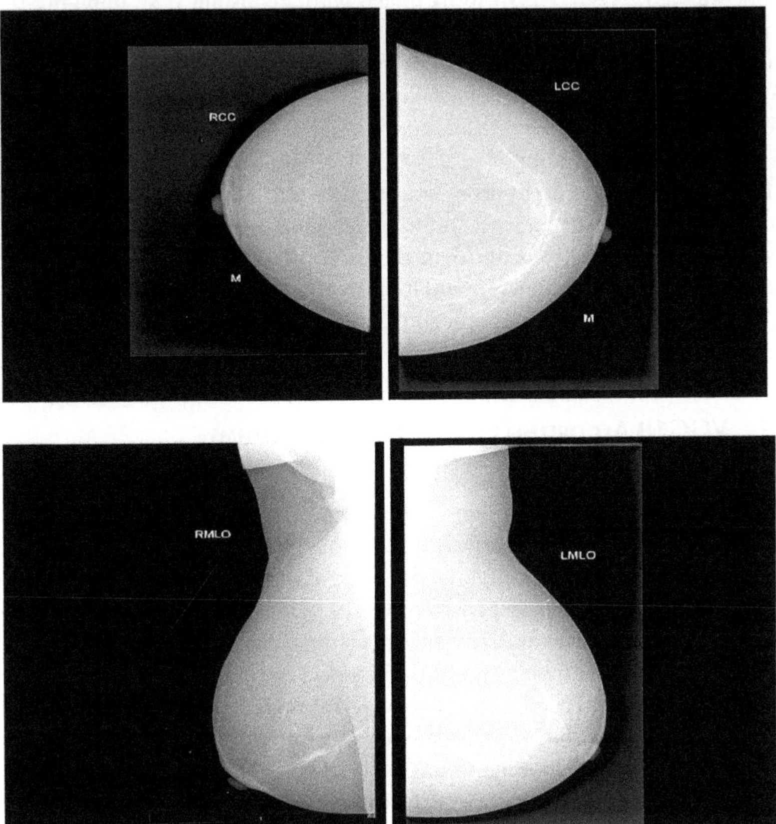

FIGURE 6.5 Sample mammogram used in this study.

The shape of the array shows that it is a three-layered matrix. The first two numbers here are length and width, and the third number (i.e., 3) is for three layers: red, green, and blue. So, if we calculate the size of an RGB image, the total size will be counted as height × width × 3. The sample image of the mammogram is displayed in Figure 6.5.

6.3.5   Splitting the Data to Train Test

Before splitting the data to train and test, I converted the DICOM file to. png format and resized the data into 224 × 224. Further, the data is converted into grayscale, and the image is normalized.

6.3.6   VGG19 Architecture

VGG19 is a VGG model variation that consists of 19 layers (16 convolution layers, 3 fully linked layers, 5 MaxPool layers, and 1 SoftMax layer). Other VGG variations include VGG11, VGG16, and more. VGG19 has a total of 19.6 billion FLOPs. This network was fed a (224 * 224) RGB image with a constant size, implying that the matrix had the structure (224,224,3). The only preprocessing done was to subtract the mean RGB value from each pixel over the whole training set. They used kernels of (3 * 3) size with a stride size of 1 pixel to cover the entire visual concept.

To keep the image's spatial resolution, spatial padding was applied. Stride 2 was used to conduct max pooling over a 2 * 2-pixel window. This was followed by the rectified linear unit (ReLu) to bring nonlinearity into the model to enhance classification and computing speed, while prior models employed tanh or sigmoid functions, which proved significantly better than those. Three fully linked layers were implemented, the first two of which were 4096 in size, followed by a layer with 1000 channels for 1000-way ILSVRC classification, and the final layer is a softmax function. The detailed architecture of the VGG 19 is displayed in Table 6.4.

6.3.7   VGG19 Algorithm versus Other Algorithms

Though many other algorithms have been used by researchers for similar studies, we choose to use VGG19 for the following reasons. Compared to other deep learning models, VGG19 has a straightforward design that makes it simple to understand and implement. However, it has a high number of parameters, which can make training and using it computationally expensive. VGG19 produced cutting-edge results on the ImageNet dataset and was utilized as a benchmark model for image classification tasks. VGG19 comes with many pretrained models, making it ideal for transfer learning in other computer vision tasks. The VGG19 model learns to extract rich visual features that can be used in various computer vision tasks.

Due to the employment of numerous 3 × 3 filters in each convolutional layer, it is a particularly common method for picture classification. This demonstrates that 16 convolutional layers are utilized for feature extraction, and the following three layers are used for classification. The feature extraction layers are divided into five groups, each of which is followed by a max-pooling layer. This model receives an image of size 224 × 224 and outputs the label of the object in the image. A pretrained VGG19 model is used to extract features in this experimental study.

6.3.8   Modelling the Data

Modelling data for VGG19 involves several steps, starting with data preprocessing. This is an important step because it ensures that the input images are in the correct

TABLE 6.4

Detailed Architecture of the VGG 19 Algorithm

ConvNet Configuration					
A	A-LRN	B	C	D	E
11 Weight layers	11 Weight layers	13 Weight layers	16 Weight layers	16 Weight layers	19 Weight layers
Input (224×224 RGB image)					
conv3–64	conv3–64	conv3–64	conv3–64	conv3–64	conv3–64
	LRN	**conv3–64**	**conv3–64**	**conv3–64**	**conv3–64**
maxpool					
conv3–128	conv3–128	conv3–128	conv3–128	conv3–128	conv3–128
		conv3–128	**conv3–128**	**conv3–128**	**conv3–128**
maxpool					
conv3–256	conv3–256	conv3–256	conv3–256	conv3–256	conv3–256
conv3–256	conv3–256	**conv3–256**	conv3–256	conv3–256	conv3–256
			conv1–256	**conv3–256**	conv3–256
					conv3–256
maxpool					
conv3–512	conv3–512	conv3–512	conv3–512	conv3–512	conv3–512
conv3–512	conv3–512	conv3–512	conv3–512	conv3–512	conv3–512
			conv1–512	**conv3–512**	conv3–512
					conv3–512
maxpool					
conv3–512	conv3–512	conv3–512	conv3–512	conv3–512	conv3–512
conv3–512	conv3–512	conv3–512	conv3–512	conv3–512	conv3–512
			conv1–512	**conv3–512**	conv3–512
					conv3–512
maxpool					
FC-4096					
FC-4096					
FC-1000					
softmax					

format for the VGG19 model, which is a CNN that is 16 layers deep. Typically, the images are resized to a fixed size (224×224), and the pixel values are normalized. In addition, the images may need to be converted to the appropriate colour space, such as RGB or BGR, depending on the implementation of the VGG19 model.

Once the images have been preprocessed, the next step is to load the VGG19 model architecture. This can be done using a pretrained model provided by a deep learning library such as TensorFlow, Keras, or PyTorch. The VGG19 model is a convolutional neural network that is trained on a large dataset of images, and it has been shown to be effective for a wide range of image recognition tasks.

After loading the VGG19 model, the next step is to apply data augmentation techniques to the input images. Data augmentation can help to improve the robustness and generalization of the VGG19 model by generating new training data from the original images. Techniques such as random cropping, flipping, and rotation can be used to create variations of the input images.

At this point, the VGG19 model can be used for feature extraction. This involves removing the final fully connected layers of the network and using the output of the last convolutional layer as input to another classifier or regression model. This is a powerful technique because the VGG19 model has already learned to recognize a wide range of features in images, so the output of the last convolutional layer can be used as a general-purpose feature representation for a wide range of image recognition tasks.

Finally, it is also possible to fine-tune the VGG19 model on a specific dataset. This involves training the last few layers of the network on the new dataset while keeping the earlier layers fixed. This can improve the accuracy of the model on the specific task, but it requires a large amount of training data and can be computationally expensive.

6.4 FINDINGS

In this confusion matrix, each row represents the actual class, while each column represents the predicted class. The diagonal entries represent the true positive (TP) instances, while the off-diagonal entries represent the false positive (FP) instances. The values in each row sum up to the total number of instances for each actual class, while the values in each column sum up to the total number of instances predicted for each predicted class. Based on the confusion matrix, we can calculate various evaluation metrics, such as accuracy, precision, recall, and F1 score for each class, represented in Table 6.5. The overall accuracy of the model is 90%, which means that out of all the instances, 90% were predicted correctly.

In addition to the true positives (TP), false positives (FP), true negatives (TN), and false negatives (FN), we can calculate several other metrics: Precision: the proportion of true positives among all predicted positives, calculated as TP/(TP + FP). Recall: the proportion of true positives among all actual positives, calculated as TP/(TP + FN). F1 score: the harmonic mean of precision and recall, calculated as 2 * (precision * recall)/(precision + recall). Accuracy: the proportion of correct predictions among all predictions, calculated as (TP + TN)/(TP + TN + FP + FN). The precision, recall, and F1 score for each class can be calculated and are represented in Table 6.6.

The various parameters that were used in this experimental study were also compared with other works that have been done earlier in the domain and are presented

TABLE 6.5

Confusion Matrix

Predicted Class	Class1	Class2	Class3	Class4
Class 1 (n = 72)	65	5	1	1
Class 2 (n = 256	10	230	14	1
Class 3 (n = 57)	0	8	48	1
Class 4 (n = 20)	0	0	1	19

TABLE 6.6

Confusion Matrix

Predicted Class	Precision	Recall	F1-Score
Class 1 (n = 72)	0.897	0.903	0.900
Class 2 (n = 256	10	230	14
Class 3 (n = 57)	0	8	48
Class 4 (n = 20)	0	0	1

TABLE 6.7

Comparison with Other Work in the Field

Study	Data Source	Algorithm	Total Data Points	Data Types	Format	Class	Class Type	Accuracy
Yadavendra and Chand (2020)	Open source	Logistic regression, random forest, support vector classifier AdaBoost, Bagging Voting and Xception	277,524	PET scan	.png	2	Benign Malignant	90%
Tsochatzidis et al. (2019)	Open source	Convolutional neural network (CNN)	10,239	CBIS-DDSM	DDSM-400	3	Normal Benign Malignant	80%
Benbrahim et al. (2020)	Open source	Neural network	569	FNA	Numeric	2	Malignant Benign	92%
Zheng et al. (2020)	Open source	Mask R-CNN	5,000	Mammogram	.png	3	Benign Malignant Normal	96.4%
Present study	Father Muller Medical College	VGG-19	1,620	Mammogram	DICOM	4	Incomplete Normal Benign Malignant	90%

in Table 6.7. This work was done on the real-time dataset that was collected from the Health Care Institution and gave us good accuracy. One more advantage of this model is that it is built over the DICOM format, which is the global standard.

6.5 CONCLUSION

This AI model can be used to assist radiologists when interpreting mammography images. With an overall precision of 90%, this prediction model produces results that can be used by radiologists. The use of this approach for mammography analysis and report writing will also save time for the radiologist. To improve the model's precision, more mammography images from different medical facilities and other sites can be included in the training dataset.

ACKNOWLEDGMENTS

The investigation was conducted in a lab supported by the Karnataka government's VGST, K- FIST(L2)-545, and data were obtained from Father Muller Hospital using a protocol number (FMMCIEC/CCM/2165/2021).

CONFLICT OF INTEREST

On behalf of all authors, the corresponding author states that there is no conflict of interest.

REFERENCES

Abeje, S., Seme, A., and Tibelt, A. 2019. Factors associated with breast cancer screening awareness and practices of women in Addis Ababa, Ethiopia, *BMC Women's Health* 1–12.

Agide, F.D., Sadeghi, R., Garmaroudi, G., and Tigabu, B.M. 2018. A systematic review of health promotion interventions to increase breast cancer screening uptake: From the last 12 years, *European Journal of Public Health* 1149–1155.

Antony, M.P., Surakutty, B., Vasu, T.A., and Chishti, M. 2018. Risk factors for breast cancer among Indian women: A case-control study, *Nigerian Journal of Clinical Practice* 436–442.

Benbrahim, H., Hachimi, H., and Amine, A. 2020. Comparative study of machine learning algorithms using the breast cancer dataset, *Advanced Intelligent Systems for Sustainable Development (AI2SD'2019)*. Chicago: IEEE, 1–12.

Chen, Y., Sun, J., Huang, L.C., Xu, H., and Zhao, Z. 2015. Classification of cancer primary sites using machine learning and somatic mutations, *BioMed Research International* 1–9.

Dang Vu, Q. et al. 2019. Methods for segmentation and classification of digital microscopy tissue images, *Frontiers in Bioengineering and Biotechnology* 1–12.

Ebrahimpour, M., and Vosooghifard, H. 2015. Applying grey wolf optimizer-based decision tree classifier for cancer classification on gene expression data, *5th International Conference on Computer and Knowledge Engineering (ICKE)*. Mashhad: IEEE, 147–151.

Gravina, M., Marrone, S., Sansone, M., and Sansone, C. 2021. DAECNN: Exploiting and disentangling contrast agent effects for breast lesions classification in DCE-MRI, *Pattern Recognition Letters* 145.

Kumar, P., Srivastava, S., Mishra, R.K., and Sai, Y.P. 2020. End-to-end improved convolutional neural network model for breast cancer detection using mammographic data, *The Journal of Defense Modeling and Simulation* 375–384.

Mehmood, M. et al. 2021. Machine learning enabled early detection of breast cancer by structural analysis of mammograms, *Computers, Materials and Continua* 641–657.

Murtaza, G. et al. 2020. Deep learning-based breast cancer classification through medical imaging modalities: State of the art and research challenges, *Artificial Intelligence Review* 1655–1720.

National Cancer Society. December 10, 2021. Accessed February 15, 2023. www.cancer.gov/types/breast/patient/breast-prevention-pdq.

Ping, Q., Yang, C.C., Marshall, S.A., Avis, N.E., and Ip, E.H. 2019. Breast cancer symptom clusters derived from social media and research study data using improved K-Medoid clustering, *IEEE Transactions on Computational Social Systems* 63–72.

Prusty, R.K. et al. 2020. Knowledge of symptoms and risk factors of breast cancer among women: A community-based study in a low socio-economic area of Mumbai, India, *BMC Women's Health* 1–12.

Ramadan, S.Z. 2020. Using a convolutional neural network with a cheat sheet and data augmentation to detect breast cancer in mammograms, *Computational and Mathematical Methods in Medicine* 1–9.

Rodriguez-Ruiz, A., Lång, K., Gubern-Merida, A., Broeders, M., Gennaro, G., Clauser, P., Helbich, T.H., Chevalier, M., Tan, T., Mertelmeier, T., Wallis, M.G., Andersson, I., Zackrisson, S., Mann, R.M., and Sechopoulos, I. 2019a. Stand-alone artificial intelligence for breast cancer detection in mammography: Comparison with 101 radiologists, *Journal of the National Cancer Institute* 916–922.

Rodriguez-Ruiz, A., Krupinski, E., Mordang, J.-J., Schilling, K., Heywang-Köbrunner, S.H., Sechopoulos, I., Mann, R.M. 2019b. Detection of breast cancer with mammography: Effect of an artificial intelligence support system, *Radiology* 1–10.

Sathwara, J., Balasubramaniam, G., Bobdey, S., Jain, A., and Saoba, S. 2017. Sociodemographic factors and late-stage diagnosis of breast cancer in India: A hospital-based study, *Indian Journal of Medical Pediatric Oncology* 277–281.

Tripathi, P., Tyagi, S., and Natha, M. 2019. A comparative analysis of segmentation techniques for lung cancer detection, *Pattern Recognition and Image Analysis* 167–173.

Tsochatzidis, L., Costaridou, L., and Pratikakis, I. 2019. Deep learning for breast cancer diagnosis from mammograms—a comparative study, *Journal of Imaging* 1–10.

Vishwakarma, G. et al. 2019. Reproductive factors and breast cancer risk: A meta-analysis of case-control studies in Indian women, *South Asian Journal of Cancer* 80–84.

Wahab, N., and Khan, A. 2020. Multifaceted fused-CNN based scoring of breast cancer whole-slide histopathology images, *Applied Soft Computing* 1–12.

Wang, S., Chen, F., Gu, J., and Fang, J. 2015. Cancer classification using collaborative representation classifier based on non-convex LP-norm and novel decision rule, *Seventh International Conference on Advanced Computational Intelligence (ICACI)*. Wuyi: IEEE, 189–198.

World Health Organization. March 26, 2021. Accessed January 13, 2023. www.who.int/news-room/fact-sheets/detail/breast-cancer.

Yadavendra, and Chand, S. 2020. A comparative study of breast cancer tumor classification by classical machine learning methods and deep learning method, *Machine Vision and Applications* 1–14.

Yap, M.H. et al. 2018. Automated breast ultrasound lesions detection using convolutional neural networks, *IEEE Journal of Biomedical and Health Informatics* 1218–1226.

Zhang, Y. et al. 2020. Automatic detection and segmentation of breast cancer on MRI using mask R-CNN trained on non–fat-sat images and tested on fat-sat images, *Academic Radiology* 1–10.

Zheng, J., Lin, D., Gao, Z., Wang, S., He, M., and Fan, J. 2020. Deep learning assisted efficient AdaBoost algorithm for breast cancer detection and early diagnosis, *IEEE Access* 1–20.

7 Optimization Techniques and Their Applications in Prenatal Congenital Heart Defects

A Survey

D. Kavitha and R. Geetha

7.1 INTRODUCTION

The Cardiological Society of India has found 1 child in every 10% births still has prenatal disease related to the heart. A transvaginal sonography study found that the likelihood of successfully visualizing the four-chamber view, left ventricular outflow view, pulmonary trunk with three-vessel view (3VV) and crossover of the major arteries with gestational age, rose from 21% in week 12 to 90% in week 15. A study proved that a high frequency linear transducer had a 94.1% accuracy in finding congenital heart defects (CHDs) in women who had chorionic villus sampling between 12 and 15 weeks of pregnancy. Congenital heart disease is a common disease found among foetuses, the defects vary from mild to severe based on the type of anomalies present in the foetus. The presence of a minute hole in the heart comes under the category of mild defect and the absence or underdevelopment of portions of the heart chambers belong to the category of severe congenital heart anomaly. There are 18 different forms of congenital heart disease: aortic stenosis, atrial septal defect, atrioventricular septal defect, coarctation of aorta, pulmonary atresia, pulmonary stenosis, patent ductus arteriosus, transposition of great arteries, tricuspid atresia, double outlet right ventricle, hypoplastic left heart syndrome, ventricular septal defect, truncus arteriosus, tetralogy of Fallot, aortic valve stenosis, complete atrioventricular canal defect, Ebstein's anomaly and single ventricle defect. Several ultrasound visualization planes are used by radiologists to screen images; each plane will infer different details of the image plane. Some of the commonly used visualization planes are the 3VV, four-chamber view (4CV), five-chamber view, outflow view and trachea view. Ogge et al. (2006) found that prenatal CHD screening using the 4CV plane leads to poor sensitivity, resulting in flawed prediction of illness. According to Chaoui (2003) description, insufficient investigation frequently leads to misreporting a normal heart, even when there is a heart anomaly, because cardiac aberration is invisible in 4CV during real-time scanning.

According to a recent survey, image processing and artificial intelligence algorithms play a vital part in the analysis of the image to overcome the constraints. The inherent

DOI: 10.1201/9781003405436-9

speckle noise in the ultrasound images masks the medical diagnosis. This type of noise will reduce the image contrast and resolution. Hence, diagnosing the image with speckle noise is still a challenging task. Basic image processing algorithms (Sridevi and Nirmala 2013) help in denoising the image, segmenting the particular biomarker of the image, image localization and edge detection. Images that have been distorted by impulse noise, Gaussian noise and minor blurring components are subjected to the efficient median filter (Qiu 1996). Peak signal-to-noise ratio (PSNR) findings of several filters were evaluated, and the suggested median filter outperforms other existing filters, as demonstrated by Elnakib et al. (2011). To increase the accuracy and efficiency as iteration increases, different kernel sizes are used by Rubel et al. (2021) for processing the image. The nonlocal weighted estimating technique (Xing et al. 2017) also aids in lowering noise levels and improving accuracy. The results are compared both qualitatively and quantitatively using the Lee, Gamma, and MAP filters. Arif Anaqi Abang Isa et al. (2021) suggested several filters, including the median filter, the hybrid median filter, the Frost filter, the Lee-Diffusion filter, and the Kuan filter (Diller et al. 2019).

Tian et al. (2014) contrast the manual segmentation strategy with the automatic segmentation technique. For segmentation, they suggested using the Markov decision process, which deep reinforcement learning was used to resolve. With the use of both methods, the data can be trained to separate the region of interest (RoI) in medical images. A unique segmentation algorithm for deep learning was put forth by Dozen et al. (2020). This technique aids in the ventricular segmentation in fatal cardiac ultrasound imaging. The forementioned analysis employs various categorization techniques. A superior sparse representation classification approach is presented by Liu et al. (2020) which improves the accuracy of image classification when compared to conventional classification algorithms.

7.2 SIGNIFICANCE OF IMAGE DESPECKLING ALGORITHMS

There are numerous despeckling algorithms in the literature that can be used for both the spatial and frequency domains. The two primary categories of spatial domain filters were those based on statistical methods and those based on speckle noise modelling. A well-known method for minimizing the cost function is described by Qiu (1996). In comparison to the traditional median filter, a novel recursive median filter is intended to reduce mean square error. Images that have been distorted by impulse noise, gaussian noise, and minor blurring components are subjected to the efficient median filter (Baccouche et al. 2020). The PSNR results from several filters were evaluated, and the proposed filter performs better when compared to the current filters (Senel et al. (2001). A revised version of the median filter was utilized to eliminate impulse noise in the images introduced. Anderson et al. (2013) claims that ultrasound contains noise that is of a random and distracting pattern that arises when coherent imaging techniques are used to acquire an image. By accurately interfering the ultrasound images, effective despeckling can be accomplished. An advanced despeckling filter for image smoothing is described by Nirmala and Sridevi (2014).

Mathematical equations that describe a corrupted image are given by

$$\mathrm{Ut} = \mathrm{K} * \Delta u = \mathrm{K}\frac{\partial^2 u}{\partial x^2} + \frac{\partial^2 u}{\partial y^2} \tag{7.1}$$

$$U(x,y,t) = \theta\left(x,y\right) \tag{7.2}$$

where

$\theta(x, y)$ states corrupted image with speckle noise, with K = 1.
x and y are the coordinates of the current pixel to be filtered;
(u, v) are the deformed images,
K is the constant,
Δ denotes the laplacian function.
$U(x,y,t)$ controls the rate of diffusion, to preserve the edges of the image.

Özkan and İnal (2014) discussed the overall picture of different modules with a fuzzy inference rule-based system and adaptive neuro fuzzy inference system that are employed to differentiate hypoplastic left heart syndrome and hypoplastic right heart syndrome. In addition, they also explain the application of morphological algorithms such as dilation and erosion in the hidden neurons. The next generation four-dimensional image analysis and its significance in the field of CHD, as well as the comparison of four-dimensional and three-dimensional imaging are well narrated by this author.

Sridevi and Nirmala (2016) state that 33% to 54% of CHD is still life threatening. Scanning techniques used by sonographers are two-dimensional echocardiography, color Doppler and sonography, which are the standard imaging modalities. However, a concern is the high-level skill sets; harmless ultrasound imaging is highly preferred to other modalities. Empirical and modelling-based techniques are widely used compared to statistical filters. To envision the smoothing process in the image region, Lee first modelled an equivalent number of glances based on statistical features. Although the Lee filter seeks to despeckle the image, it does not offer process flexibility. In order to achieve a trade-off between mean filtering and all filtering (Rubel et al. 2021) used an exponential-shaped kernel for the modelling of statistical adaptive weighted mean filter.

The primary impression of the probabilistic patch-based weighted maximum likelihood estimation (PPWMLE) filter is that it successfully strikes a balance between protecting small details at the edge of the image and denoising the image. Comparing the qualitative outcomes of several advance filters, the PPWMLE filter is shown to be more effective, as stated by Sridevi and Nirmala (2013). According to Elnakib et al. (2011), the random noise can be eliminated using a combination of morphological operators and anisotropic filters. Utilizing these methods reduces the appearance of blurring while maintaining edge information. We therefore conclude that the type of the problem statement and the type of images will determine how well filters will function. By identifying the quality of actual edges in relation to noise, the suggested filter regulates the rate of diffusion. The brightness variations are smoothed out while keeping the contrast in the edges as such. The challenge is to distinguish the image with respect to its noise and its background as the noise level increases for every iteration. Thus, the true value of edges is assessed by taking into account contrast and brightness throughout each iteration. The original image and the preprocessed image using a Perona-Malik filter are shown in Figure 7.1.

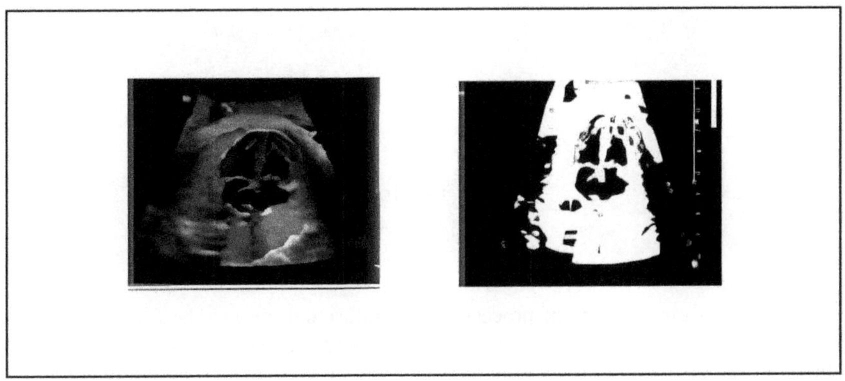

FIGURE 7.1	Comparison of a preprocessed ultrasound image with a denoised foetal image.

The novel anisotropic diffusion filtering technique is proposed to filter the noise inherent in the image. An anisotropic diffusion filter is proven to give better smoothness in conserving the minute details of the image and thereby the uniqueness of the image will never get disturbed or changed. Thus, this filter possesses the same features as that of the original image.

## 7.3	EXPLORATION OF SEGMENTATION ALGORITHMS AND CHALLENGES

This segment deals with the survey of various segmentation algorithms and their importance in ultrasound images. The biomarkers of the images can be easily extracted by using segmentation algorithms as it divides the images into meaningful parts like edges, gradient values, colour, texture and pixel brightness, etc. According to the literature survey by Baccouche et al. (2020), it is one of the most common methods among medical images to divide the image from its background and also texture features. This part is still a challenging part as the similarity between the pixels is very small and it's not easy to segment without missing any details contained in an image. Sundaresan and Sridevi (2013) suggested segregating the images into several sections and then each section is separated into various regions based on the different properties such as grey colour and texture, etc.

It is important to diagnostically predict the region of interest for the prenatal images and which is also a critical task as stated by Elnakib et al. (2011). Zhiqiang Tian et al. (2020) proved the results by comparing manual segmentation with segmentation using optimization algorithms. They used the Markov random process algorithm for extracting the RoI of an image. Arif Anaqi Abang Isa et al. (2021) suggested a combined methodology named K-means clustering with pseudo-color conversion for dividing the image contained intensity similarities like red, blue and green channels. The image with lesions can automatically be diagnosed based on the distinct K means clusters. The aim is to transform every image into useful information, making it simple to conduct further analysis.

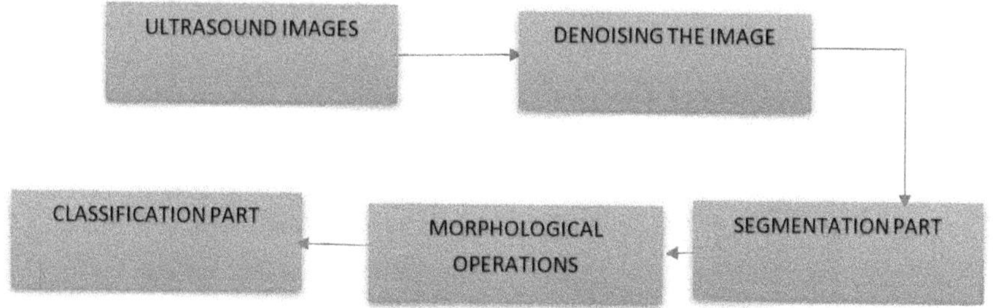

FIGURE 7.2   Various stages of processing the ultrasound image. The four-chamber view ultrasound image is taken as the input; as it has more speckle noise, it is preprocessed using filtering techniques.

The K-means clustering algorithm is described by Dindoyal et al. (2007) as being more straightforward and computationally quicker than the hierarchical type of clustering algorithms. The crucial aspect of this algorithm is the creation of various clusters and initializing the number of centroids. Hence morphological algorithms are carried out to bring out the biomarkers of ultrasound images. The novel K-means clustering algorithm sounds good in describing the structure of any images. The main part is to assign the data points to its center and updating the centroids. Different stages of processing an image are shown in Figure 7.2.

An additional K-means clustering segmentation algorithm is proposed to segment the biomarkers of the CHD images. The combined effect of preprocessing and segmentation algorithms will help the radiologist to easily delineate the biomarkers present in an image, thereby allowing a conclusion as to whether the image is normal or abnormal.

The various steps of the K-means algorithm are as follows:

Step I: The greyscale image is obtained by converting the ultrasound images.

Step II: The ultrasound image is segmented into various clusters, specified as "k," and then the centroid values for each cluster are determined based on the potential value of each pixel in the image.

Step III: The initial cluster should be the one with the maximum potential points from step II.

Step IV: Using the Euclidean distance formula, the mean value of each cluster is calculated and making use of that data, distance between each pixel and the centres of the clusters was measured.

Step V: According to the centroid, if the pixel is adjacent to the same cluster, move to that cluster and continue the process there until that specific cluster is segmented; otherwise, if the pixel is not adjacent to the same cluster, move to the next cluster.

Step VI: Repeat the steps, until all the clusters are segmented.

There was quite a difference between the manually segmenting an image and the image treated by automatic segmentation. Figure 7.3 shows a comparison of both methods.

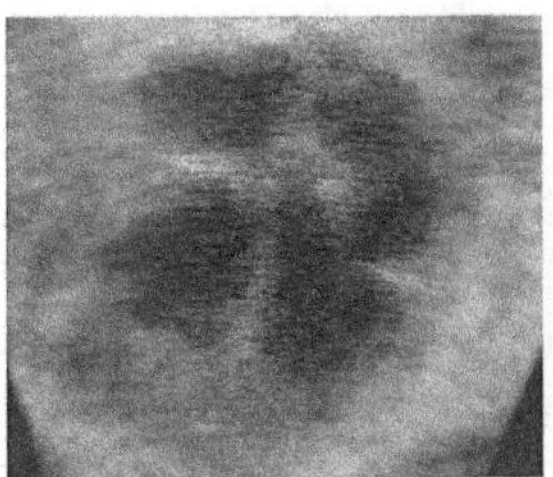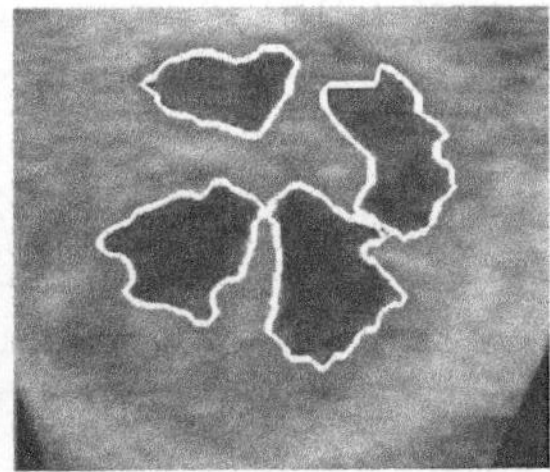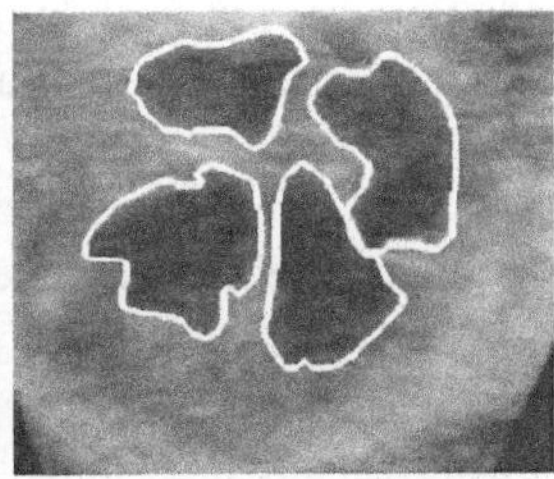

FIGURE 7.3 The qualitative outcome of a manually segmented image compared to automatic segmentation. Although there is a huge variation among the three images, when compared to manual segmentation, automatic segmentation performs better in extracting the region of interest and fine details of an image.

Morphological image processing aids in the performance of binary object analysis using the connected component characteristics. The procedure of extracting the biological sonographic biomarkers of the foetal heart that are diagnostically significant is facilitated by morphological processing.

7.4 CLASSIFICATION ALGORITHMS AND THEIR SIGNIFICANCE IN ULTRASOUND IMAGES

This section discusses different methods presently in use for classifying ultrasound images. The features are divided into numerous classes by this image categorization. Properly chosen features are needed for a good image categorization system. Diller et al. (2019) suggested a dense deep learning algorithm to analyze CHD in adults. This deep learning model was created using raw image data to classify the diagnostic groups based on the disease severity. This convolutional neural network (CNN) method is simply scaled down to improve the accuracy for numerous datasets, and eventually acts as an online decision-making tool. Diller et al. (2019) use a novel CNN for categorizing patients with or without transposition of great arteries (TGA). To study the minimum classification error (MCE), Xing et al. (2017) proposed an approach based on error back propagation. An improved result is obtained by combining the cross entropy and MCE. The aforementioned analysis employs various categorization techniques. The optimal sparse representation classification approach developed by Liu et al. (2020) enhances the accuracy of image classification when compared to traditional classification methods. The effectiveness of approach in diagnosing heart disease is determined using a proportionally high number of classifier algorithms in accordance with the methods recommended by Bharti et al. (2021). A deep learning classifier approach, which reduces the error between the actual and learned output, is frequently employed for large datasets. To diagnose CHD using 4D magnetic resonance images, Zhang (2007) studied a case-based reasoning approach based on neural networks.

According to Abdulshahed et al. (2015), ANFIS—adaptive neuro fuzzy inference system—is a hybrid strategy that combines the fuzzy logic approach based on linguistic visibility with decision-making for the learning capabilities of an artificial neural network (ANN). A novel feature selection algorithm for heart disease prediction was put forth by Li et al. (2020). Through this effort, the diagnostic system's processing time is shortened while also improving classification accuracy. Weighted random forest (WRF), support vector machine (SVM), ANN, and naive Bayes are four learning classifiers that Jakka et al. (2020) generalized. ANN is one of the more effective classifiers in identifying prenatal CHDs. A radial basis function neural network (RBFNN), proposed by Murugesan and Thilagamani (2020), has drawn the attention of many researchers to be incorporated in many engineering applications due to its benefits such as high accuracy, enhanced trustworthiness and convergence rate. RBFNN has been used in the multilayer perceptron case instead of sigmoid functions, but the main issue is that it does not perform well for noisy images and requires a lot of computation. The multilayer back propagation neural network model, the most widely applied neural network model, was created by Rumelhart et al. (1994). It reduces the difference between the actual and predicted results. The different algorithms being used for delineation are given in Table 7.1.

This chapter discusses the importance of using trained or deep learning classifiers to evaluate whether a 2D ultrasound image has an abnormality. To recognize the prenatal ventricular septal abnormalities in the provided images Sridevi and Nirmala (2016) proposed an ANFIS classifier. Because it takes time to train the system, machine learning algorithms are not used. Deep learning algorithms are therefore highly recommended because they will automatically train the system. To train their machine using the bidirectional long short-term memory (BiLSTM) ensemble classification technique, Baccouche et al. (2020) used an unbalanced dataset for heart disease. They demonstrated that BiLSTM achieves a high accuracy when compared to machine learning techniques with an F1 score of 96%. According to Murugesan and Thilagamani (2020), deep learning approaches are frequently utilized to overcome the limits of redundancy, dimensionality reduction and accuracy. Employing the confusion matrix to categorize the images and accurate disease prediction at the proper moment can save the life of the foetus.

7.5   CONCLUSION

The primary basis for ultrasound prenatal CHD screening is clinical experience, and it is subject to clinical practise. One of the frontier areas in prenatal medical diagnostics is the detection of CHD from ultrasound imaging. Congenital heart disease is characterized by anatomical and functional abnormalities of the heart during pregnancy. Some congenital heart disease requires catheterization, surgery or long-term treatment and monitoring since they are very critical and they can lead to miscarriage. The most popular method used in clinical practise for detecting foetal cardiac problems is manual evaluation of ultrasound images. Although prenatal cardiac abnormalities screening and gestational ultrasound cardiac screening have improved, CHD is still the most common birth defect linked to life-threatening illness and mortality. This is due to the fact that a wide variety of screening abilities and diagnostic skills are required for the ultrasound clinical diagnosis of prenatal heart disease. However, because of the inherent speckle noises, the ultrasound modality catches the biological internal heart structures with an incorrect border. To correct all the errors in the ultrasound images, various preprocessing, segmentation and classifier

TABLE 7.1

Performance Analysis of Existing Methods

Type	Author	Method	Dimension	User Interaction	Performance Measures	Comparison
Curvelet transform	Qiu (1996)	Improved Median Filtering	Two-dimensional (2D) ultrasound	Manual	Denoising rate is high	Manual delineation
Computer aided diagnosis (CAD)	Zuluaga et al. (2015)	Atlas-based analysis method	2D MRI	Manual	–	Manual delineation
Different preprocessing methods	Nirmala and Sridevi (2013)	Rayleigh maximum likelihood ilter using fuzzy rules	2D ultrasound	Manual	High peak signal-to-noise ratio, feature similarity index measure and structural similarity index measure	Manual delineation of different methods
Markov random field	Nirmala and Sridevi (2016)	Pre-processing PPBMLE, segmentation MRF	2D ultrasound	Automatic	Segmentation error = 9.83%	Manual delineation
ANFIS	Sridevi and Nirmala (2016)	ANFIS, RBFNN, modified back propagation neural network (MBPNN), KNN classifier	2D ultrasound	Automatic	Average error = 0.013%	Automatic delineation
Different preprocessing methods	Sridevi and Nirmala (2013)	Daubechies wavelet transform, Haar transform, symlet transform and coiflet wavelet transform	2D ultrasound	Manual	Maximum signal-to-noise ratio and minimum root mean square error	Manual delineation of different methods

ANFIS = adaptive neuro fuzzy inference system; KNN = K-nearest neighbour; MBPNN = multi-task back propagation neural network; MRF = Markov random field; PPBMLE = probabilistic patch based maximum likelihood estimation; RBFNN = radial basis function neural network.

algorithms are needed. Work in this area by different authors was discussed in this chapter.

7.6 FUTURE SCOPE

Based on the survey, optimized algorithms in the field of image processing under various categories are identified and advanced deep learning algorithms for processing number of images can be chosen and all these modules are integrated into one design for diagnosing the congenital heart defects. This can diagnose the disease more accurately with less time and can act as a secondary tool for the radiologist and doctors in the future.

REFERENCES

Abdulshahed, A.M., Longstaff, A.P. and Fletcher, S. 2015. The Application of ANFIS Prediction Models for Thermal Error Compensation on CNC Machine Tools, *Applied Soft Computing*, vol. 27, pp. 158–168.

Anderson, M.E. and Trahey, G.E. 2013. *A Beginner's Guide to Speckle, a Seminar on k-Space Applied to Medical Ultrasound*. Retrieved January 4 from http://dukemil.bme.duke.edu/Ultrasound/k-space/node.

Arif Anaqi Abang Isa, A.M., Kipli, K., Jobli, A.T., Mahmood, M.H., Sahari, S.K., Hernowo, A.T. and Hamdan, S. 2021. Pseudo-Colour with K-Means Clustering Algorithm for Acute Ischemic Stroke Lesion Segmentation in Brain MRI, *Pertanika Journal of Science and Technology*, vol. 29, no. 2, pp. 743–757, DOI: 10.47836/pjst.29.2.03.

Baccouche, A., Garcia-Zapirain, B., Castillo Olea, C. and Elmaghraby, A. 2020. Ensemble Deep Learning Models for Heart Disease Classification: A Case Study from Mexico, *Information*, vol. 11, p. 207, DOI: 10.3390/info11040207.

Bharti, R., Khamparia, A., Shabaz, M., Dhiman, G., Pande, S. and Singh, P. 2021. Prediction of Heart Disease Using a Combination of Machine Learning and Deep Learning, *Computational Intelligence and Neuroscience*, vol. 2021, Article ID 8387680, July, DOI: 10.1155/2021/8387680.

Chaoui, R. 2003. The Four-Chamber View: Four Reasons Why It Seems to Fail in Screening Forcardiac Abnormalities and Suggestions to Improve Detection Rate, *Ultrasound in Obstetrics & Gynecology*, vol. 22, no. 1, pp. 3–10.

Diller, G.P., Lammers, A.E., Babu-Narayan, S., Li, W., Radke, R.M., Baumgartner, H., Gatzoulis, M.A. and Orwat, S. 2019. Denoising and Artefact Removal for Transthoracic Echocardiographic Imaging in Congenital Heart Disease: Utility of Diagnosis Specific Deep Learning Algorithms, *The International Journal of Cardiovascular Imaging*, vol. 35, July, DOI: 10.1007/s10554-019-01671-0.

Dindoyal, I., Lambrou, T., Deng, J. and Todd-Pokropek, A. 2007. Level Set Snake Algorithms on the Fetal Heart, *IEEE International Symposium Proceedings on Biomedical Imaging: From Nano to Macro*, pp. 864–867.

Dozen, A., Komatsu, M., Sakai, A., Komatsu, R., Shozu, K., Machino, H., Yasutomi, S., Arakaki, T., Asada, K., Kaneko, S., Matsuoka, R., Aoki, D., Sekizawa, A. and Hamamoto, R. 2020. Image Segmentation of the Ventricular Septum in Fetal Cardiac Ultrasound Videos Based on Deep Learning Using Time-Series Information, *Biomolecules*, vol. 10, DOI: 10.3390/biom10111526.

Elnakib, A., Gimel'farb, G., Suri, J.S. and El-Baz, A. 2011. Medical Image Segmentation: A Brief Survey, in *Multi-Modality State-of-the-Art Medical Image Segmentation and Registration Methodologies*, Springer, New York, pp. 1–39.

Jakka, A. and Vakula Rani, J. 2020. Diagnosis of Progressive Optic Neuropathy Disorder Using Machine Learning Classifiers, *International Journal of Advanced Science and Technology*, vol. 29, no. 5, pp. 7489–7500.

Li, J.P., Haq, A.U., Din, S.U., Khan, J., Khan, A. and Saboor, A. 2020. Heart Disease Identification Method Using Machine Learning Classification in E- Healthcare, *IEEE Access*, vol. 8, June, DOI: 10.1109/ACCESS.2020.3001149.

Liu, J.-E. and An, F.-P. 2020. Image Classification Algorithm Based on Deep Learning-Kernel Function, *Hindawi Scientific Programming*, vol. 2020, pp. 1–14, DOI: 10.1155/2020/7607612.

Murugesan, M. and Thilagamani, S. 2020. Efficient Anomaly Detection in Surveillance Videos Based on Multi-Layer Perception Recurrent Neural Network, *Embedded Hardware Design*, vol. 79, November, DOI: 10.1016/j.micpro.2020.103303.

Nirmala, S. and Sridevi, S. 2013. Modified Rayleigh Maximum Likelihood Despeckling Filter Using Fuzzy Rules, *IEEE Proceedings of International Conference on Information Communication and Embedded Systems*, pp. 755–760.

Nirmala, S. and Sridevi, S. 2014. Fuzzy Connectedness Based Segmentation of Fetal Heart from Clinical Ultrasound Images, *Advanced Computing, Networking and Informatics*, vol. 1, pp. 329–337.

Nirmala, S. and Sridevi, S. 2016. Markov Random Field Segmentation Based Sonographic Identification of Prenatal Ventricular Septal Defect, *Procedia Computer Science*, no. 79, pp. 344–350.

Ogge, G., Gaglioti, P., Maccanti, S., Faggiano, F. and Todros, T. 2006. Prenatal Screening for Congenital Heart Disease with Four-chamber and Outflow-tract Views: A Multicenter Study, *Ultrasound in Obstetrics and Gynecology: The Official Journal of the International Society of Ultrasound in Obstetrics and Gynecology*, vol. 28, no. 6, pp. 779–784.

Özkan, G. and İnal, M. 2014. Comparison of Neural Network Application for Fuzzy and ANFIS Approaches for Multi-Criteria Decision-Making Problems, *Applied Soft Computing*, vol. 24, pp. 232–238, November, DOI: 10.1016/j.asoc.2014.06.032.

Qiu, G. 1996. An Improved Recursive Median Filtering Scheme for Image Processing, *IEEE Transactions on Image Processing*, vol. 5, no. 4, pp. 646–648, April.

Rubel, O., Lukin, V., Rubel, A. and Egiazarian, K. 2021. Selection of Lee Filter Window Size Based on Despeckling Efficiency Prediction for Sentinel SAR Images, *Remote Sensing*, vol. 13, p. 1887, May, DOI: 10.3390/rs13101887.

Rumelhart, D.E., Widrow, B. and Lehr, M.A. 1994. The Basic Ideas in Neural Networks, *Communications of the ACM*, vol. 37, no. 3, pp. 87–93.

Senel, H.G., Peters II, R.A. and Dawant, B. 2001. Topological Median Filter, *IEEE Transactions on Image Processing*, vol. 10, no. 12, pp. 1–16, December.

Sridevi, S. and Nirmala, S. 2013. Generalization of Rayleigh Maximum Likelihood Despeckling Filter Using Quadrilateral Kernels, *ICTACT Journal on Image and Video Processing*, vol. 3, no. 3, pp. 559–564.

Sridevi, S. and Nirmala, S. 2016. ANFIS Based Decision Support System for Prenatal Detection of Truncus Arteriosus Congenital Heart Defect, *Applied Soft Computing*, vol. 46, pp. 577–587.

Sundaresan, M. and Sridevi, S. 2013. Survey of Image Segmentation Algorithms on Ultrasound Medical Images, *Prime*, pp. 215–220, February, DOI:10.1109/ICPRIME.2013.6496475.

Tian, Y., Chen, Q., Wang, W., Peng, Y., Wang, Q., Duan, F., Wu, Z. and Zhou, M. 2014. A Vessel Active Contour Model for Vascular Segmentation, *Journal of Biomedicine and Biotechnology*, vol. 6, DOI: 10.1155/2014/106490.

Tian, Z., Si, X., Zheng, Y., Chen, Z. and Li, X. 2020. Multi-step Medical Image Segmentation Based on Reinforcement Learning, *Journal of Ambient Intelligence and Humanized Computing*, March, DOI: 10.1007/s12652-020-01905.

Xing, X., Chen, Q., Yang, S. and Liu, X. 2017. Feature-Based Nonlocal Polarimetric SAR Filtering, *Remote Sensing*, vol. 9, p. 1043, October, DOI:10.3390/rs9101043.

Zhang, H. 2007. *Segmentation and Computer-Aided Diagnosis of Cardiac MR Images Using 4-D Active Appearance Models*, vol. 154, Theses and Dissertations, University of Iowa, Iowa, DOI: 10.17077/etd.jso8hznb.

Zuluaga, M.A., Burgos, N., Mendelson, A.F., Taylor, A.M. and Ourselin, S. 2015. Voxel Wise Atlas Rating for Computer Assisted Diagnosis: Application to Congenital Heart Diseases of the Great Arteries, *Medical Image Analysis*, vol. 26, no. 1, pp. 185–194.

8 Analyzing Aortic Stenosis Diagnosis and Medication with Artificial Intelligence

Densil Raj V. Francis and L. R. Aravind Babu

8.1 INTRODUCTION

Aortic stenosis is the most common and serious valve disease. It is serious because of the progressive nature of narrowing the aortic valve, which reduces the flexibility to fully open and close, shown in Figure 8.1. Aortic stenosis mainly affects people aged 65 and older due to scarring and calcium deposits on the valve leaflets. Although aortic stenosis is age-related and begins after age 60 the sad part is that it will not develop any symptoms for years. The European Society of Cardiology (ESC)/European Association for Cardio-Thoracic Surgery (EACTS)/American College of Cardiology (ACC) has formulated certain guidelines for the management of valvular heart disease that helps in the study of various cardiovascular diseases (CVDs; Nishimura et al. 2017). These guidelines provide complete, evidence-based suggestions for the management of patients with aortic stenosis. In real-world scenarios, we have practical challenges in the management of patients with aortic stenosis due to a lack of evidence-based information. This study aimed to review the current guidelines and recommendations to highlight the important gap that prevails in the management of patients with aortic stenosis.

The fibrotic changes in the aortic valve leaflets eventually progress to extensive calcification, further obstructing left ventricular outflow. The pathobiology of aortic stenosis is not attributed solely to aging but involves a complex inflammatory process. This inflammatory process is characterized by endothelial damage due to lipid accumulation, oxidative stress, angiogenesis, and genetic factors.

The inflammatory process involved in aortic stenosis leads to fibrosis and thickening of the valve leaflets. These changes ultimately result in calcification of the valve leaflets, further narrowing the aortic valve opening and causing a pressure gradient across the valve. The pathobiology of aortic stenosis involves progressive narrowing of the aortic valve and subsequent changes in the myocardium (Baumgartner et al. 2017). In response to the increased afterload caused by aortic valve narrowing, the left ventricle undergoes hypertrophy to maintain wall stress and cardiac function. This cascade of compensatory responses leads to alterations in the myocardial cellular structure, resulting in fibrosis. Furthermore, cellular changes in the aortic

DOI: 10.1201/9781003405436-10

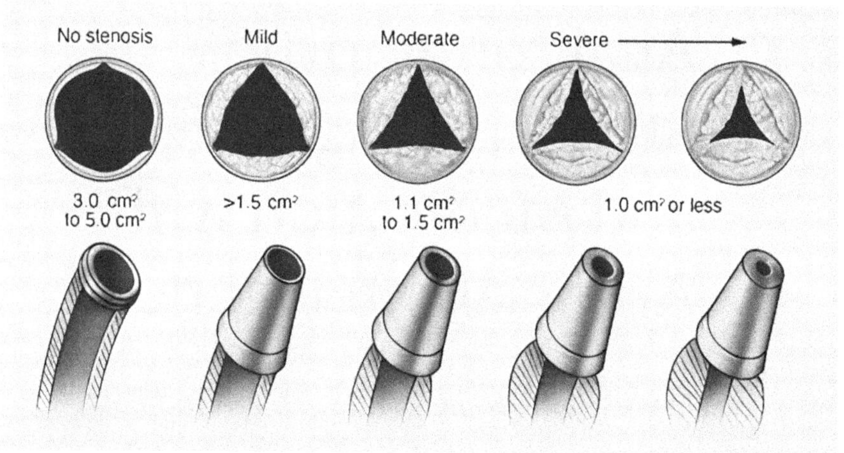

FIGURE 8.1 Stages in aortic stenosis (www.mayoclinic.org/diseases-conditions/aortic-stenosis/symptoms-causes/syc-20353139).

valve leaflets lead to fibrotic thickening and eventually extensive calcification. These changes ultimately lead to significant aortic valve stenosis and left ventricular outflow obstruction, necessitating surgical valve replacement as the only viable treatment option.

The calcification of the aortic valve leaflets further narrows the aortic valve opening, leading to increased leaflet stiffness and the development of a pressure gradient across the valve (Otto et al. 2021).

8.2   INTERPRETATION

Each valve has flaps, which are sometimes called cusps or leaflets, that open and close during the heartbeat. Sometimes these valves will not open and close properly, which leads to minimal blood flow. The valve is made of three leaflets of tissue that swing open when blood pushes against them. The leaflets will no longer fully open or close properly when they get stiffened and lose their flexibility. This will lead to the narrowing of the valve opening that causes aortic stenosis. Due to the narrowing of the valve, the heart needs to work harder. Further, this reduces the blood flow fully or partially from the heart to the aorta and other parts of the body.

8.3   DIAGNOSIS

8.3.1   BENCHMARK

The identification of aortic stenosis is done through a physical examination that reveals a systolic murmur. Patients with undetermined systolic murmur are advised to undergo a transthoracic echocardiogram (TTE), which records the heart sound for a single second and is used to reveal the history of bicuspid aortic valve or other symptoms due to aortic stenosis. The diagnosis is based on the presence of an abnormal valve, whether calcified or congenital or thickened, with restricted opening and closing. When aortic stenosis is diagnosed in asymptomatic patients there is a need for repeated TTE to monitor the progression of the disease, which ranges from mild to moderate to severe. As a general guideline, patients with mild aortic stenosis should udergo TTE at least every 3 to 5 years; those patients with moderate aortic stenosis should undergo TTE every 1 to 2 years; and patients with severe aortic stenosis should undergo TTE every 6 to 12 months, unless a second opinion is required due to clinical signs or symptoms.

8.3.2 COGNIZANCE GAPS

Most of the time detecting the systolic murmur is unreliable in the initial diagnosis of aortic stenosis, as well as the disease progression. A detailed study finds that there is poor sensitivity and specificity in diagnosing valvular heart disease with the help of auscultation by both general physicians (specificity 69% and sensitivity 44%), as well as cardiologists (specificity 81% and sensitivity 31%). When it comes to the diagnosis of overweight patients through physical examination it shows a very poor accuracy. It has been proposed to perform the screening with office-based TTE as well as point-of-care ultrasound as the progression rate of aortic stenosis increases in the elderly population based on age factor. The patient groups to be screened, the clinical outcomes, and the financial implications of echocardiogram screening of aortic stenosis remain to be established.

8.4 MEDICAL THERAPY

8.4.1 BENCHMARK

Most people suffering from aortic stenosis may not complain of any symptoms. The reason for this is that aortic stenosis is a slow, progressive condition that cannot be indicated until late in the progress of the disease. A kind of medical intervention is required which may result in slowing down the disease progression or preventing severe aortic stenosis. Prior studies have revealed that there is a strong relationship between traditional cardiovascular risk factors such as hypertension, diabetes, anxiety, and certain genetic transformations in the development of calcific valve disease (Yan et al. 2017). But so far, no medical treatment is said to prove its beneficial part in either preventing or treating aortic stenosis. In adult patients with asymptomatic mild to severe aortic stenosis, randomized trials of statin therapy do not provide any efficient outcome. However, these trials helped with intervention only after a valve obstruction was identified. Overall, it seems that the optimal control of cardiovascular risk factors would reduce an individual's lifetime or after the progression of aortic stenosis.

The latest guidelines mainly focus on the management of cardiac comorbidity conditions that reduce the capacity of working and standard cardiovascular risk reduction in adult patients with calcific aortic stenosis along with the optimal management of hypertension. Angiotensin-converting enzyme (ACE) inhibitors play a vital role in preventing an enzyme in the body from making angiotensin 2, a substance that narrows down the blood vessels. Also, the angiotensin receptor blockers (ARB) are used to treat heart failures in order to decrease the left ventricular (LV) fibrosis, which in turn reduces the pressure of aortic stenosis (Shah et al. 2023). In general, medical therapies do not give active outcomes but still with the optimization of cardiovascular risk factors, the overall cardiovascular outcomes may be improved.

8.4.2 Cognizance Gaps

There is no medical evidence that general medication is effective in treating Aortic Stenosis. Still, there is some hope, based on recent scientific research, that cellular and molecular basis of calcific valve disease may be reduced with this targeted medical therapy in the future. To identify the people with more risk due to calcific valvular disease, epidemiologic and genetic studies can be used. Mendelian randomization studies can be used to look at whether levels of a substance found naturally in the body are associated with the development of calcific valvular disease.

8.4.3 Timing of Medical Intervention

It is difficult to decide when to intervene in the progress of aortic stenosis as multiple factors need to be considered. Even though medical technology develops it is very difficult to categorize symptomatic and asymptomatic aortic stenosis. Sometimes asymptomatic aortic stenosis will lead to sudden cardiac arrest, which will lead to death (Banović et al. 2022). However, there are significant advances in the safety measures and effectiveness of both surgical and transcatheter aortic valve replacement (TAVR; Foroutan et al. 2016) but each has its own risk.

8.5 TRANSCATHETER AORTIC VALVE IMPLANTATION

Transcatheter aortic valve replacement (TAVR), commonly referred to as transcatheter aortic valve implantation (TAVI), is a medical procedure in which a specific valve is introduced into the compromised valve. TAVI offers a less intrusive alternative to conventional open-heart surgery.

Preceding the TAVI procedure, the patient is administered either a general or local anaesthetic, and the entire process usually spans between 1 and 2 hours. A catheter, possessing a balloon at its extremity, is carefully inserted into an artery located in either the groin or under the collarbone. The catheter is then guided into the heart, where the fresh valve is meticulously positioned within the aortic valve.

8.5.1 Balloon Valvuloplasty

Balloon valvuloplasty is a medical procedure that is intended to enlarge the valve, allowing for enhanced blood flow into the aorta. Through the utilization of a catheter, introduced into a blood vessel within the groin region, the apparatus is carefully maneuvered towards the vicinity of the heart. Subsequently, the catheter's terminal end is positioned within the aortic valve, followed by the inflation of a balloon

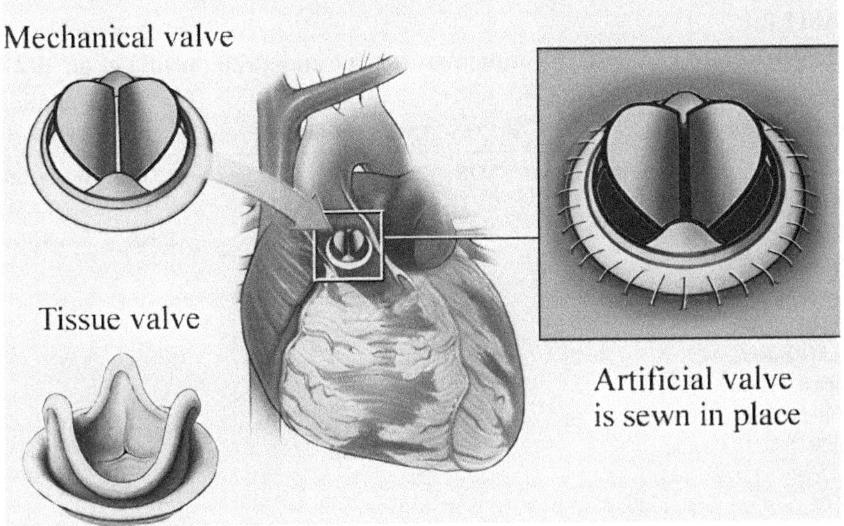

FIGURE 8.2   Types of artificial valve.

to stretch the valve. After the completion of the procedure, the balloon is deflated and both it and the catheter are subsequently extracted. Generally, balloon valvuloplasty is employed as a transitory solution or as a means of alleviating symptoms in instances where alternate approaches are unavailable (Williams and Hildick-Smith 2020). It is plausible that additional surgical interventions or procedures may be required at a later stage in the patient's life.

Following the decision to perform aortic valve replacement the type of valve to be used needs to be carefully chosen. There are two types of valves: bioprosthetic and mechanical (Figure 8.2).

The choice of valve depends on various factors, including patient preference, age, life expectancy, metabolic factors, anticoagulation, previous infection, and re-operation risk. Both types have their drawbacks. Bioprosthetic valves have a finite period of usefulness and the clock starts at the time of valve implantation itself. Mechanical valves increase the risk of bleeding. The most advantageous moment to intervene in a medical condition can be most accurately characterized as the specific juncture within the progression of the ailment at which the advantages of replacing the affected valve surpass the potential dangers associated with the inherent malfunction of the valve in question. In adults with valvular aortic stenosis, the progression of the disease can be divided into multiple stages based on various factors, including the presence of clinical symptoms, the examination of the valve anatomy, the assessment of the severity of outflow obstruction, and the evaluation of the left ventricular (LV) function. High-gradient severe aortic stenosis, also known as Stage D1, which is characterized by a mean gradient exceeding 40 mm Hg and a maximum velocity surpassing 4.0 m/s, is the prevailing phenotype observed in symptomatic patients and is considered the most straightforward to diagnose. Table 8.1 represents the general guidelines for aortic valve replacement based on clinical outcomes.

TABLE 8.1

Comparison of ESA/EACTS Guidelines in 2017 and 2020 (Sevilla et al. 2021)

Indications for AVR in Asymptomatic AS	2017 ESC/EACTS Guidelines			2020 AHA/ ACC Guidelines		
	Class	LOE	Intervention	Class	LOE	Intervention
LVEF <50%	I	C	SAVR	I	B-NR	SAVR, TAVI
Undergoing other cardiac surgery	I	C	SAVR	I	B-NR	SAVR, TAVI
Symptoms on exercise tests related to AS	I	C	SAVR	Considered as symptomatic		
Exercise-induced fall in blood pressure	IIa	C	SAVR	IIa	B-NR	SAVR
Aortic velocity >5.5 m/s (ESC/EACTS); aortic velocity >5.0 m/s (AHA/ACC) and low surgical risk	IIa	C	SAVR	IIa	B-NR	SAVR
Rate of peak transvalvular velocity progression >0.3 m/s per year and low surgical risk	IIa	C	SAVR	IIa	B-NR	SAVR
Repeated elevated (×3) BNP and low surgical risk	IIa	C	SAVR	IIa	B-NR	SAVR
Severe pulmonary hypertension without explanation	IIa	C	SAVR	–		
Progressive decrease in LVEF <60% on ≥3 serial studies	–			IIb	B-NR	SAVR

ACC = American College of Cardiology; AHA = American Heart Association; AS = aortic stenosis; AVR = aortic valve replacement; EACTS = European Association for Cardio-Thoracic Surgery; ESC = European Society of Cardiology; LOE = Level of Evidence; LVEF = left ventricular ejection fraction; SAVR = surgical aortic valve replacement; TAVI = transcatheter aortic valve implantation.

Current Guideline Recommendations for Aortic Valve Replacement in Asymptomatic Aortic Stenosis

However, the presence of severe AS (aortic valve area [AVA] <1.0 cm^2) can be concealed by a low transaortic volume flow rate, despite a small valve area, resulting in a low velocity (or gradient). According to the American Heart Association (AHA)/ American College of Cardiology (ACC) guidelines, low-gradient severe AS is categorized into two stages: low-flow, low-gradient (LF-LG) severe AS with a reduced LV ejection fraction (EF <50%) (Stage D2) and LF-LG severe AS with preserved LVEF (Stage D3). Both conditions can be recognized by the presence of a reduced stroke volume, which is adjusted to the body surface area and falls below 35 mL/m^2. This outcome leads to a calculated aortic valve area (AVA) of less than 1.0 cm^2, despite the relatively low aortic valve gradients. A calcium score exceeding 1200 AU in women and surpassing 2000 AU in men has been established to exhibit a strong correlation with severe AS and serve as a reliable predictor of unfavourable clinical outcomes. Based on the above guidelines as per clinical examination the data is trained and tested with different machine learning and deep learning models.

The evidence that supports the appropriate timing of aortic valve replacement is most compelling for patients who exhibit symptoms as a result of severe high-gradient aortic stenosis. Nevertheless, it is difficult to emphasize that the recommendations for aortic valve replacement do not apply to any of the three variants of aortic stenosis, including Stages D1, D2, and D3. Patients may adapt to their physical limitations without recognizing overt symptoms, as the insidious onset of symptoms makes it difficult to identify them. Consequently, it is crucial to acquire a comprehensive patient history to elicit symptoms and recognize any lifestyle changes that might have occurred as a result of adaptation. Treadmill stress testing can be a valuable tool in cases where patients have equivocal symptoms or are asymptomatic according to their history, as it can help elicit symptoms, measure aerobic capacity, and assess the risk of patients who do not exhibit overt symptoms. If a patient experiences symptoms or has reduced aerobic capacity during stress testing without any other obvious explanation, it is important to consider the possibility of symptomatic valve disease. Furthermore, patients with diminished aerobic capacity or irregular hemodynamic responses to physical activity (such as a rise in blood pressure of less than 20 mm Hg or a hypotensive reaction) exhibit elevated rates of unfavourable occurrences and necessitate assessment for aortic valve replacement.

One promising avenue for the early detection of aortic stenosis is the use of biomarkers. Biomarkers are measurable indicators, such as molecules or substances, that can provide information about the presence or progression of a disease. Several studies have been conducted to identify potential biomarkers for detecting aortic stenosis.

One study, the Early Valve Replacement Guided by Biomarkers of Left Ventricular Decompensation in Asymptomatic Patients with Severe Aortic Stenosis, aimed to identify biomarkers that could predict left ventricular decompensation in asymptomatic patients with severe aortic stenosis (Shah et al. 2023). The study utilized emerging imaging modalities, such as positron emission tomography (PET) and magnetic resonance imaging (MRI), to evaluate the potential of these biomarkers in predicting ventricular decompensation.

The study found that certain biomarkers, such as the uptake of 18-fluorodeoxyglucose (18F-FDG) and 18-sodium fluoride (18F-NaF) on PET/MRI scans, correlated with left ventricular decompensation in asymptomatic patients. These findings highlight the potential role of biomarkers in detecting aortic stenosis and predicting disease progression (Peeters et al. 2019). Furthermore, The Early Valve Replacement in Severe Asymptomatic Aortic Stenosis Study also investigated the use of biomarkers for early detection.

We aimed to determine whether biomarkers could guide the decision for early valve replacement in patients with severe asymptomatic aortic stenosis.

The study utilized various biomarkers, including N-terminal pro-B-type natriuretic peptide, high-sensitivity troponin assays, and imaging biomarkers, to assess their ability to predict adverse outcomes in patients with severe asymptomatic aortic stenosis (Nakatsuma et al. 2018). These biomarkers were found to have predictive value in identifying patients at higher risk of adverse outcomes, leading to potential early intervention and improved patient management. In conclusion, the use of biomarkers holds promise in detecting aortic stenosis and predicting disease progression.

Biomarkers have emerged as valuable tools in the early detection of aortic stenosis and in guiding patient management.

They provide critical information about the presence and progression of the disease by analyzing through various deep learning models, allowing for timely intervention and improved patient outcomes. Moreover, the development of new imaging modalities and clinical trials, such as the ones mentioned above, are further advancing our understanding of biomarkers and their role in detecting aortic stenosis. Collectively, these studies emphasize the importance of biomarkers in detecting aortic stenosis and predicting disease progression. In summary, biomarkers play a significant role in detecting aortic stenosis and predicting disease progression (Redfors et al. 2017). The use of biomarkers in predicting ventricular decompensation in patients with aortic stenosis shows promise. These biomarkers, such as the uptake of 18-fluorodeoxyglucose (18F-FDG) and 18-sodium fluoride PET/MRI scans have shown potential in detecting aortic stenosis and predicting disease progression. In addition to predicting disease progression, biomarkers in aortic stenosis have also been explored for their potential to guide the optimal timing of valve intervention. Using biomarkers in the management of aortic stenosis has been a topic of debate, but recent studies have shown their potential for early detection and prediction of adverse outcomes in patients with severe asymptomatic aortic stenosis. Several types of biomarkers have been studied in patients with aortic valve stenosis, including cardiac troponins and natriuretic peptides. These biomarkers have been shown to correlate with disease severity and patient mortality, making them valuable in risk prediction and prognosis assessment. Furthermore, biomarkers such as B-type natriuretic peptides have been used to predict symptom-free survival and postoperative outcomes in patients with aortic stenosis.

In the context of transcatheter aortic valve implantation, biomarkers have also been investigated for risk prediction.

Biomarkers, such as natriuretic peptides, have shown elevated values in aortic stenosis and have been used to predict symptom-free survival, prognosis, and postoperative outcomes in patients undergoing transcatheter aortic valve implantation. The use of biomarkers in detecting aortic stenosis and predicting disease progression is a topic of great interest in cardiovascular research. The significance of accurate biomarkers in detecting aortic stenosis and predicting disease progression cannot be understated.

8.5.2　MEDICAL ADVANCEMENT

Recent advancements in medical treatment for aortic stenosis have provided alternative options for high-risk patients who may not be suitable candidates for aortic valve replacement. These advancements have allowed for improved management and outcomes in patients with severe aortic stenosis. However, neither of these treatment options has been shown to reduce mortality.

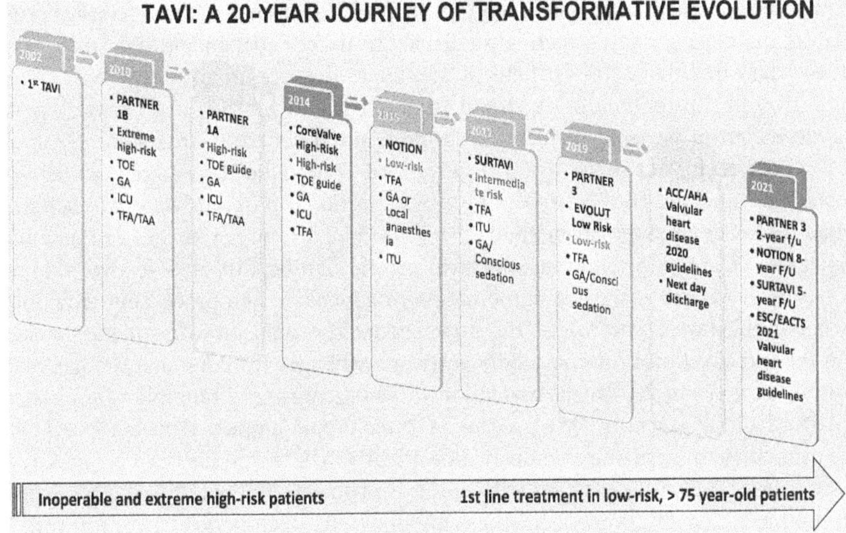

FIGURE 8.3 A transformative model of the transcatheter aortic valve implantation procedure from 2002 to 2021. (Reproduced with permission from Kalogeropoulos et al. 2022.)

Another emerging treatment option for high-risk patients with severe aortic stenosis is transcatheter aortic valve replacement. Transcatheter aortic valve replacement is a less invasive procedure that has shown promise in reducing mortality rates similar to surgical aortic valve replacement aortic valve replacement. To date, aortic valve replacement remains the only definitive therapy for aortic stenosis. In summary, recent medical treatment for aortic stenosis has focused on the early detection of ventricular decompensation, as well as providing alternative options for high-risk patients who are not suitable candidates for conventional surgical aortic valve replacement. Current treatment options for severe aortic stenosis include surgical aortic valve replacement, transcatheter aortic valve replacement, medical treatment, and percutaneous balloon aortic valvuloplasty. In recent years, transcatheter aortic valve replacement has emerged as a less invasive therapeutic option for high-risk patients with severe aortic stenosis, offering comparable reductions in mortality rates to surgical aortic valve replacement (see Figure 8.3).

These advancements have allowed for improved management and outcomes in patients with severe aortic stenosis. Patients with severe aortic stenosis who are not eligible for conventional surgical valve replacement due to high surgical risk or comorbidities now have alternative treatment options available.

8.6 TERMINOLOGY AND TECHNIQUES

Artificial intelligence (AI) is defined as a computational program capable of executing tasks that resembles human intelligence. These tasks encompass pattern recognition, planning, language comprehension, object identification and sound recognition, as well as problem-solving. In practical terms, AI can be conceptualized as the capacity of a machine or device to make independent decisions based on the data it gathers. In the field of medicine, this commonly involves utilizing data from health records or extracted from images to forecast a probable diagnosis, detect novel diseases, or determine the optimal course of treatment (Szolovits et al. 1988; Darcy et al. 2016). The advent of AI can be traced back to the pioneers who emerged in the 1950s; however, substantial progress was only achieved when robust computational methods were developed to organize and evaluate data (Szolovits et al. 1988) and datasets have grown significantly in size over the past 25 years. In recent years, the combination of advanced data processing techniques through multilayered networks, the unprecedented expansion of available datasets, and the introduction of user-friendly software packages for data analysis has made the prospect of an AI-driven revolution in healthcare more tangible (Darcy et al. 2016). Machine learning (ML) serves as a technique employed to equip AI with the capability to learn autonomously. Specifically, ML techniques can progressively learn rules and identify patterns from extensive datasets without the need for explicit programming or any preconceived assumptions. ML techniques have proven to be highly effective in prediction and intelligent decision-making across various aspects of everyday life, including internet search engines, personalized advertising, spam email filtering, character recognition and language processing, financial trends, and robotics (Obermeyer and Emanuel 2016; Deo 2015). Two fundamental prerequisites for the functioning of ML are (1) the availability of relevant and detailed data that can adequately address the question at hand and (2) the utilization of an appropriate computational ML technique that matches the type, quantity, and complexity of the available data. Finally, it is crucial to validate the outputs generated by ML and demonstrate their utility in clinical practice (Figure 8.4).

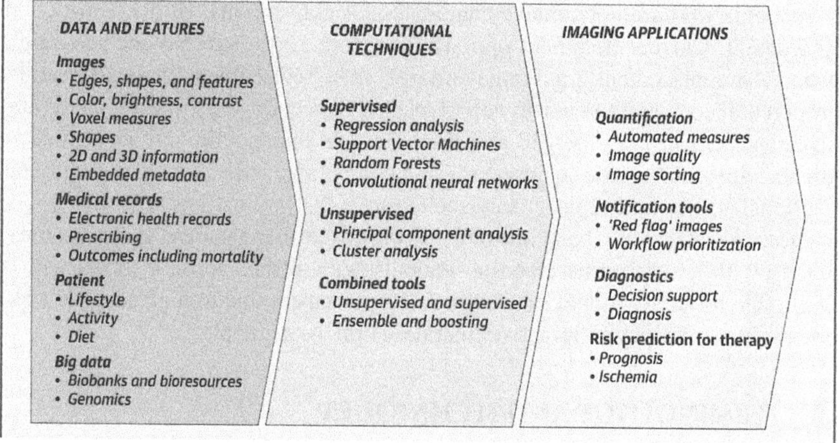

FIGURE 8.4   Image analysis using artificial intelligence. (Reproduced with permission from Dey et al. 2019.)

8.7 COMPUTATIONAL APPROACHES AND TOOLS (ARTIFICIAL INTELLIGENCE, MACHINE LEARNING, DEEP LEARNING)

Aortic stenosis is a progressive disease that can result in severe complications, including irreversible myocardial damage and increased morbidity and mortality. Accurate and timely identification of aortic stenosis is crucial for appropriate risk stratification and management of patients.

TTE is currently the first-line imaging modality for diagnosing aortic stenosis (Krishna et al. 2023).

In emergency rooms or acute inpatient settings, access to sonographers and cardiologists may be limited, resulting in delays in diagnosis and management. To overcome these challenges, researchers have turned to AI, ML, and deep learning techniques to improve the diagnosis of aortic stenosis. AI has shown great potential in transforming echocardiography by providing accurate and rapid diagnostic information. AI, ML, and deep learning techniques have been applied in the diagnosis of aortic stenosis to improve accuracy and efficiency (Pighi et al. 2021). Researchers have developed and validated deep learning models for detecting aortic stenosis using electrocardiography (Kwon et al. 2021).

These models utilize complex algorithms that can analyze electrocardiogram data and identify patterns associated with aortic stenosis.

By training these models on large datasets of electrocardiograms from patients with and without aortic stenosis, the algorithms can learn to recognize subtle changes in the ECG waveform that are indicative of aortic stenosis. The deep learning–based algorithm developed by Kwon et al. (2021) demonstrated high accuracy in detecting aortic stenosis using electrocardiography.

EuroSCORE II was based upon an updated worldwide dataset that included one-third of patients who underwent aortic valve replacement. The society of thoracic surgery (STS) score is derived from a US dataset and presents a dedicated submodel for six specific surgical procedures, including surgical isolated aortic valve replacement. Risk scores were developed based on datasets from heterogeneous groups of patients who underwent cardiac surgery; therefore, they present some inherited limitations when applied to TAVI patients.

Another study by Adedinsewo et al. (2020) also demonstrated the potential of AI-enabled ECG algorithms in identifying patients with left ventricular systolic dysfunction, which can be a common complication of aortic stenosis (Pighi et al. 2021).

Furthermore, Ko et al. (2020) utilized a convolutional neural network–enabled electrocardiogram to detect hypertrophic cardiomyopathy, which can be associated with aortic stenosis (Elias et al. 2022). The network is shown in Figure 8.5. These studies collectively highlight the promising role of AI, ML, and deep learning in improving the diagnosis of aortic stenosis (Pighi et al. 2021). In addition to electrocardiography, AI has also been applied in other imaging modalities such as echocardiography and cardiac magnetic resonance imaging to aid in the diagnosis of aortic stenosis. For example, Sánchez-Puente et al. (2023) demonstrated the potential of artificial intelligence in optimizing the appropriate timing of follow-up echocardiography for patients with aortic stenosis, which can help reduce the volume of unnecessary follow-up imaging studies (Tao et al. 2023). Furthermore, the application of

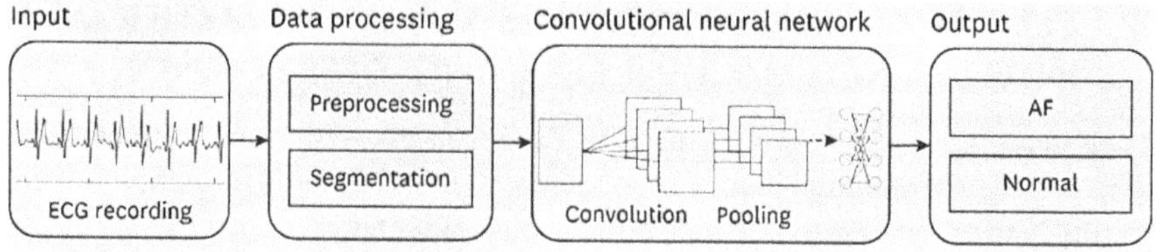

FIGURE 8.5 Convolutional neural network for atrial fibrillation prediction. (Reproduced with permission from Erdenebayar et al. 2019.)

AI in echocardiography can provide accurate and rapid diagnostic information for aortic stenosis (Pighi et al. 2021).

The utilization of AI and deep learning techniques for the diagnosis of aortic stenosis holds great potential in improving accuracy, efficiency, and patient outcomes when it comes to detecting this condition. These advanced technologies can harness the power of complex algorithms to analyze various data sources, including electrocardiograms, echocardiography results, and patient clinical data, to make accurate diagnoses of aortic stenosis. Furthermore, the use of AI and deep learning algorithms can overcome some of the limitations of traditional diagnostic methods for aortic stenosis.

For example, traditional diagnostic methods such as auscultation and physical examination may be subjective and rely on the expertise and experience of the clinician. In contrast, AI algorithms can analyze large amounts of data and identify patterns that may not be easily discernible to the human eye. Additionally, these algorithms can continuously learn and improve over time, adapting to new patterns and variations in data. This can lead to more accurate and consistent diagnoses of aortic stenosis, reducing the risk of misdiagnosis or delayed treatment. Moreover, AI algorithms have the potential to aid in the early detection and prediction of aortic stenosis. By analyzing historical patient data and identifying subtle changes in electrocardiograms, echocardiography images, and other relevant clinical information, AI algorithms can identify patients at risk of developing aortic stenosis before the onset of symptoms. This early detection can allow for earlier intervention and management, potentially improving patient outcomes and reducing the burden on healthcare resources. One study conducted by Sánchez-Puente et al. (2023) found that applying AI techniques to echocardiography data resulted in a high accuracy rate (86.6%) for the diagnosis of aortic stenosis (Busnatu et al. 2022). Furthermore, the combination of AI and ML techniques has shown promising results in enhancing the accuracy of diagnosing aortic stenosis.

For example, Yang et al. (2020) successfully utilized a feature analysis framework combined with continuous wavelet transform to reduce features extracted from cardio-mechanical signals, resulting in improved classification of aortic stenosis. AI, ML, and deep learning algorithms have the potential to revolutionize the diagnosis of aortic stenosis (Pighi et al. 2021). These algorithms can analyze various types of medical data, such as electrocardiograms, echocardiography images, and patient demographic information, to identify patterns and correlations that may not be apparent to human clinicians. By utilizing these advanced algorithms, clinicians can not only improve the accuracy and consistency of their diagnoses but also benefit from enhanced early detection capabilities. Furthermore, the integration of deep learning techniques can provide even more powerful diagnostic capabilities for aortic stenosis. Deep learning algorithms, based on convolutional neural networks, have been applied to electrocardiogram analysis for the precocious identification of moderate or severe cases of aortic stenosis. These algorithms can learn and adapt to patterns in the data, making them capable of detecting subtle changes that may indicate the presence of aortic stenosis before symptoms manifest.

The application of AI, ML, and deep learning algorithms in the diagnosis of aortic stenosis has shown promise as a tool for early detection and accurate diagnosis. Implementing these technologies has the potential to improve patient outcomes by enabling timely interventions and appropriate management strategies. In addition to improving diagnostic accuracy, the utilization of AI and ML algorithms can also enhance efficiency in healthcare settings. By automating the analysis process, these algorithms can rapidly assess large volumes of medical data and provide clinicians with actionable information in a timely manner.

The incorporation of these methodologies in the diagnosis of aortic stenosis represents a significant advancement in medical practice. The technologies have the potential to revolutionize the field of cardiology by providing clinicians with powerful tools for accurate and efficient diagnosis. Their potential in the diagnosis of aortic stenosis has been explored through various studies. One such study, conducted by Kwon et al. (2021), developed a deep learning–based algorithm for detecting aortic stenosis using electrocardiography data (Elias et al. 2022). The algorithm demonstrated high accuracy in detecting aortic stenosis, showcasing the potential of AI in improving diagnostic capabilities for this condition.

Another study by Yang et al. (2020) utilized a feature analysis framework and continuous wavelet transform to classify aortic stenosis based on cardio-mechanical signals (Busnatu et al. 2022). Their research demonstrated the usefulness of AI in accurately classifying aortic stenosis cases, further highlighting the potential of these technologies in enhancing diagnostic accuracy and clinical decision-making for aortic stenosis. Furthermore, research conducted by Dr. Sengupta's group and the Artificial Intelligence for Aortic Stenosis at Risk International Consortium focused on the application of ML and topological data analysis to diagnose and grade the severity of aortic stenosis using echocardiographic data (Ito and Oh 2022). This research utilized an ML-based ensemble classifier to identify low- and high-severity disease groups based on echocardiographic data, showcasing the potential of these technologies to enhance the accuracy of diagnosis and grading in aortic stenosis.

The application of AI, ML, and deep learning methodologies in the diagnosis of aortic stenosis offers several key advantages (Pighi et al. 2021). First, these technologies have the capability to analyze large amounts of data quickly and accurately. This allows for more efficient and timely diagnosis, saving valuable time for both the patient and the clinician. Second, AI algorithms can learn and adapt from new data, enhancing their diagnostic capabilities over time.

This adaptability and continuous learning ability can lead to improved accuracy and a reduction in misdiagnosis rates. Moreover, the use of AI, ML, and deep learning in the diagnosis of aortic stenosis can help overcome some limitations of traditional diagnostic methods.

For example, traditional methods such as echocardiography and electrocardiography rely heavily on the interpretation skills of the clinician, which can be subjective and prone to human error. By leveraging AI and ML algorithms, the diagnosis of aortic stenosis can become more objective and standardized. Additionally, the use of AI and ML can also help in detecting subtle patterns and relationships in the data that may not be easily recognizable to human interpretation, thereby leading to improved diagnostic accuracy.

The integration of artificial intelligence, machine learning, and deep learning methodologies in the diagnosis of aortic stenosis has shown promising results. Dr. Sengupta's group, along with the Artificial Intelligence for Aortic Stenosis at Risk International Consortium, have successfully developed and employed these advanced technologies to enhance the diagnosis and grading of aortic stenosis.

Their approach utilized topological data analysis, which employed a machine learning framework to cluster patient similarity networks for phenotypic recognition of the pattern of left ventricular responses in patients with aortic stenosis (Ito and Oh 2022).

This approach allowed for the identification of low- and high-severity disease groups based on echocardiographic data, which were then used to train a machine-learning-based ensemble classifier. This classifier was able to accurately identify these groups of patients, thus augmenting the echocardiographic grading of aortic stenosis severity. Additionally, another recent study applied deep learning–based algorithms and convolutional neural networks to analyze electrocardiograms for the early identification of moderate or severe aortic stenosis patients (Pighi et al. 2021). The results of this study demonstrated the potential of artificial intelligence in improving the diagnosis of aortic stenosis by leveraging deep learning algorithms and convolutional neural networks. The use of artificial intelligence, machine learning, and deep learning in the diagnosis of aortic stenosis holds great promise for improving accuracy and standardization, and ultimately leading to better patient outcomes.

8.8   CONCLUSION

In conclusion, the detection and treatment of aortic stenosis are crucial components of cardiovascular care that can significantly impact a patient's quality of life and overall health. Aortic stenosis, characterized by the narrowing of the aortic valve opening, poses a serious threat if left untreated. Fortunately, advances in medical technology and our understanding of the condition have led to improved detection methods and treatment options.

Echocardiography and other imaging techniques play a vital role in diagnosing the condition, allowing healthcare providers to assess the severity and plan appropriate treatment strategies. Routine cardiac check-ups, especially for individuals at higher risk, are instrumental in identifying aortic stenosis in its early stages.

The treatment of aortic stenosis depends on its severity and the patient's overall health. Lifestyle modifications, such as blood pressure control and heart-healthy diets, can help manage mild cases. However, for moderate and severe cases, surgical interventions are often necessary. The two main approaches for treatment are valve replacement and valve repair. Surgical options include traditional open-heart surgery and minimally invasive procedures, such as transcatheter aortic valve replacement (TAVR). The choice of treatment depends on various factors, including the patient's age, comorbidities, and the extent of valve damage.

In recent years, TAVR has gained prominence as a less invasive and highly effective alternative to open-heart surgery, particularly for elderly or high-risk patients. It offers quicker recovery times and reduced hospital stays. As medical science continues to advance, we can anticipate further refinements in techniques and the development of even less invasive procedures.

Regular check-ups and advancements in surgical techniques provide patients with better health and a more active life. With ongoing research and innovation, we can look forward to further improving the management of aortic stenosis, ultimately enhancing the prognosis and quality of life for those affected by this condition.

In conclusion, the integration of artificial intelligence, machine learning, and deep learning methodologies in the diagnosis of aortic stenosis has shown significant potential for improving diagnostic accuracy and patient outcomes.

REFERENCES

Adedinsewo, D., Carter, R. E., Attia, Z., Johnson, P., Kashou, A. H., Dugan, J. L., Albus, M., Sheele, J. M., Bellolio, F., Friedman, P. A., & Lopez-Jimenez, F. 2020. Artificial intelligence-enabled ECG algorithm to identify patients with left ventricular systolic dysfunction presenting to the emergency department with dyspnea. *Circulation: Arrhythmia and Electrophysiology*, 13(8). https://doi.org/10.1161/circep.120.008437.

Banović, M., Putnik, S., Pěnička, M., Doros, G., Deja, M., Kočková, R., Kotrč, M., Glaveckaitė, S., Gašparović, H., Pavlović, N., Velicki, L., Salizzoni, S., Wojakowski, W., Van Camp, G., Nikolic, S., Iung, B., & Bartúnek, J. 2022. Aortic valve replacement versus conservative treatment in asymptomatic severe aortic stenosis: The AVATAR trial. *Circulation*, 145(9), 648–658. https://doi.org/10.1161/circulationaha.121.057639

Baumgartner, H., Falk, V., Bax, J. J., De Bonis, M., Hamm, C. W., Holm, P. J., Iung, B., Lancellotti, P., Lansac, E., Muñoz, D., Rosenhek, R., Sjögren, J., Mas, P. T., Vahanian, A., Walther, T., Wendler, O., Windecker, S., & Zamorano, J. L. 2017. ESC/EACTS guidelines for the management of valvular heart disease. *European Heart Journal*, 38(36), 2739–2791. https://doi.org/10.1093/eurheartj/ehx391

Busnatu, S., Niculescu, A., Bolocan, A., Petrescu, G., Păduraru, D. N., Năstasă, I., Lupusoru, M., Geanta, M., Andronic, O., Grumezescu, V., & Martins, H. 2022. *Clinical applications of artificial intelligence—an updated overview.* https://scite.ai/reports/10.3390/jcm11082265

Darcy, A. M., Louie, A. K., & Roberts, L. W. 2016. Machine learning and the profession of medicine. *JAMA*, 315(6), 551. https://doi.org/10.1001/jama.2015.18421

Deo, R. C. 2015. Machine learning in medicine. *Circulation*, 132(20), 1920–1930. https://doi.org/10.1161/circulationaha.115.001593

Dey, D., Slomka, P. J., Leeson, P., Comaniciu, D., Shrestha, S., Sengupta, P. P., & Marwick, T. H. 2019. Artificial intelligence in cardiovascular imaging. *Journal of the American College of Cardiology*, 73(11), 1317–1335. https://doi.org/10.1016/j.jacc.2018.12.054

Elias, P., Poterucha, T. J., Rajaram, V., Moller, L. M., Rodriguez, V. A., Bhave, S., Hahn, R. T., Tison, G. H., Abreau, S., Barrios, J., Torres, J. N., Hughes, J. W., Perez, M., Finer, J., Kodali, S., Khalique, O., Hamid, N., Schwartz, A., Homma, S., Perotte, A. J. et al. 2022.

Deep learning electrocardiographic analysis for detection of left-sided valvular heart disease. *Journal of the American College of Cardiology*, 80(6), 613–626. https://doi.org/10.1016/j.jacc.2022.05.029

Erdenebayar, U., Kim, H., Park, J. U., Kang, D., & Lee, K. J. 2019. Automatic prediction of atrial fibrillation based on convolutional neural network using a short-term normal electrocardiogram signal. *Journal of Korean Medical Science*, 34(7). https://doi.org/10.3346/jkms.2019.34.e64.

Foroutan, F., Guyatt, G. H., O'Brien, K. et al. 2016. Prognosis after surgical replacement with a bioprosthetic aortic valve in patients with severe symptomatic aortic stenosis: Systematic review of observational studies. *BMJ*, 354, i5065.

Ito, S., & Oh, J. K. 2022. *Aortic stenosis: New insights in diagnosis, treatment, and prevention.* https://scite.ai/reports/10.4070/kcj.2022.0234

Kalogeropoulos, A. S., Redwood, S., Allen, C., Hurrell, H., Chehab, O., Rajani, R., Prendergast, B., & Patterson, T. 2022. A 20-year journey in transcatheter aortic valve implantation: Evolution to current eminence. *Frontiers in Cardiovascular Medicine*, 9. https://doi.org/10.3389/fcvm.2022.971762

Ko, W. Y., Siontis, K. C., Attia, Z. I., Carter, R. E., Kapa, S., Ommen, S. R., Demuth, S. J., Ackerman, M. J., Gersh, B. J., Arruda-Olson, A. M., & Geske, J. B. 2020. Detection of hypertrophic cardiomyopathy using a convolutional neural network-enabled electrocardiogram. *Journal of the American College of Cardiology*, 75(7), 722–733. https://doi.org/10.1016/j.jacc.2019.12.030.

Krishna, H., Desai, K., Slostad, B., Bhayani, S., Arnold, J. H., Ouwerkerk, W., Hummel, Y., Lam, C. S., Ezekowitz, J., Frost, M., & Jiang, Z. 2023. Fully automated artificial intelligence assessment of aortic stenosis by echocardiography. *Journal of the American Society of Echocardiography*, 36(7), 769–777. https://doi.org/10.1016/j.echo.2023.03.008.

Kwon, J., Jo, Y., Lee, S., & Kim, K. 2021. *Artificial intelligence using electrocardiography: Strengths and pitfalls.* https://scite.ai/reports/10.1093/eurheartj/ehab090

Nakatsuma, K., Taniguchi, T., Morimoto, T., Shiomi, H., Ando, K., Kanamori, N., Murata, K., Kitai, T., Kawase, Y., Izumi, C., Miyake, M., Mitsuoka, H., Kato, M., Hirano, Y., Matsuda, S., Inada, T., Nagao, K., Mikami, H., Takeuchi, Y., Kimura, T. et al. 2018. B-type natriuretic peptide in patients with asymptomatic severe aortic stenosis. *Heart*, 105. https://doi.org/10.1136/heartjnl-2018-313746

Nishimura, R. A., Otto, C. M., Bonow, R. O., Carabello, B. A., Erwin, J. P., Fleisher, L. A., Jneid, H., Mack, M. J., McLeod, C. J., O'Gara, P. T., Rigolin, V. H., Sundt, T. M., & Thompson, A. 2017. AHA/ACC focused update of the 2014 AHA/ACC guideline for the management of patients with valvular heart disease: A report of the American college of cardiology/American heart association task force on clinical practice guidelines. *Circulation*, 135(25). https://doi.org/10.1161/cir.0000000000000503

Obermeyer, Z., & Emanuel, E. J. 2016. Predicting the future—big data, machine learning, and clinical medicine. *The New England Journal of Medicine*, 375(13), 1216–1219. https://doi.org/10.1056/nejmp1606181

Otto, C. M., Nishimura, R. A., Bonow, R. O. et al. 2021. ACC/AHA guideline for the management of patients with valvular heart disease: A report of the American college of cardiology/American heart association joint committee on clinical practice guidelines. *Journal of the American College of Cardiology*, 77, e25–e197. https://doi.org/10.1161/CIR.0000000000000923

Peeters, F., Kietselaer, B., Hilderink, J. M., Linden, N. V. D., Niens, M., Crijns, H. J., & Meex, S. J. 2019. *Biological variation of cardiac markers in patients with aortic valve stenosis.* https://scite.ai/reports/10.1136/openhrt-2019-001040

Pighi, M., Giovannini, D., Scarsini, R., & Piazza, N. 2021. *Diagnostic work-up of the aortic patient: An integrated approach toward the best therapeutic option.* https://scite.ai/reports/10.3390/jcm10215120

Redfors, B., Furer, A., Lindman, B. R., Burkhoff, D., Marquis-Gravel, G., Francese, D. P., Ben-Yehuda, O., Pîbarot, P., Gillam, L. D., Leon, M. B., & Généreux, P. 2017. Biomarkers in aortic stenosis: A systematic review. *Structural Heart*, 1(1–2), 18–30. https://doi.org/10.1080/24748706.2017.1329959

Sánchez-Puente, A., Dorado-Díaz, P. I., Sampedro-Gómez, J., Bermejo, J., Martinez-Legazpi, P., Fernández-Avilés, F., Sánchez-González, J., Pérez del Villar, C., Vicente-Palacios, V., & Sanchez, P. L. 2023. Machine learning to optimize the echocardiographic follow-up of aortic stenosis. *JACC: Cardiovascular Imaging*, 16(6), 733–744. https://doi.org/10.1016/j.jcmg.2022.12.008.

Shah, S. M., Shah, J., Lakey, S. M., Garg, P., & Ripley, D. P. 2023. *Pathophysiology, emerging techniques for the assessment and novel treatment of aortic stenosis.* https://scite.ai/reports/10.1136/openhrt-2022-002244

Szolovits, P., Patil, R. S., & Schwartz, W. B. 1988. Artificial intelligence in medical diagnosis. *Annals of Internal Medicine*, 108(1), 80. https://doi.org/10.7326/0003-4819-108-1-80

Tao, M., Al-Sadawi, M., Ahmed, N., Dianati-Maleki, N., Mann, N., & Kort, S. 2023. *The use of quality improvement interventions in reducing rarely appropriate echocardiograms: A systematic review and meta-analysis.* https://scite.ai/reports/10.1111/echo.15653

Williams, T., & Hildick-Smith, D. 2020. Balloon aortic valvuloplasty: Indications, patient eligibility, technique and contemporary outcomes. *Heart*, 106(14), 1102–1110. https://doi.org/10.1136/heartjnl-2019-315904

Yan, A. T., Koh, M., Chan, K. K., Guo, H., Alter, D. A., Austin, P. C., . . . & Ko, D. T. (2017). Association between cardiovascular risk factors and aortic stenosis: The CANHEART aortic stenosis study. *Journal of the American College of Cardiology*, 69(12), 1523–1532.

Yang, C., Ojha, B. D., Aranoff, N. D., Green, P., & Tavassolian, N. 2020. Classification of aortic stenosis using conventional machine learning and deep learning methods based on multi-dimensional cardio-mechanical signals. *Scientific Reports*, 10(1). https://doi.org/10.1038/s41598-020-74519-6.

9 A Study of the Role of Artificial Intelligence in Monitoring Environmental and Health Issues in the Post-COVID-19 Pandemic Era for Sustainable Living

Mahua Basu and Mausumi Das Nath

9.1 INTRODUCTION

The unanimous adoption of 'Transforming Our World: The 2030 Agenda for Sustainable Development' in the UN Sustainable Development Summit 2015, created 17 Sustainable Development Goals (SDGs) for ending poverty and promoting peace, thereby protecting our planet through protecting human rights and dignity. Sustainable development in the field of environment is one of the dimensions warranted by this agenda. An enormous amount of effort was given to support and capacity building on issues such as water, energy, climate change, transport, wildlife, health, and well-being targeting several of the SDGs. In the contemporary world, numerous issues have produced a gargantuan impact on the environment. Issues like global warming and climate change, environmental pollution, lack of waste management, and health and hygiene standards, have given rise to several other controversial issues. Tackling these issues confronts industrialists, academicians, researchers, and medical practitioners with a mammoth challenge. Air, land, and water pollution issues such as ground-level ozone, carbon monoxide, NO_x, SO_x particulate matter, volatile organic compounds, etc., have contributed significantly to severe health complications. This is causing a growth spike in hospitalization and frequent visits to medical practitioners. With the onset of COVID-19, attention was temporarily drawn away from these grave issues. But the problem exists in reality. Falling immunity post-COVID-19 coupled with increasing environmental pollution increases people's vulnerability, hence the need for smart environmental management.

The COVID-19 pandemic has significantly affected the world, resulting in severe consequences for the environment and human health, bringing to light the urgent

DOI: 10.1201/9781003405436-11

need for sustainable living, as well as the critical role of technology in achieving this goal. It has become more essential than ever to focus on sustainable development to address various challenges, including climate change, pollution, waste management, water management, biodiversity management, and health and hygiene standards. Environment, economy, and society are three perspectives of sustainable development that are related to each other. Artificial Intelligence (AI) introduction in common environmental issues makes it simple to address what otherwise is a complex sustainability challenge and can put people's interest into action (Stein 2020). With smarter and smarter AI that boosts human efficiency, it is now imperative to extend their implication beyond industries and business, that is, the planet Earth. This notion makes a paradigm shift from a human-friendly AI to an Earth-friendly AI. AI has the potential to change environmental science by helping to analyze vast amounts of data, improving predictive models for climate change, and assisting in the development of sustainable solutions for environmental challenges. In this study, we explore the application of AI in monitoring environmental and health issues in the post-COVID-19 pandemic era to further sustainable living.

9.2 APPLICATION OF ARTIFICIAL INTELLIGENCE AND PROSPECTS

9.2.1 Environmental Issues and AI Application

In the post-COVID-19 era, AI can be used for monitoring numerous environmental issues. AI-powered sensors can be used to monitor air quality, water quality, soil quality, wildlife, waste scenarios, etc. Such AI-generated models will help identify pollution hotspots and enable governments and environmental agencies to take action to reduce pollution levels. Such predictive models can also be used to monitor deforestation and forest degradation, which are major drivers of climate change. In this section we outline the various environmental issues where AI can make a positive impact.

9.2.1.1 Climate Change Issues and AI Applications

9.2.1.1.1 Global Scenario

Climate change is the long-term shifts in temperatures and weather patterns. Despite natural shifts, anthropogenic activities such as fossil fuel combustion mainly drive climate change, making the planet 1.1°C warmer compared to the late 1880s. The change results in wildfires, droughts, floods, intense storms, melting ice caps, rising sea level, wildlife extinction, water and food scarcity, displacement, poverty, and increased health risks, to name a few. A universally agreed limit for the global temperature increase to less than 1.5°C could make the world still sustainable. But the existing policies will contribute to a 2.8°C rise at the end of this century. Sixty-eight percent of the global emission comes from the 10 largest emitter nations. Beginning to act, the coalition of nations committed to net zero emissions by 2050, cutting 50% of the emission by 2030, keeping temperature rise below 1.5°C with a pledge to reduce fossil fuel production by 6% annually between 2020 and 2030. SDG 13 is about taking action to deal with the effects of climate change and protect ourselves from natural disasters caused by it. Achieving this goal can also help us make progress towards other important goals, such as promoting good health (SDG 3), using clean energy (SDG 7), supporting economic growth (SDG 8), building better cities (SDG 11), promoting innovation, industrial and infrastructural development (SDG 9), and using resources responsibly (SDG 12).

9.2.1.1.2 AI Application

At its core, AI involves using algorithms and computational models to enable machines to perform tasks that normally require human intelligence, such as learning, reasoning, problem-solving, and perception. Man provides data and states a few key parameters. At each passing the algorithm makes a skilled speculation regarding the kind of information to look for. Subsequent guesses are updated based on how well the preceding guess has worked (Stein 2020). Hence learning-based AI diagnoses problems while interacting with the problem itself (Nicholas 2019) without human involvement but by making hypotheses, reassessing models, and reevaluating the data. Previously AI was used to track climate change impacts by analyzing the images of shallow water reefs (Jeffery 2019). The use of AI to track forest health by gathering data on temperature, humidity, and CO_2 is known. These techniques can help predict extreme weather events, such as hurricanes, floods, and droughts, allowing for better planning and response. The AI for Earth program by Microsoft uses AI to tackle various environmental issues, including climate change (Joppa 2017). The FarmBeats project is an Internet of Things (IoT) platform for optimizing agricultural practices. The AI uses data from sensors and satellite imagery for precision agriculture (Vasisht et al. 2017). The Climate Informatics research program funded by the US National Science Foundation for developing models for predicting climate patterns with high accuracy aims to develop machine learning tools and methodologies to better understand and predict climate patterns. Machine learning techniques were used to analyze data from a variety of sources, such as satellite images, weather station measurements, and ocean buoy data, to develop models that can predict climate patterns with high accuracy (Monteleoni et al. 2013). Indigenous knowledge can be assimilated by AI, for example, utilizing Natural Language Processing (NLP) (Nost and Colven 2022). Enormous amounts of data on various dissimilar variables, such as temperature and humidity, are required for meaningful climate science but the handling of such massive data presents a challenge (Faghmous and Kumar 2014). Machine learning techniques, such as deep learning (DL), artificial neural networks (ANNs), and support vector machines (SVMs), can help predict and analyze climate patterns. Since 1960, complex, nonlinear combinations of signal and noise in ANN indicator patterns are identified in simulated and observed surface temperature maps, so that novel insights into climate variability and climate change are exposed (Barnes et al. 2019). DL technique application seems to suit the broad field of earth system science. Diverse sources starting from sophisticated Earth Observation (EO) satellites to enormously used inexpensive crowd-sourcing sensors provide constantly escalating amounts of earth system data. The basis of weather forecasts and climate monitoring is observations that involve steps such as the retrieval of information, quality control, bias rectification, and data fusion/assimilation. The amount of precipitation was estimated by combining Meteosat Spinning Enhanced Visible and Infrared (SEVIRI) geostationary satellite images with rain gauges (Dewitte et al.

2021). Learning-based AI can diagnose problems without human involvement by making hypotheses, reassessing models, and reevaluating data, which has been used to track climate change impacts and forest health by gathering data on temperature, humidity, and CO_2, and can help predict extreme weather events for better planning and response. Several initiatives, such as the AI for Earth program by Microsoft and the Climate Informatics research program funded by the US National Science Foundation, aim to use machine learning to better understand and predict climate patterns. DL, ANNs, and SVMs can help predict and analyze climate patterns, using diverse sources of earth system data. Combining satellite images with rain gauges can estimate the amount of precipitation. Indigenous knowledge can be assimilated by AI through NLP. However, handling massive amounts of data remains a challenge.

Electricity accounts for 25% of global greenhouse gas (GHG) emissions. And with the transition to electric vehicles (EVs), the demand for electricity is expected to increase further. Hence the power sector, comprising power generation, transmission, distribution, and consumption, has immense opportunities for AI application. AI can influence minimizing energy losses in electricity transmission and distribution (Bughin et al. 2017).

AI may speed up the development of clean energy technologies, improve forecasting of power demand, strengthen optimization and management of systems, and improve system monitoring (Rolnick et al. 2022). Through micro-grid development, AI can democratize electricity, rendering reasonable access and facilitating off-grid zero-carbon electrification. AI can assist in predicting the source of GHG emissions. This might help policymakers and financers to know where to regulate and where to invest in energy production (Stein 2020). Reduction of GHG emissions from the power sector can be achieved through AI-assisted optimization of grid assets, enhancing energy efficiency and increased reliability and resiliency. AI can help to discover new substances to be used in batteries for energy storage or for absorbing atmospheric CO_2 (Rolnick et al. 2022). AI, by integrating data from hazards like wildfires and severe storms, can adjust grid operations accordingly to make a safer, more efficient, and more reliable grid (Victor 2019).

AI can bring transformation by improving EV charge scheduling, managing congestion, vehicle-to-grid algorithm, and battery energy management. They can also contribute to the research and development of EV batteries (Vollrath 2020). AI in advanced distribution management systems can locate the fault, manage peak demand, and support micro-grids and EVs (Stein 2020). In Europe, operators analyze voltage, load, and grid topology data with AI for assessing available capacity on the system ad also planning future needs (Wei et al. 2019).

AI can enhance the predictability of intermittent renewable energy to afford more incorporation into the grid, better manage its intermittency, improve storage, and tackle power fluctuations. For example, AI is used to adjust the propeller blades of wind turbines in keeping with changing wind directions. This in turn reduces the intermittency issues of wind turbines. Machine learning is used to predict wind power output 36 hours in advance of real power generation by Google and DeepMind (Carter 2019). Improvement in prolonged lifespan and maximizing the energy efficiency of large batteries with a small number of issues can be achieved through AI (Saustania 2019). Table 9.1 summarizes the applications of machine learning in tackling climate change.

TABLE 9.1

Sectoral Application of Machine Learning to Tackle Climate Change

Sector	Role of ML	Examples/Case Studies
Electricity/power systems	Forecasting supply and demand	Google's DeepMind used machine learning (ML) to optimize energy consumption in data centers, reducing electricity usage by 40% (Gao 2014).
	Power system optimization and scheduling	The California Independent System Operator (CAISO) uses ML algorithms to optimize power scheduling, resulting in improved grid stability (Yang et al. 2022).
	Identification and management of geothermal energy sites	Using advanced ML, deep learning (DL), and cloud computing, the GeoThermalCloud tool is used to discover, explore, and develop hidden geothermal resources (Mudunuru et al. 2023; Vesselinov et al. 2022).
		Stanford University researchers used ML to identify optimal sites for geothermal energy extraction, increasing the efficiency of energy production (Vesselinov et al. 2020).
	Maintenance of nuclear fission reactors	Researchers at Texas A&M University employed ML techniques for predictive maintenance of nuclear reactors, improving safety and efficiency (Revels 2023).
Transportation	Improving vehicle efficiency and demand forecasting	Uber employs ML algorithms to optimize its ride-sharing platform, reducing empty trips and improving overall vehicle efficiency (Tyagi 2020).
	Optimizing vehicle routing and reducing vehicle miles traveled	United Parcel Service (UPS) utilizes ML-based routing algorithms to optimize package delivery routes, reducing fuel consumption and miles traveled (eLogii 2020).
	Traffic pattern analysis and road traffic forecasting	The City of Chicago implemented DL models for short-term forecasting of traffic patterns and optimizing travel routes, traffic signal timing, and forecasting congestion (Polson and Sokolov 2017).
	Enhancing operational efficiency in aviation	Boeing uses ML algorithms to analyze flight data and improve aircraft fuel efficiency, leading to reduced emissions (Thollander et al. 2020).
	Predicting runway demand and reducing fuel consumption	Frankfurt Airport implemented an ML-based runway demand prediction system, optimizing air traffic and reducing fuel consumption (Thollander et al. 2020).

Buildings and cities	Enhancing energy efficiency in buildings	Honeywell developed an autonomous building solution to study energy consumption and then optimize energy settings keeping in tune with occupant's comfort levels (Chin 2020).
	Building load forecasting and energy management	Cornell University's ML-based load forecasting system improved energy management in its buildings, reducing electricity costs (https://fcs.cornell.edu/departments/energy-sustainability/energy-management-overview).
	Building diagnostics and maintenance	DeepMind collaborated with the U.K.'s National Grid to develop an ML system that identifies energy inefficiencies in buildings for targeted maintenance (Madhumita Murgia 2017).
	District heating and cooling systems	Helsinki's district heating provider, Helen along with Silo AI, employs ML for optimizing its heating network, resulting in energy savings and reduced emissions (Alanen 2020).
	Predicting energy consumption and ranking building efficiency	The US Environmental Protection Agency's ENERGY STAR's Portfolio Manager uses ML algorithms to predict energy consumption, water use, wastes, greenhouse gas emission and rank the efficiency of commercial buildings (Star 2023).
	Smart city applications and data processing	Singapore implemented a smart city project using AI and ML (Trends 2022).
Climate modeling	ML algorithms improve climate modeling accuracy and simulate complex climate processes	The Intergovernmental Panel on Climate Change (IPCC), Climate Risk Prediction uses information on climate models from 20 labs worldwide (Monteleoni et al. 2011).
Climate risk prediction	ML models analyze historical data to predict and assess climate and climate-related risks	The National Center for Atmospheric Research (NCAR) uses ML algorithms to improve weather predictions, forecast hazards such as hailstorms, hurricanes, blizzards (Gagne 2020). —Terrafuse AI, USA provides climate and weather risk forecasting products and hyperlocal climate risk scores powered by ML and develops wildfire risk models using physics-informed ML (AI).

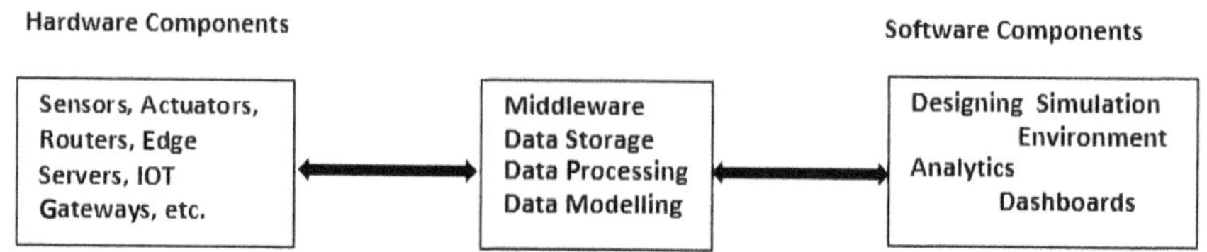

FIGURE 9.1 Diagram of a digital twin system. Various types of data, such as satellite, meteorological, soil moisture, environmental, climate change data, etc., are collected through sensors, and network devices (e.g., routers, gateways, servers). They are then fed into a system for further processing and modeling. The data output is fed into a simulator for several runs, which in turn again is used on the original physical object to understand the impact on the environment.

An emerging technology that has experienced a surge recently for predictive analysis is the digital twins, which have the ability to have a deep insight into the inner operations of a system, any interactions between the various components of the system and the upcoming activities of their physical counterpart. It has immense application in various fields such as smart cities, automotive industries, medicine, and engineering fields. It is a virtual depiction of a physical object or system that covers its lifecycle. The decision-making is facilitated by the reorganization of real-time data, the use of simulation, machine learning, and reasoning. Let's consider the object to be a wind turbine—a range of sensors are fitted interrelated to fundamental functional areas. In return, the sensors would generate data about various aspects of the performance of physical objects such as energy harvest, temperature, weather conditions, and more. This output would then be fed into a processing system followed by application to the digital copy. Once such data is informed, simulations could be run, performance issues could be studied, and potential improvements could be generated on the virtual model (Figure 9.1). The helpful insights can then be used on the original physical object (How does a digital twin work? From https://www.ibm.com/topics/what-is-a-digital-twin).

9.2.1.2 Air Quality Issues and AI applications

9.2.1.2.1 Global Scenario

According to Polk (2019), atmospheric pollution stood as the fourth leading risk factor for untimely death worldwide, with 6.67 million deaths attributable to pollution. Ninety percent of the global population experienced an annual average of fine particulate matter (PM 2.5) concentration that exceeds the World Health Organization's (WHO's) Air Quality Guideline of 10 micrograms/m^3. Asia, Africa, and the Middle East experienced the highest exposure. India exhibits the highest exposures. Ozone levels rose between 30% and 70% compared to levels a century earlier. This indicates the production and release of chemicals such as NO_x and volatile organic compound (VOC) from industries and automobiles that trigger tropospheric ozone formation. This ground-level ozone affects cropland, foods, and human health. The combustion of coal, charcoal, wood, agricultural residues, dung, and kerosene for heating and cooking contributes to domestic pollution. Open fires and cookstoves producing black carbon, CO, and particulates worsen the situation with poor ventilation. SDG target 3.9.1 calls for substantially reducing deaths and ailments from atmospheric pollution, while SDG objective 7.1.2 aims for access to clean energy to curb indoor air pollution.

9.2.1.2.2 AI Application

Air pollution is a major environmental problem that is caused by emissions from various sources, such as transportation, industry, and agriculture. The quality of air around us poses serious health risks to humans and other living organisms. In preventing and controlling air pollution, geospatial methods such as Aerosol Optical Depth (AOD), satellite data, and Land Use Regression (LUR) were often used. But AI applications in the field of weather and climate science, disaster management, vehicular pollution control, and energy use optimization play a very crucial role. Monitoring air quality is essential for identifying sources of pollution and developing strategies to reduce it. Many of the applications are discussed in the discussion of climate change earlier in the chapter. IBM helps China in its efforts to predict air quality levels and reduce pollutants through its Green Horizon initiative that uses IoT. A startup, Dositracker, helps battle radon pollution in Romania with its RadonAir project (Aayush et al. 2020).

AI offers better management of complicated and nonlinear interactions between the spatiotemporal factors and hence holds better potential for the prediction and forecasting of air pollution. Of the three forecasting models (physical, statistical, and AI), AI models are found to outperform the others. The most commonly used AI techniques are back propagation neural network (BPNN), extreme learning machine (ELM), long short-term memory (LSTM), the generalized regression neural network (GRNN), gated recurrent unit (GRU), the wavelet neural network (WNN), fuzzy logic and SVM (Subramaniam et al. 2022), genetic algorithms (GAs), and swarm intelligence models, as well as combination ones such as adaptive neuro-fuzzy interference system (ANFIS) (Oprea et al. 2017). Of these, the most intelligent ones are ANN, fuzzy logic, SVM, and deep neural network (DNN). For the past two decades, SVM in association with other ML approaches is used to predict CO, SO_2, O_3, and PM (Masood and Ahmad 2021). Integrated SVM with the partial least squares-based model (PLS-SVM) offered better and satisfactory results (Qader et al. 2021). Accurate monitoring of CO levels in urban areas with the neuro-fuzzy model was reported in Delhi (Mishra and Goyal 2015). In the case of forecasting, the ANN model predicted CO concentrations every day more precisely than the integrated ANN and fuzzy interference systems (FIS), known as the ANFIS model (P. Wang et al. 2015). Again when compared, ANFIS performs better than data mining (decision tree) though it takes a long time in the training phase (Oprea et al. 2017). A hybrid model incorporating ANN, such as GA-ANN, fuzzy neural network (FNN) (see Figure 9.2), and ANFIS, is often preferred for its easy implementation, its ability to overcome specific shortcomings and provide synergistic advantages for solving complicating issues (Ye et al. 2020).

In training different air quality indexes for NO_2 and CO employing fuzzy logic with simulated annealing (SA) and particle swarm optimization (PSO), ANFIS-PSO is found to work better (Ly et al. 2019). In Delhi, satisfactory results were obtained in CO prediction using wavelet decomposition and ANFIS (Mittal and Bhardwaj 2011).

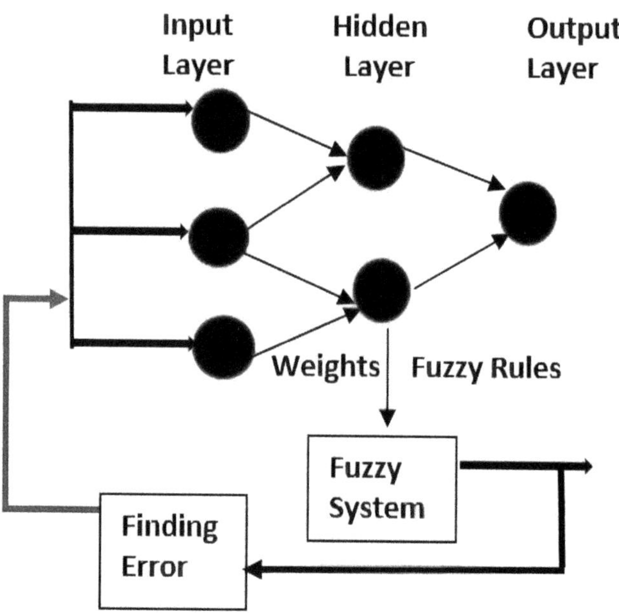

FIGURE 9.2 The neural network and fuzzy system do not need to communicate with each other. Both online and offline learning can occur. The fuzzy set is used as a weights input variable, output variable, and rules are models as neurons that can be included or excluded depending upon the situation. Last, the fuzzy rule base is represented by network neurons. Thus, this hybrid model gives better results.

The PM concentration can be measured by applying the DL technique on outdoor images where a convolutional neural network (CNN) was used to analyze the images. A vigorously pretrained deep recurrent neural network (DRNN) was proposed for predicting the time-series level of PM2.5 in Japan. Environmental monitoring data used in this process was obtained from physical sensors (Ong et al. 2016). Air pollution can also be predicted by spatiotemporal deep learning (Li et al. 2016). In a district of Turkey, SO_2, CO, and PM10 were estimated using geographic forecasting models and neural networks (GFM_NN) (Kurt and Oktay 2010).

AI can help in monitoring air quality by providing accurate and real-time data that can be used to identify sources of pollution and take corrective measures. AI can assist in analyzing data from sensors and satellite imagery, allowing for targeted interventions to improve air quality. One example of an AI application for air quality is the BreezoMeter project, which uses AI to provide real-time air quality data for various locations worldwide. The system uses ML techniques to analyze data from various sources, including satellite imagery and air quality sensors, to provide accurate and reliable air quality information (Silva et al. 2019). ML is a technique used in AI that permits machines to learn from data without unambiguous programming. It entails the development of algorithms that can study patterns in data and make forecasts based on those patterns. ML is used in air quality monitoring to identify sources of pollution and predict air quality. One of the most commonly used ML algorithms in air quality monitoring is the random forest algorithm. It is a type of supervised learning algorithm that can be exploited to predict air quality by analyzing data from sensors. Random forest algorithms use decision trees to classify data based on the features of the data. The algorithm then combines the predictions of the decision trees to make a final prediction (Zimmerman et al. 2018).

DL, a subset of ML, applies ANNs to model and resolve complex problems. It is a more advanced and sophisticated technique that allows machines to learn from large amounts of data and make more accurate predictions. DL is used in air quality monitoring to identify patterns and predict air quality. CNNs are one of the most commonly used DL algorithms in air quality monitoring. CNNs are used to analyze data from air quality sensors and identify patterns in the data. The algorithm learns the features of the data and uses them to make predictions about air quality. Reinforcement learning is a technique used in AI where machines learn by interacting with their environment and receiving feedback based on their actions. It involves developing algorithms that can learn from trial and error and improve over time. Reinforcement learning is used in air quality monitoring to optimize air quality control systems by identifying the most effective control measures to reduce pollution. The algorithms learn from data collected from air quality sensors and optimize the control measures to reduce pollution levels. The IoT is a network of physical devices, sensors, and other objects that are connected to the internet and can exchange data with each other. IoT is used in air quality monitoring to collect statistical data from air quality sensors and other devices and transmit it to a central database. The data can then be analyzed using AI techniques such as ML and DL to predict air quality and identify sources of pollution.

9.2.1.3   Water Quality Issues and AI Applications

9.2.1.3.1   Global Scenario

In the Earth, freshwater is only 2.5%, of which merely a little more than 1.2% is surface water, 30% is underground, and the remaining freshwater is found in ice sheets and glaciers. Of the 1.2% that is surface water, 0.49% is river water (Gautam and Dahal 2020). Water management, climate change, and natural cycles lead to the world's wetter landscapes becoming wetter and dry areas becoming drier. Overdrafting groundwater for farming use is crucial for freshwater depletion. Between 2007 and 2015 groundwater recharge from rain and snow was decreased by severe drought in California's central valley. Farmers pumped out more water as a result. From 2002 to 2016, the region lost 6.1 gigatons per year of stored groundwater. In contrast, between 2002 and 2016, water storage increased on average by 29 gigatons every year in the western Zambezi basin and Okavango delta in Africa.

India receives around 4% of total rainfall, very meager for such a densely populated nation, and ranks 133 in terms of water availability per capita per year. Both 'blue water' (riverine, lacustrine, and aquifer water) and 'green water' (used by plants) are released into the atmosphere. Just to mention that the water cycle comprises precipitation, evaporation from water bodies and soil, transpiration from plants, infiltration into the soil, groundwater discharge, etc. In 2021 around 2.3 billion people reside in water-stressed areas (Nations 2021). Annually 3.4 million deaths occur from scarce and contaminated water. Even today 2.1 billion people are without access to clean and safe, potable water. On a daily average, women and children have to walk 3.7 miles for procuring water, spending almost 3 to 6 hours a day that could otherwise be utilized meaningfully. The average yearly per-person water availability fell from 1,545 m^3 in 2011 to around 1,486 m^3 in 2021. The situation is grim, calling for immediate action and proper management. SDG 6 aims for access to clean water and sanitation for all.

9.2.1.3.2 AI Application

AI techniques have the potential to revolutionize the way freshwater is monitored, distributed, and managed. AI can be used to monitor the water quality in real-time, detecting contaminants and other issues that may affect the health of consumers. Freshwater management is a critical issue as freshwater is becoming increasingly scarce due to population growth, climate change, and pollution. ML algorithms can be used for predicting water quality parameters, such as temperature, pH, dissolved oxygen (DO), and nutrient concentrations in freshwater bodies to identify patterns, detect changes in freshwater ecosystems, and predict future trends. Remote sensing aids in data collection. Hydraulic modeling involves the use of computer simulations to analyze the movement of water in freshwater systems, aiding in identifying areas of high flow velocity or areas where water is stagnating, which in turn can help to optimize freshwater management strategies such as dam operations or water withdrawal. The method can increase efficiency and sustainability, and reduce the risk of water scarcity.

Current methods for assessment of water quality are both expensive and time-consuming. Assessment comprises sample collection, transportation, and calculation involving both laboratory and statistical analysis, which is quite difficult (Hmoud Al-Adhaileh and Waselallah Alsaade 2021). Advanced AI methods can be advantageous and cost-effective. The delivery of quality water infrastructure requires resolving challenges such as systemic analysis of raw water, disposal systems, and monitoring issues. Advanced computing with AI techniques can develop modeling of water quality. ANN helps in monitoring water quality systems by predicting changes in water quality (Gomolka et al. 2017). Classification of water quality with SVM, NNs, DNN, and K-nearest neighbors (KNNs) techniques employing data from the Pakistan Council of Research in Water Resources was reported by Ahmed et al. (2019) A hybrid CNN-long short-term memory (LSTM) is employed for predicting water quality in terms of total nitrogen, total phosphorus, and total organic carbon (Baek et al. 2020). Prediction of drinking water quality in the Yangtze River basin was carried out with LSTM using pH, DO, COD, and NH3-N (Liu et al. 2019). The sewage effluent water quality was predicted by Zheng et al. with immune particle swarm optimization (PSO) that used a neural network with a hidden layer (Zheng et al. 2010). To guarantee water security in the South to North Water Transfer Project of China, a generalized regression neural network (GA-GRNN) was proposed (Z. Wang et al. 2015). Deep learning techniques and ANFIS showed better performance in the prediction of water quality index in comparison to traditional ML methods (Hmoud Al-Adhaileh and Waselallah Alsaade 2021). GAs, inspired by biological functions, can be used as an optimization technique for minimizing or maximizing an objective task. GAs can yield excellent solutions in cases of highly complex,

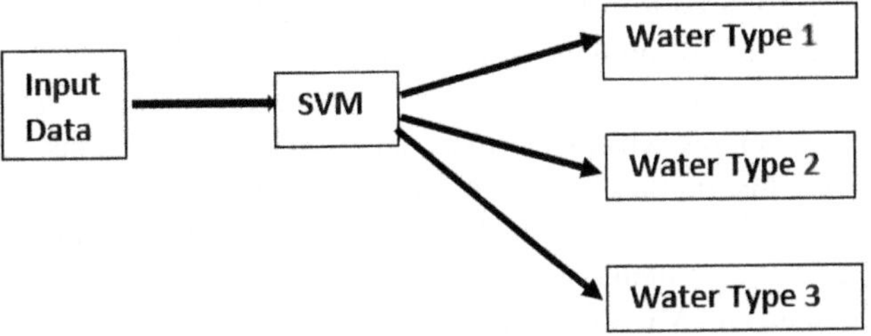

FIGURE 9.3 The support vector machine (SVM) classifier is provided with inputs for multiclass classification in the context of water quality assessment.

multi-parameter problems. For example, GA was used for the optimization of regional wastewater treatment in a river water quality management model and also of energy usage in water treatment plants. ANN was used to optimize watershed management to uphold an equilibrium between water quality demand and resulting limitations for the agricultural industry (Chau 2006). The improved Water Distribution Network (WDN) with sensors, networks, and integrated operations permits water utility companies to supervise and regulate water supply in real time, making it affordable and sustainable. A Smart Water Grid (SWG) can reduce water issues without compromising sustainability but increasing the efficiency of the WDN by integrating information and communications technology (ICT) and the usual water management system. Despite high installation costs, Singapore, Australia, the European Union, the United States, and South Korea took the lead in implementing SWG technology in smart water management (Koo et al. 2021). AI algorithms that can analyze data from sensors in water distribution systems to predict when maintenance is needed, such as when pipes are at risk of breaking or when equipment is about to fail, help proactively address issues before they turn into major problems, reducing both downtime and saving money. Likewise, leaks can also be detected to reduce water loss and minimize the need for repairs and predict water demand with the rising population. Due to various manpower and expertise issues, the classification of water quality is done using the SVM technique, which is easy, fast, and very accurate. Samples of water are fed and then trained through an SVM classifier. The output determines the water quality based on the samples collected. The efficiency obtained by this model is also high (see Figure 9.3).

Marine water management is a critical issue on account of increasing pressures from climate change, pollution, and overfishing. AI can help in marine biodiversity surveillance and conservation by analyzing data from sensors and satellite imagery, allowing for the identification of areas that require conservation efforts. AI can also assist in deep-sea resource modeling and predicting sea ice and climate, allowing for better planning and management of marine resources. One example of an AI application for marine research is the Ocean Health Index project. The project uses machine learning to analyze data from various sources, including satellite imagery and ocean sensors, to evaluate the health of the world's oceans to identify areas that require conservation efforts, allowing for targeted interventions to protect marine biodiversity. Information from remote sensors such as satellites, drones, and other types of sensors are used to collect data about ocean temperature, salinity, currents, and more to be used to monitor ocean ecosystems, track changes over time, and identify areas that need attention. ML algorithms can analyze vast amounts of data collected through remote sensing or other sources to identify patterns, detect changes in ocean ecosystems, predict future trends, predict the location of fish populations, identify areas at risk of harmful algal blooms, and forecast the impacts of climate change on marine ecosystems, which can aid in ocean zoning. Predictive analytics uses historical data and ML algorithms to predict future events or trends. NLP algorithms can be used to analyze social media posts, news articles, and other sources of information to identify issues and concerns related to both freshwater and marine water management. Robots can work continuously to automate various tasks, such as underwater surveys, monitoring marine life, and cleaning up ocean pollution, and can work in hazardous or difficult environments, such as deep sea or polluted areas, which are unsafe for human workers.

Water pollution is a major ecological problem that can have grave consequences on public health, the environment, and the economy. Computer vision acting as the eyes of the AI system enables the machines to see and observe visual inputs. It involves the use of cameras and other sensors to collect images and videos of water bodies. These images can be analyzed using computer vision algorithms to identify potential issues such as oil spills or litter. Oil spillage can be detected using an unmanned aerial vehicle (UAV) and infrared camera (De Kerf et al. 2020). NLP can improve public awareness and engagement, and allocate resources more effectively. Underwater robots can be used to collect water samples from hard-to-reach areas of water bodies, reducing the need for human intervention and improving efficiency, allowing for more accurate and comprehensive monitoring of water quality, providing time series three-dimensional imagery of the sea floor and image-based habitat classification by using huge volumes of high-resolution stereo imagery and multibeam sonar (Dunbabin and Marques 2012). A bluetooth-operated robot can remove waste debris, such as plastic, papers, metals, and garbage, from water bodies to reduce fuel-operated garbage collection, thus saving aquatic life (Preeti and Gadgay 2018).

9.2.1.4 Sanitation Issues and AI Applications

9.2.1.4.1 *Global Scenario*

Approximately 3.6 billion of the global population lack access to safe and managed sanitation, with nearly 8% of people still practicing open defecation, worldwide. Between 2000 and 2020, 2.4 billion of the global population got access to improved toilets. Despite sanitation being one of the Millennium Development Goals (MDGs), 1.7 billion people lack basic services even today. Sixty-six percent of them reside in rural areas and 50% of them in sub-Saharan Africa. Sanitation is an important aspect of public health, the inadequacy of which poses significant health threats, with more than 700 children dying daily from diarrhea. Poor sanitation costs billions of dollars in many countries and holds back economic growth. It accounted for 6.3% of the gross domestic product (GDP) in Bangladesh (2007) and 6.4% of the GDP in India (2006). Premature deaths, expensive health care, loss of productive time due to ailments and treatments, etc., all contribute to economic loss. Therefore, in a report titled 'State of the World's Sanitation', the United Nations Children's Fund and WHO (2020) have called urgently for transformed sanitation for all by 2030.

9.2.1.4.2 AI Applications

AI is increasingly being used in sanitation to help improve the efficiency and effectiveness of sanitation systems. ML is one of the most commonly used AI techniques in sanitation. ML algorithms can be used to analyze data collected from sensors and other sources to identify patterns and predict future trends. For example, ML algorithms can be used to analyze sewage data to identify potential outbreaks of diseases such as cholera (Midani et al. 2018) and typhoid fever. This information can be used to alert public health officials and help prevent the spread of disease. Moreover, machine learning techniques have been applied to optimize sensor selection and classify sanitation-related malodors. This approach enables the discrimination between non-offensive odors and the accurate classification of distinct malodor types, specifically distinguishing between urine and feces (Zhou et al. 2020). The other popular techniques, computer vision, robotics, predictive analytics, and NLP, have also been applied in sanitation.

Computer vision involves the use of cameras and other sensors to collect images and videos of sanitation systems. These images can be analyzed further using computer vision algorithms to identify potential issues such as blockages or leaks in sewage pipes (Midani et al. 2018; Su and Yang 2014) thus preventing sewage overflows and contamination of water sources. Robots can be used to inspect and clean sanitation systems, reducing the need for human intervention and improving efficiency, which in turn helps prevent human workers having to enter hazardous environments. One such instance is the EARTHBOT (Krithiga 2019). Further, predictive analytics involves the use of data analysis and ML algorithms to predict future events or trends. It can be used to predict when a sewer system is likely to fail, enabling proactive maintenance and reducing the risk of system failures (Malek Mohammadi et al. 2020; Thiyagarajan et al. 2017).

NLP algorithms can be used to analyze data from social media, news articles, and other sources to identify issues and concerns related to sanitation (Leeson et al. 2019), inform sanitation policies and strategies, improve public awareness and engagement, and allocate resources more effectively. AI is also being used to optimize sanitation systems. Optimization algorithms can be used to identify the most efficient and cost-effective ways to manage sanitation systems, taking into account factors such as population density, water availability, and waste disposal methods.

9.2.1.5 Waste Management Issues and AI Application

9.2.1.5.1 Global Scenario

By 2059, global waste generation is likely to increase to 3.40 billion tones. This is twice the population growth for the same period. On average, every day, the per capita waste generation is 0.74 kg, varying between 0.11 and 4.54 kg. Thirty-three percent of the annual 2.01 billion tonnes of municipal solid waste generated every year, fails to be managed in an environmentally safe manner. The Food and Agriculture Organization (FAO and UNICEF 2020) estimated that 720 to 811 million people suffered from hunger in 2020. Globally and annually 5 trillion plastic bags are used while 1 million plastic bottles are purchased every minute. Fifty percent of all plastics are single-use items. Huge amounts of waste are dumped in the form of landfill. Out of the 37% landfill waste, only 8% is dumped in sanitary landfills with landfill gas collection structures. Open dumping amounts to 31% of waste, of which 11% goes to incinerators and 19% could be recovered for recycling and composting. Plastics, including microplastics, are ubiquitous and have acquired the name of 'plastisphere' for novel marine microhabitats and are likely to become an Anthropocene marker in the fossil records.

9.2.1.5.2 AI Applications

Waste management is a crucial aspect of modern society, as waste production continues to increase due to population growth and lifestyle changes. With the advent of AI, waste management is becoming more efficient and sustainable. Waste-related service features in SDG Targets 11 and 12 contribute to sustainable cities and communities, as well as catering to responsible consumption and production. One example of an AI application for waste management is the Zero Waste Management project by the city of San Francisco. The city uses AI-based systems to identify and sort different types of waste, enabling the development of efficient waste collection and disposal methods. The system uses ML techniques to analyze data from sensors and cameras installed in waste collection vehicles and at waste disposal sites. AI can help optimize waste management by identifying areas that require attention, enabling the development of efficient waste collection and disposal methods. AI-based systems can also help in recycling by identifying and sorting different types of waste, making the process more efficient and cost-effective.

ML algorithms can be used to identify and classify waste based on different categories such as organic, inorganic, hazardous, or recyclable. Waste sorting is an essential part of waste management because it enables efficient disposal, reduces the amount of waste going to landfills, and enables effective recycling. By using ML algorithms, waste sorting can be automated, resulting in faster and more accurate sorting. Waste management facilities can use ML algorithms to train machines to recognize different types of waste, which results in fewer errors in waste sorting. The application of forthcoming IoT and ML technologies targets efficient waste management in every aspect in the field of domestic waste management for making a green smart society. These technologies emphasize smart waste collection and decomposition to maximize benefits and minimize actual waste efficiently. Waste segregation is facilitated on two levels: at the individual house level in the society and society level. Biodegradable waste is recycled into compost. KNN is employed to produce alert messages for different combinations of three sensor values, such as levels of biodegradable and nonbiodegradable waste, poisonous gas concentration, etc. Hence the technology enables pollutant reduction, conservation, resourcing, and energy reuse and caters to uplifting green technology (Dubey et al. 2020).

NLP enables machines to understand and analyze human language. Waste management authorities can use NLP algorithms to analyze social media posts, news articles, and other sources of information to identify waste-related issues and concerns in a particular area and take prompt action to address such issues, thus preventing them from escalating. The real-time data collected can be used to inform waste management policies and strategies, improve public awareness and engagement, and allocate resources more effectively.

A conceptual verification of the municipal waste management system was planned to make it more cost-effective. Deep learning enabled classifiers and cloud computing was employed for precise classification of wastes at the onset of waste collection. Waste was categorized into plastics, glass, paper or cardboard, metals, fabric, and other recyclables to facilitate waste disposal. Garbage classification was realized by applying CNN. Seven state-of-the-art CNNs and data preprocessing methods for waste categorization were analyzed. The precision of nine were the categories that ranged from 91.9% to 94.6% in the validation set. MobileNetV3 has 94.26% classification accuracy, involving a little storage space of 49.5 MB and a shortest running time of 261.7 ms. Additionally, IoT not only facilitated information exchange between the waste containers and waste management centers but also the total waste generation in an area and the operating state of any waste container with the help of sensors. Based on the surveillance, the waste management center can plan the operation and maintenance of adaptive equipment, collection of wastes, and route plan of vehicles. This serves as the basis of a successful municipal waste management system (C. Wang et al. 2021).

Robots automate various tasks, such as sorting, processing, and recycling waste. AI-powered robots use sensors and cameras to identify and classify waste, and then use mechanical arms to sort and recycle it. This process is faster, more efficient, and more accurate than traditional manual waste sorting processes. Moreover, robots can work continuously without rest, which is an advantage in managing large quantities of waste. Robots can also work in hazardous or dirty environments that are not safe for human workers (Chen et al. 2022; Coffey et al. 2021; Sarc et al. 2019). By using robots, waste management authorities can improve efficiency, safety, and sustainability in waste management. Predictive analytics can be used to forecast the amount and types of waste generated in a particular area. This information can be used to plan and allocate resources more effectively, such as waste collection routes, recycling facilities, and disposal sites. Predictive analytics can also be used to optimize waste management processes by predicting failures or identifying areas of improvement.

9.2.1.6 Forestry, Biodiversity, and Crop Management Issues and AI Applications

9.2.1.6.1 Global Scenario

From oceans to peat lands to deserts, the entire planet is a sequence of coupled ecosystems. All the parts of an ecosystem are interdependent, much like a jigsaw puzzle. Globally, the land area under forest cover is nearly 30%. Forests are dominant terrestrial ecosystems that harbor 90% of all terrestrial biodiversity (Victor 2019). Forests provide essential ecological services such as breathable air, carbon sequestration, food security, nutrient and water cycles, and other productive services. They trap around 30% of the global CO_2 emissions every year (Bughin et al. 2017). Worldwide, 1.6 million people are dependent on the forest for their livelihoods (Andrae and Edler 2015). A change in the temperature of an ecosystem will have knock-on effects on other things, like what plants and animals can grow and live there. About 1.74 million species have been databased out of an estimated 2 million to 1 trillion of earth's current species. Biodiversity loss is mostly due to human activity, including deforestation, over-fishing, and over-hunting, as well as to pollution, climate change, agriculture, and overdevelopment that place the global populations of animals and plants under huge threat. Throughout the world, in the past 25 years, the loss of forest land amounted to 129 million hectares resulting in a drop of global carbon stock by 17.4 Gt (Nost and Colven 2022). According to the 2019 Intergovernmental Platform on Biodiversity and Ecosystem Services' Global Assessment Report, one million floral and faunal species face the threat of extinction. The Living Planet Report 2020 by the World Wildlife Federation reports an average decline of 68% in the mammalian, piscine, reptilian, and avian and amphibian populations since 1970, globally.

Biodiversity is vital. It plays a huge role in the integrity of the forest, grasslands, and marine ecosystems, provisions important adaptation functions; it buffers from extremes of climate conditions, regulates the water cycle, protects soil, regulates temperature in urban areas, reduces food insecurity, and provides an option for economic diversification, predominantly during times when the impact of climate change cuts agricultural yield. All these functions are important, not just for the climate, but for life on earth. Often, distant ecosystems are interdependent in unforeseen ways. For example, every year nearly 22,000 tonnes of dust transported by the wind from the Sahara Desert replenishes the Amazon Rainforest with phosphorus separated by thousands of miles. Just one species removal due to climate change or pollution or habitat loss or any other reason can lead to a domino effect affecting an entire ecosystem.

9.2.1.6.2 AI Applications

Even now in several countries, conventional pen and paper are used to carry out forest inventory. Such methods are linked to drawbacks such as the slow pace of data collection and analysis, lack of scalability of approaching, and individual biases. Rapid proliferation in agriculture, urbanization, and booming development jeopardizes the rich biodiversity (Rolnick et al. 2022). Biodiversity management is crucial, as the loss of biodiversity can have significant ecological, economic, and social consequences. SDG 14 and SDG 15 call for protecting life in water and on land, respectively. Additionally, biodiversity contributes to SDG 2 (zero hunger), SDG 1 (no poverty), and SDG 3(good health and well-being). Switching to AI applications, especially progression in data science together with satellite and digital revolution, potentiates improved monitoring, management, and conservation in forestry and wildlife sectors. Companies like sagarobotics.com and FarmBot employ robots for accurate de-weeding and exact application of fertilizer and pesticides, resulting in increased yield per acre (Rolnick et al. 2022).

AI is increasingly being used in biodiversity management efforts to help researchers and conservationists better understand and protect biodiversity but the use of AI still lags behind other fields. The survival and conservation of both marine and terrestrial animals are critical on account of threats. The development and application of AI require improved access to big data related to forest and biodiversity, IoT network infrastructure, advanced and digital and satellite technology (high-resolution cameras, satellite technology, sensors, drones, and UAVs), and computational space and storage, cloud computing, all of which can help improve adoption of AI technology in India (Rolnick et al. 2022). ML and NLP help to predict ecosystem services (Stein 2020). AI is also being used to design and optimize protected areas. Optimization algorithms can be used to identify the best locations and sizes for protected areas, taking into account factors such as species richness, habitat quality, and connectivity. Though technology cannot solve all problems, it can certainly prevent illegal practices, and offer transparency. These algorithms can also be used to evaluate trade-offs between conservation objectives and other goals such as economic development. Globally, AI-based startups and nonprofit organizations aim for forest digitization, fight rising CO_2 levels, protect wildlife trafficking, prevent unlawful wildlife trade, and assist floral and faunal classification and identification, their monitoring and census thus improving management (Nost and Colven 2022). The startup SilviaTerra integrates high-resolution satellite images (15 m × 15 m resolution) from remote sensing, field survey data from the United States Forest Department, and cloud computing for developing a predictive model and estimating forest conditions to solve forest inventory problems (Parisa and Nova 2020). The Chesapeake Bay Conservancy teamed with Esri GIS mapping software and Azure cloud services of Microsoft to develop a highly exhaustive land cover map of that region with the aid of ML libraries. A Portugal-based startup, 20tree. AI combined remote sensing, big data, cloud computing, and AI to monitor forest inventory in real time. Likewise, a German and Finland-based company, CollectiveCrunch, started an AI platform called 'Linda Forest' that exploits several sources. Accurate prediction of the target area regarding wood mass, wood species and wood quality can be obtained (Shivaprakash et al. 2022).

GainForest, based in Switzerland, for monitoring and forecasting deforestation, employs massive amounts of satellite images, game theory, video prediction model, and ML-based measurement, reporting, and verification. This helps them in designing carbon payment schemes (Nicholas 2019). Accurate estimation of individual tree size, volume, and carbon density is estimated by a tech company, Pachama, using ML with satellite, drone, and LIDAR images. Several nonprofit organizations, such as the Erol Foundation, the Center for Global Discovery and Conservation Science (GDCS) at Arizona State University, and Planet.Inc, combine computer vision models with LIDAR (Light Detection and Ranging) to map carbon stock and emissions in a cost-effective and automated way (Shivaprakash et al. 2022). An AI application can promote reforestation and afforestation by planting over a trillion trees that could sequester hundreds of gigatons of carbon emissions. Three startup companies, Droneseed, Dendra, and Land Life are on their way to addressing this issue. Droneseed devised a seed vessel to carry desirable seed species; once planted it assists to safeguard and promote faster germination. Droneseed sent out a Federal Aviation Administration-certified cluster of four to five drones and drone swarms bearing about 57 lb. of weight. First, the study area is scanned, the suitability for planting (such as moisture) is recognized, and the seed vessels are dropped. The Nature Conservancy, Oregon, partnered with Droneseed to restore Oregon's rangeland that is distressed by invasive species.

Technologies such as optical character recognition and NLP can digitize large quantities of data obtained from previous monitoring in paper format for record keeping. These data can then be fed into various analytical algorithms for further analysis. Real-time information on temperature and moisture can be obtained by installing

IoT devices in the forest to get insight into forest health and impending threats. A Canadian startup, Terrafuse, employs AI models to comprehend climate-related risks like wildfire. For this purpose, it uses historic wildfire information, numerical simulations, and satellite images on Microsoft Azure. An Estonian entity, Timbeter, fights illegitimate logging and timber wood trafficking through the online tracking of roundwood by an integrated AI and database on photometric measurement. Likewise, the German-based Xylene tracks real-time wood supply chain combining space technology, blockchain, supply chain mapping, IoT device-gathered automated data, and Earth Observation.

Anthropogenic sounds like illegal tree felling can use acoustic systems with DNN and send alerts to the protected area managers (Azzi et al. 2020). Rainforest Connection uses bio-acoustic monitoring to prevent unlawful tree felling and address deforestation issues. They use old discarded mobile phones, enable them with solar power, and fix them on treetops to record the sounds of a chainsaw in the forest. The recorded data is transmitted to the base stations via the cell phone towers. Google's machine learning library, Tensorflow, is used to recognize and perceive chainsaw sound. Subsequently, the location info is shared with forest officials for immediate control measures. Real-time forest mapping monitoring is also carried out by Terramonitor, Global Forest Watch, and Future Forest Map using open source satellite data and AI. Outland analytics use bio-acoustic identification not only to detect chainsaw sound but also sounds of unauthorized vehicles in real-time to prevent environmental crime. The World Resources Institute partnered with the Central Africa Regional Program for the Environment to comprehend the responsible factors and predict forest loss in the Democratic Republic of Congo (Shivaprakash et al. 2022).

ML algorithms can analyze hefty amounts of data collected through sensors, drones, and other sources. These algorithms can identify patterns and behaviors of species, detect changes in their habitats, and predict future trends. For example, ML algorithms can be used to track the movement of marine animals such as whales, dolphins, and sharks, and to identify areas where they are at risk of encountering human activities such as shipping or fishing. This information can help conservationists to plan management strategies and protect the animals from harm. ML algorithms can also be used to identify and track the movements of migratory species such as birds and marine animals. This information can help conservationists to plan management strategies and protect these species from harm. For example, ML algorithms can be used to track the movement of endangered species such as elephants, tigers, and gorillas, and to identify areas where they are at risk of encountering human activities such as farming or mining. This information can help conservationists to plan management strategies and protect the animals from harm.

AI applications can assist in species classification accurately. A community of 48 species was classified in the Serengeti ecosystem. Individual animals can also be identified using criteria. Individual tigers (*Panthera tigris*) have been recognized based on stripe patterns. On this basis, a CNN-enabled identification of individual tigers was developed by Shi et al. (2020). This ability to recognize animals at both species and individual levels has significant effects not only in wildlife monitoring but also averting man-wildlife conflict. This application combined with image analysis-enabled automated cameras can also warn farmers about the entry of elephants into crop fields and villages.

Deep artificial neural networks (or deep learning), a cascading set of multiple layers of digital neurons, are a subset of ML. This approach is used for the automated classification of a dataset comprising more than 3.2 million images and 48 potential species in a large camera trap. The DNN was previously trained to differentiate images with animals from those without animals with the aid of a two-stage workflow. Hence the empty images comprising more than 75% of the data were filtered out first. The prediction accuracy was 96.8%, indicating that the DL model could suitably classify a prelabelled 'test set' of 105,000 images. Next, a trained network used labeled imagery of more than 48 species to procure confidence scores or probabilities per species for the rest of the imagery. Confidence thresholds were placed on these probabilities in such a way that the model might match the prediction accuracy of human volunteers at 96.6% (Lamb et al. 2019).

Computer vision is another popular AI technique used in biodiversity management. Computer vision involves the use of cameras and other sensors to collect images and videos of species. These images can be analyzed using computer vision algorithms to identify species, count individuals, and track their movements. For example, computer vision algorithms can be used to identify and track invasive species, which can help to prevent their spread and reduce their impact on native biodiversity. For example, computer vision algorithms can be used to identify and track sea turtles and their nests, which can help to protect them from predators and human activities; it can also identify and track large herbivores such as elephants and rhinoceroses, which can help to protect them from poachers.

Acoustic monitoring is another AI technique used in biodiversity management. Acoustic sensors can be used to monitor the sounds that species make, such as the vocalizations of birds, primates, and marine animals. These sounds can be analyzed using ML algorithms to identify species and track their movements. For example, acoustic monitoring can be used to track the migration patterns of birds and the distribution of marine animals such as whales and dolphins. Acoustic sensors can be used to monitor the sounds that marine animals make, such as the vocalizations of whales and dolphins. These sounds can be analyzed using ML algorithms to identify species and track their movements. The conservationists can plan protection strategies and reduce the risk of habitat loss and also reduce the risk of collisions with ships.

NLP is another AI technique used in biodiversity management. NLP algorithms can be used to analyze social media posts, news articles, and other sources of information to identify issues and concerns related to biodiversity management. This information can be used to inform conservation policies and strategies, improve public awareness and engagement, and allocate resources more effectively. AI is also being used to design and optimize protected areas. Optimization algorithms can be used to identify the best locations and sizes for protected areas, taking into account factors such as species richness, habitat quality, and connectivity. These algorithms can also be used to evaluate trade-offs between conservation objectives and other goals such as economic development.

9.2.1.7 Resource Management and AI Applications

AI techniques can also be applied to identify potential mineral deposits, optimize exploration strategies, and improve the efficiency and safety of extraction operations.

AI algorithms can be used to analyze geological and geospatial data to identify potential mineral deposits. This can help companies target exploration efforts more effectively and reduce the risk of exploration failures. Predictive models can be created about mineral deposits based on historical data and other factors helping in the optimization of extraction strategies accordingly. Robots can be used to automate mining operations, including drilling, blasting, and material handling, thus improving efficiency and reducing the risk of accidents.

AI-enabled environmental monitoring helps companies comply with environmental regulations and reduce their environmental footprint. AI algorithms can be used to predict when mining equipment is likely to require maintenance, helping to prevent costly downtime and extend the lifespan of the equipment.

9.2.2 HEALTH ISSUES AND AI APPLICATIONS

9.2.2.1 Global Scenario

Health is the key to all three international initiatives—the SDGs, the Paris Agreement under the United Nations Framework Convention on Climate Change (UNFCCC), and the Sendai Framework for Disaster Risk Reduction. The COVID-19 pandemic highlighted the centrality of effective public health and healthcare systems. Lack of protective equipment, erroneous diagnostic tests, overworked doctors, inequitable access to healthcare services, skyrocketing expenses, lack of transparency, and lack of information exchange exposed the flaws in healthcare systems (Shaheen 2021). The key environmental factors affecting our health are air quality, hazardous chemicals, climate change and disasters, microbial diseases, poor quality of water, lack of access to healthcare, infrastructural issues, and other global issues. Climate change is estimated to cause an additional 250,000 deaths from malnutrition, malaria, dengue, diarrhea, and heat stress between 2030 and 2050. Chemicals, key to economic development, managed improperly can pose major health risks and cause disorders such as allergies, cancer, cardiac, pulmonary, neurological, congenital, urinary, and reproductive ailments. Different chemicals affect human health in various ways; hence chemical safety is of utmost importance. WHO reported more than 1.5 million deaths in 2016 from exposure to selected chemicals (CHE 2016). With increasing dependency on plastics, toxic additives, such as bisphenol A (BPA), plasticizers, and flame retardants that are often added to plastic to improve its properties, get released into the environment under various atmospheric conditions. In 2018, WHO reported the presence of microplastics in 90% of the tested bottled water (only 17 free out of 259 tested) (Readfearn 2018).

9.2.2.2 AI Applications

The recent developments in AI are revolutionizing and made us speculate whether AI is going to replace physicians in the coming days! Certainly, AI tools can help doctors to accomplish improved results in healthcare management and the medical field in terms of improved efficiency, accuracy, and quality of healthcare services.

TABLE 9.2

Medical Applications of Different Types of Machine Learning

Type of Machine Learning	Definition	Examples in Medical Applications
Supervised learning	Learning from labeled data to make predictions or decisions	Predicting disease outcomes, diagnosis from medical imaging
Unsupervised learning	Learning from unlabeled data to discover patterns and relationships	Identifying disease subtypes, discovering biomarkers
Semisupervised learning	Learning from both labeled and unlabeled data	Combining clinical and genomic data for diagnosis, drug discovery
Reinforcement learning	Learning by interacting with an environment to maximize a reward signal	Optimization of clinical trials, personalized treatment planning
Deep learning	Using neural networks with multiple layers to learn complex representations of data	Automated medical image analysis, drug discovery
Transfer learning	Leveraging knowledge learned from one task to improve performance on a different but related task	Predicting disease outcomes using electronic health records, transfer learning for medical image analysis
Online learning	Learning in real-time as data arrives sequentially	Real-time monitoring of patient vital signs, personalized treatment recommendations

The spectrum of AI techniques, such as neural networks, SVM, decision trees, etc., is being used to diagnose diverse ailments. ANN has shown more precision in categorizing diabetes and cardiovascular disease (CVD) than other methods (Eren et al. 2008).

ML, a subset of AI, applies computational algorithms (Ramkumar et al. 2021). Algorithms are built up to tutor datasets for statistical applications to permit data processing correctly and these principles underlie ML. Computers thus use past experiences to make successful predictions (Eren et al. 2008). In the ML technique, we prepare models with preexisting data. A brief account of the various ML techniques is summarized in Table 9.2. ML will recognize the test input based on prelearning when anyone feeds the data used for testing. An ML software library has been taught for detecting alteration in Parkinson's disease by DaTscan image analysis (Eren et al. 2008). With medical imagery like CT scans, MRI, X-rays, and ultrasound, AI has the aptitude to identify signs of disease more precisely and quickly. It assists patients with the rapid detection of disease and more precise treatment options. IBM Watson is used by Pfizer. It caught excellent media attention owing to its skill to spotlight precision medicine, particularly cancer diagnosis and immune-oncology treatment (Manne and Kantheti 2021; Shaheen 2021). Roche subsidiary Genentech is depending on an AI system from GNS Healthcare in Cambridge, Massachusetts, while Sanofi opted to use Exscientia's AI platform to try metabolic disease medications (Shaheen 2021). AI enables comprehensive analysis of massive quantities of data, resulting in speedy and precise decision-making (Ellahham et al. 2020). The AI-assisted care program Partnership in AI-Assisted Care (PAC) by Stanford University uses multiple sensors. Its intelligent well-being support system for seniors and smart intensive care units (ICUs) can sense any change in the behavior of aged people staying alone and ICU patients, respectively. PAC also extends intelligent hand hygiene support and healthcare conversational agents with depth sensors that refine computer vision technology to accomplish perfect hand hygiene for clinicians and nursing staff, thus reducing hospital-acquired infections (Malik et al. 2019). ML algorithms can be used to analyze patient data and identify patterns that can be used to predict disease outcomes, optimize treatment plans, and improve patient outcomes. For example, ML algorithms can be used to predict the likelihood of readmission or complications after surgery, allowing healthcare providers to take preventive measures. NLP can be used to analyze medical notes, patient records, and other sources of unstructured data to extract meaningful insights. For example, NLP algorithms can be used to identify patients at risk of sepsis, a life-threatening condition caused by infection, by analyzing the signs and symptoms documented in their medical notes. Computer vision is another AI technique used in healthcare management. Computer vision algorithms can be employed for the analysis of medical images, such as X-rays and CT scans, to spot abnormality and support diagnosis. For example, computer vision can be used to identify breast cancer in mammograms, allowing for earlier detection and treatment.

Though drug discovery and development is a time-consuming process, the discovery time can be reduced greatly with the help of ML techniques (Manne 2021). Billions of dollars can be saved as it can reduce the time of drug discovery and development (Erguzel and Ozekes 2014). Repetitive work is greatly reduced thus streamlining the process of drug discovery (Chan et al. 2019). Information on drug interaction and probable side effects from the medical literature can be extracted using AI algorithms. ML is one of the most commonly used AI techniques in drug delivery. ML algorithms can be used to analyze large datasets and identify patterns that can be used to optimize drug delivery. For example, ML algorithms can be used to predict the pharmacokinetics of a drug, which refers to how the drug is absorbed, distributed, metabolized, and excreted by the body. This information can be used to optimize the drug dosage and delivery schedule. Another AI technique used in drug delivery is virtual screening. Virtual screening involves using computer simulations to identify potential drug candidates that could be effective against a specific disease target. AI algorithms can be used to analyze large databases of chemical compounds and predict which compounds are most likely to be effective. This information can be used to design more effective drug delivery systems. AI is also being used to optimize drug delivery systems. Optimization algorithms can be used to identify the most efficient and effective drug delivery methods, taking into account factors such as the drug properties, the target tissue, and the desired therapeutic effect. For example, optimization algorithms can be used to design drug delivery systems that target specific cells or tissues, reducing side effects and improving drug efficacy. Computer vision is also being used in drug delivery. Computer vision algorithms can be used to analyze images of drug delivery systems, such as nanoparticles or liposomes, and predict how the system will interact with the body. This information can be used to optimize the design of drug delivery systems and improve their effectiveness.

The prediction here is defined as "taking the information you have, often called 'data', and using it to generate information you don't have". Predictive analytics is very intuitive that uses advanced computer algorithms to analyze large existing datasets to identify trends and patterns that can be used to predict disease outcomes and inform treatment plans (Shaw et al. 2019). For example, predictive analytics can be used to predict the risk of heart attack or stroke in patients with diabetes, allowing healthcare providers to intervene before these events occur.

ANNs are a network of artificial neurons comprising three layers; the input layer, the hidden layer that gives training on the dataset fed as input, and the output layer. Deep learning is of more interest than other ML techniques. A DNN is an ANN with numerous hidden layers and hence increases the accuracy and performance (Manne and Kantheti 2021). ANN algorithms consist of interconnected nodes that process and transmit information, allowing the machine to learn and make predictions based on patterns in the data. ANN is used in various fields, such as computer vision, NLP, and robotics. CNNs are a type of deep learning algorithm commonly used in computer vision applications, such as image and video recognition. CNNs are designed to learn spatial hierarchies of features from input data automatically and adaptively by using convolution operations and pooling layers to extract and compress information. This makes them highly effective for tasks such as object detection, face recognition, and image classification. CNN classifies visual content and recognizes objects that are fed as input. It relies on connections and weights across units that are followed by sub-sampling. Virtual AI comprises informatics from deep learning applications like electronic health records and image processing. It helps physicians to diagnose and manage disease. Physical AI involves mechanical advances, such as the use of robotics in surgery and physical rehabilitation (Ellahham et al. 2020). AI also has the potential to shorten the span of clinical trial durations, maintaining improved productivity. Biopharma businesses procure huge volumes of scientific and research information from a large number of sources known as real-world data (Marr 2017).

For any diagnosis, first, a pathological test is conducted (e.g., cancer). Then the pathologist would collect and analyze the images from the procedure conducted. This classification and analysis are done by ANN (see Figure 9.4). Their method is based on the determination of nuclei regions on the images and then using these regions in the algorithm that would perform the classification. It can then find out whether the regions are cancerous or noncancerous. With ANN, classification accuracies are more than 90% and even exceed 99% in some cases. Hence, ANNs have substantial potential in CVD diagnosis too.

Robotics has numerous applications in medical science, including surgical procedures, diagnosis, and rehabilitation. Surgical robots can be used to perform minimally invasive surgeries with increased precision, reduced risk of complications, and improved patient outcomes. Robots can also be used for medical imaging and diagnosis, such as in the case of robot-assisted ultrasound. Additionally, robotics can be used in physical therapy and rehabilitation to help patients recover from injuries or neurological conditions, such as stroke. Overall, robotics has the potential to improve patient outcomes and reduce healthcare costs. Robots made with AI technologies

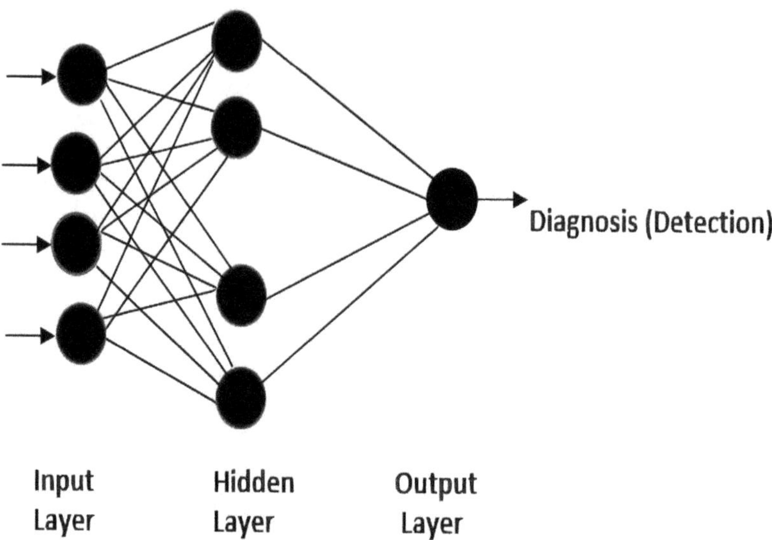

FIGURE 9.4 Data inputs are fed into an artificial neural network (ANN), which proceeds to classify the output utilizing the input layer, hidden layers, and output layer.

can carry out X-rays and CT scans more rapidly and precisely. Robots can be used to assist with surgeries, monitor patients, and perform routine tasks such as dispensing medication. Urological and gynecological surgeries have been revolutionized by the Da Vinci robotic surgical system developed by Intuitive Surgical. The better precision robotic arms mimic the surgeon's hands. It has a three-dimensional view and magnification options to allow minute incisions (Malik et al. 2019).

Mobile health refers to the use of mobile devices, such as smart phones and tablets, for healthcare purposes. This includes a wide range of applications and technologies, such as mobile apps for health tracking, remote monitoring of patients, and telemedicine. Mobile health can be used to improve access to healthcare in remote or underserved areas, provide real-time monitoring and feedback for chronic conditions, and enable personalized healthcare through data analysis and ML algorithms. Mobile health has the potential to revolutionize the healthcare industry by making healthcare more accessible, affordable, and efficient. It requires the use of a short messaging service (SMS). These devices can monitor heart rate, blood pressure, glucose level, sleep pattern, brain activities, etc. More complicated operations like General Packet Radio Service (GPRS), Global Positioning System (GPS), and bluetooth-based technology are also used. Globally there are over 500 mobile health projects and around 40,000 medical mobile applications. Big data, both organized and unorganized, comprises clinical details of the physicians, their notes and prescriptions, lab data, drugstore documents, CT scan and MRI images, insurance files, and records from administrative operations. The data can be examined by data mining. Various fields like IoT, machine vision, driver assistance, and NLP have been employed (Khan and Alotaibi 2020). Since 2018, the collaboration continues between Buoy Health and

Boston Children's Hospital to provide parents with advice about their ailing child. They answer the questions regarding medication and if the symptoms require a visit to a doctor. The AiCure App by the National Institutes of Health (NIH) monitors the medication used by the patient with the help of a smart phone webcam (Malik et al. 2019). One example of an AI application for monitoring the spread of COVID-19 is the COVID-19 Open Research Dataset (CORD-19) project that uses AI to analyze data from various sources, including scientific papers and news articles, to identify patterns and insights related to the pandemic. The system uses ML techniques such as NLP and deep learning to identify relevant information, allowing researchers to develop effective strategies to combat the virus. AI technique-built chatbots can be used to identify depression and anxiety. Right Eye LLC innovated an AI-powered experiment for the detection of autism spectrum disorder by applying eye-tracking technology (Erguzel and Ozekes 2014).

AI can also play a crucial role in monitoring and addressing health issues, especially in the context of the COVID-19 pandemic. AI-based systems can assist in monitoring the spread of the virus, predicting outbreaks, and developing effective treatments and vaccines. AI can also help in identifying potential disease risks and developing targeted interventions to prevent the spread of infectious diseases.

AI can also help in developing effective treatments and vaccines for COVID-19. One example of AI application in this context is the BenevolentAI project, which uses AI to analyze huge amounts of data for identifying potential drug candidates for COVID-19. The system uses ML techniques such as deep learning and NLP to analyze scientific literature, clinical trial data, and other relevant information to identify potential drug candidates. Another example of an AI application for health issues is the Deep Patient project by Mount Sinai Hospital in New York. The project uses ML techniques such as deep learning and neural networks to analyze medical records to identify potential disease risks and develop personalized treatment plans for patients.

The use of AI is increasing but its application is mostly restricted to a few diseases, such as cancer, disease of the nervous system, and cardiovascular diseases (Marr 2017). Nonetheless AI has the potential to handle huge quantities of data, synthesis of insights, increased outreach, easy access to information, virtual follow-up and consultations, saving cost and time for diagnosis and disease management, hence maximizing efficiency. Sometimes AI is linked to low or inaccurate prediction. CNNs are taught and authorized using datasets in the clinical background but may not decipher well for a bigger population. Take for instance the monitoring of skin lesions in detecting skin cancer that could be additionally diversified in the general population. Lessening or minimizing risk implies safety in healthcare and holds the key to AI applications (Eren et al. 2008). ML techniques create a 'black box' phenomenon; the user can only access the inputs and outputs of an algorithm, but not the internal mechanism of the specific relations assessed by the algorithm (Ramkumar et al. 2021). AI is also being used to optimize healthcare management systems. Optimization algorithms can be used to identify the most efficient and cost-effective ways to deliver healthcare services, taking into account factors such as patient needs, resource availability, and healthcare provider expertise. For example, optimization algorithms can be used to schedule appointments and allocate resources in a way that minimizes wait times and maximizes patient satisfaction.

9.3 CHALLENGES AND LIMITATIONS

AI-based systems function effectively and are adopted widely. Despite the potential of AI and ML in monitoring environmental and health issues they have several limitations when it comes to their application in healthcare (see Table 9.3). Some of these limitations include:

1. Data bias. AI models may exhibit bias if they are trained on data that is not representative of the population they are intended to serve.
2. Lack of transparency. Some AI models are known as 'black boxes', meaning it can be difficult to understand how they arrive at their predictions or recommendations. This lack of transparency and explainability can limit the trust in and adoption of AI-based systems, especially in critical applications such as healthcare and environmental monitoring.
3. Limited data availability. In some cases, there may not be enough data available to train an effective AI model. The data used in AI-based systems must be reliable, accurate, and up to date to ensure that the system produces accurate insights and predictions.
4. Limited scope. AI models may be limited in their ability to recognize rare diseases or conditions that are not well represented in the training data.
5. Ethical considerations. The use of AI and ML in healthcare raises ethical concerns, such as privacy, security, and the potential for discrimination.

It is important to be aware of these limitations and work to address them to ensure that AI and ML are used effectively and ethically in healthcare.

AI by itself is a big electricity consumer, with data centers consuming over 2% of the global electricity (Pearce 2018). It is predicted that the consumption will increase between 8% and 25% by 2025 (Andrae and Edler 2015). Hence AI contributes to the very problem it claims to solve—the climate crisis. The term 'nubecene' was coined by Gonzalez (Nost and Colven 2022) to capture the wide-ranging impacts, promises, and challenges that many AI applications might offer. The potential to transform environmental challenges through AI and related machine learning (ML) approaches resonates with the concept of 'digital solutionism'. AI is expected to address social and environmental inequities but raises new ethical and justice-related questions by reproducing them. Extracting socioenvironmental data for reputational profit can make us apprehensive (Nost and Colven 2022).

9.4 CONCLUSION

The United Nations' Sustainable Development Goals (SDGs) are a set of 17 goals aimed at achieving a sustainable and equitable future for all. These goals cover a range of areas, such as poverty reduction, gender equality, education, health, clean energy, sustainable cities, and climate action. AI applications can play a significant role in achieving these SDGs by providing innovative solutions to complex problems. AI applications can help in creating sustainable cities by optimizing urban planning, reducing traffic congestion, and managing waste and water resources. AI-powered sensors can also monitor air and water quality to improve public health.

TABLE 9.3

Merits and Demerits of the Various AI Techniques Applied to Solve Environmental and Health Issues

Issues	Techniques Applied	Merits	Demerits
Climate change	ML-DL, ANN, SVM, NLP	Predict & analyze climate patterns; assimilation of knowledge	AI consumes 2% of the global electricity; raises new ethical issues
Air quality	AI-BPNN, ELM, LSRM, GENN, GRU, WNN, FL, SVM, GA Hybrid techniques: adaptive neuro-fuzzy interference system, SVM-ML approach, integrated SVM, PLS-SVM, PSO	Prediction is easy to implement. Adopt corrective measures, reliable data	A short dataset restricts required accuracy; a large dataset is computationally expensive. Risks such as leakage probability, measurement of economic, ecological, and social impacts, environmental perturbation strength assessment, etc.
Water quality	SVM, NNs, DNN, K-nearest neighbors, ML, NLP Hybrid techniques: CNN-LSTM, PSO-NN, GA-GRNN	Modeling of water quality, predicting changes, classification. Better prediction of drinking water quality, optimize wastewater treatment	Expensive, time-consuming. High installation costs
Sanitation	AI, ML Computer vision Predictive analytics NLP	Efficiency, and effectiveness of sanitation systems. Identify patterns & predict future trends and potential outbreaks of diseases. Identify blockages or leaks, improve efficiency. Predict the failure of sewage systems. Improve public awareness & engagement, and allocate resources effectively. Analyzes qualitative data in public health.	New strategies are to be included. Correctness and quality of detection could be improved. A comprehensive framework needs to be updated. Not an adequate alternative to human analysis.
Waste management	NLP Robotics Predictive analytics	Waste-related concerns prevent from escalating. Improve efficiency, safety, and sustainability. Plan & allocate resources effectively.	Analyzing waste-dumping behavior is a challenge. The automation level of the robot could be improved. Effectiveness could be improved.

(continued)

TABLE 9.3 *(Continued)*
Merits and Demerits of the Various AI Techniques Applied to Solve Environmental and Health Issues

Issues	Techniques Applied	Merits	Demerits
Forestry, biodiversity, and crop management	Remote sensing ML NLP Robotics Predictive analytics Optimization algorithms OCR & NLP DNN& acoustic system ML algorithms Deep learning Computer vision, NLP	Monitor forest ecosystem, track changes. Predict forest fires, insect infestation. Improve public awareness, and allocate resources more effectively. Improve efficiency, safety, and sustainability. Plan & allocate resources effectively. Best location & sizes for protected areas. Analysis of data. Send alerts for illegal tree felling. Identify patterns, predict future trends; help to plan management strategies. Classify datasets from images. Biodiversity management improves public awareness and allocates resources more effectively. Help predict ecosystem services	Enhanced capabilities need to be explored; the lack of sound and video data limits the expansion of ML applications. Analyzing from a large dataset is a challenge. Should be more cost-effective. Effectiveness could be improved. Various types of video and sound data are not available. Classifying from a large data set is a challenge. Analyzing from a large and different dataset is a challenge. Tackling different ecosystem issues is a challenge.
Health management	NN, SVM, Decision trees, etc. ANN ML algorithms NLP Computer vision	Diagnose ailments. Categorize diabetes & cardiovascular diseases Analyze patient data, and identify patterns to predict disease outcomes. Analyze patient records, and identify life-threatening conditions caused by infection. Identify breast cancer in mammograms, allowing early detection & treatment.	Without medical tests it is difficult. Machine learning needs to improve with bigger data. Different problems and different datasets could be used. Effectiveness could be improved.

AI-BPNN = artificial intelligence-back propagation neural network; ANN = artificial neural network; CNN = convolutional neural network; DL = deep learning; DNN = deep neural network; ELM = extreme learning machine; FL = fuzzy logic; GA = genetic algorithm; GENN = General Neural Network; GRU = gated recurrent unit; LSRM = linear switched reluctance machine; LSTM = long short-term memory; ML = machine learning; NLP = Natural Language Processing; NN = neural networks; OCR = optical character recognition; PLS = partial least squares; PSO = particle swarm optimization; SVM = support vector machine; WNN = wavelet neural network.

In conclusion, AI applications have the potential to contribute significantly to achieving the SDGs by providing innovative solutions to complex problems. AI-powered solutions can predict climate patterns, monitor air and water quality, optimize waste and water management, and develop effective treatments and vaccines for infectious diseases to enhance healthcare, financial inclusion, clean energy, sustainable cities, and climate action. The integration of AI with the SDGs can create a more equitable and sustainable future for all.

In conclusion, the post-COVID-19 pandemic era requires a concerted effort to address various environmental and health issues to achieve sustainable development. Overall, AI has the potential to play a significant role in monitoring environmental and health issues post-COVID-19 pandemic for sustainable living.

REFERENCES

Aayush, K., Vishal, D., Hammad, N., & Manu, K. 2020. Application of artificial intelligence in curbing air pollution: The case of India. *Asian Journal of Management, 11*(3), 285–290.

Ahmed, U., Mumtaz, R., Anwar, H., Shah, A. A., Irfan, R., & García-Nieto, J. 2019. Efficient water quality prediction using supervised machine learning. *Water, 11*(11), 2210.

AI, T. 2021. Terrafuse AI launches new platform to visualize California wildfire risk. *PR Newswire.* 21 October. https://www.prnewswire.com/news-releases/terrafuse-ai-launches-new-platform-to-visualize-california-wildfire-risk-301405367.html

Alanen, P. 2020. Helen and Silo AI bring intelligence to district heat production in Helsinki. *Silo AI.* 4 September. https://www.silo.ai/blog/helen-and-silo-ai-bring-intelligence-to-district-heating-production-in-helsinki

Andrae, A. S., & Edler, T. 2015. On global electricity usage of communication technology: Trends to 2030. *Challenges, 6*(1), 117–157.

Azzi, S., Gagnon, S., Ramirez, A., & Richards, G. 2020. Healthcare applications of artificial intelligence and analytics: A review and proposed framework. *Applied Sciences, 10*(18), 6553.

Baek, S.-S., Pyo, J., & Chun, J. A. 2020. Prediction of water level and water quality using a CNN-LSTM combined deep learning approach. *Water, 12*(12), 3399.

Barnes, E. A., Hurrell, J. W., Ebert-Uphoff, I., Anderson, C., & Anderson, D. 2019. Viewing forced climate patterns through an AI lens. *Geophysical Research Letters, 46*(22), 13389–13398.

Bughin, J., Hazan, E., Ramaswamy, S., Chui, M., Allas, T., Dahlstrom, P., & Trench, M. 2017. *Artificial intelligence: The next digital frontier?* https://www.mckinsey.com/~/media/mckinsey/industries/advanced%20electronics/our%20insights/how%20artificial%20intelligence%20can%20deliver%20real%20value%20to%20companies/mgi-artificial-intelligence-discussion-paper.ashx

Carter, R. A. 2019. Running with renewables. *Engineering and Mining Journal, 220*(3), 32–37.

Chan, H. S., Shan, H., Dahoun, T., Vogel, H., & Yuan, S. 2019. Advancing drug discovery via artificial intelligence. *Trends in Pharmacological Sciences, 40*(8), 592–604.

Chau, K.-W. 2006. A review on integration of artificial intelligence into water quality modelling. *Marine Pollution Bulletin, 52*(7), 726–733.

Chemical Safety and Health Unit (CHE). 2016. *The public health impact of chemicals: Knowns and unknowns.* 23 May. https://www.who.int/publications/i/item/WHO-FWC-PHE-EPE-16-01

Chen, X., Huang, H., Liu, Y., Li, J., & Liu, M. 2022. Robot for automatic waste sorting on construction sites. *Automation in Construction, 141,* 104387.

Chin, S. 2020. Machine learning helps create "autonomous" building. *Fierce Electronics.* 26 February. https://www.fierceelectronics.com/sensors/machine-learning-helps-create-autonomous-building

Coffey, P., Smith, N., Lennox, B., Kijne, G., Bowen, B., Davis-Johnston, A., & Martin, P. A. 2021. Robotic arm material characterisation using LIBS and Raman in a nuclear hot cell decommissioning environment. *Journal of Hazardous Materials, 412*, 125193.

De Kerf, T., Gladines, J., Sels, S., & Vanlanduit, S. 2020. Oil spill detection using machine learning and infrared images. *Remote Sensing, 12*(24), 4090.

Dewitte, S., Cornelis, J. P., Müller, R., & Munteanu, A. 2021. Artificial intelligence revolutionises weather forecast, climate monitoring and decadal prediction. *Remote Sensing, 13*(16), 3209.

Dubey, S., Singh, P., Yadav, P., & Singh, K. K. 2020. Household waste management system using IoT and machine learning. *Procedia Computer Science, 167*, 1950–1959.

Dunbabin, M., & Marques, L. 2012. Robotics for environmental monitoring [from the guest editors]. *IEEE Robotics & Automation Magazine, 19*(1), 20–23.

Ellahham, S., Ellahham, N., & Simsekler, M. C. E. 2020. Application of artificial intelligence in the health care safety context: Opportunities and challenges. *American Journal of Medical Quality, 35*(4), 341–348.

eLogii. 2020. *UPS route optimization software: Review.* 9 November. https://elogii.com/blog/ups-route-optimization-software.

Eren, A., Subasi, A., & Coskun, O. 2008. A decision support system for telemedicine through the mobile telecommunications platform. *Journal of Medical Systems, 32*, 31–35.

Erguzel, T., & Ozekes, S. 2014. Artificial intelligence approaches in psychiatric disorders. *Journal of Neurobehavioral Sciences, 1*(2), 52.

Faghmous, J. H., & Kumar, V. 2014. A big data guide to understanding climate change: The case for theory-guided data science. *Big Data, 2*(3), 155–163.

FAO, I., & UNICEF. 2020. WFP and WHO. 2020. The State of Food Security and Nutrition in the World, 619.

Gagne, D. J. 2020. *Seeing the Atmosphere through machine learning.* National Center for Atmospheric Research Explorer Series, University Corporation for Atmospheric Research. https://ncar.ucar.edu/what-we-offer/education-outreach/public/ncar-explorer-series-lectures/2020-explorer-series/machine

Gao, J. 2014. *Machine learning applications for data center optimization.* https://static.google usercontent.com/media/research.google.com/en//pubs/archive/42542.pdf

Gautam, G., & Dahal, K. R. 2020. Issues and problems of community water supply schemes with special reference to Nepal. *Journal of Civil, Construction and Environmental Engineering, 5*(5), 114.

Gomolka, Z., Twarog, B., Zeslawska, E., Lewicki, A., & Kwater, T. 2017. Using artificial neural networks to solve the problem represented by BOD and DO indicators. *Water, 10*(1), 4.

Helen and Silo AI bring intelligence to district heat production in Helsinki. *Silo AI.* https://www.silo.ai/blog/helen-and-silo-ai-bring-intelligence-to-district-heating-production-in-helsinki

Hmoud Al-Adhaileh, M., & Waselallah Alsaade, F. 2021. Modelling and prediction of water quality by using artificial intelligence. *Sustainability, 13*(8), 4259.

IBM. *How does a digital twin work?* https://www.ibm.com/topics/what-is-a-digital-twin

Jeffery, J. 2019. *8 companies utilizing AI to tackle climate change.* www.entrepreneur.com/business-news/8-companies-utilizing-ai-to-tackle-climate-change/340002

Joppa, L. N. 2017. The case for technology investments in the environment. *Nature, 552*(7685), 325–328.

Khan, Z. F., & Alotaibi, S. R. 2020. Applications of artificial intelligence and big data analytics in m-health: A healthcare system perspective. *Journal of Healthcare Engineering, 2020*, 1–15.

Koo, K.-M., Han, K.-H., Jun, K.-S., Lee, G., & Yum, K.-T. 2021. Smart water grid research group project: An introduction to the smart water grid living-lab demonstrative operation in Yeong Jong Island, Korea. *Sustainability, 13*(9), 5325.

Krithiga, R. 2019. *'EARTHBOT': The smart sanitation robot*. Paper presented at the 2019 IEEE International WIE Conference on Electrical and Computer Engineering (WIECON-ECE).

Kurt, A., & Oktay, A. B. 2010. Forecasting air pollutant indicator levels with geographic models 3 days in advance using neural networks. *Expert Systems with Applications*, *37*(12), 7986–7992.

Lamba, A., Cassey, P., Segaran, R. R., & Koh, L. P. 2019. Deep learning for environmental conservation. *Current Biology*, *29*(19), R977–R982.

Leeson, W., Resnick, A., Alexander, D., & Rovers, J. 2019. Natural language processing (NLP) in qualitative public health research: A proof of concept study. *International Journal of Qualitative Methods*, *18*. https://doi.org/10.1177/1609406919887021

Li, X., Peng, L., Hu, Y., Shao, J., & Chi, T. 2016. Deep learning architecture for air quality predictions. *Environmental Science and Pollution Research*, *23*, 22408–22417.

Liu, P., Wang, J., Sangaiah, A. K., Xie, Y., & Yin, X. 2019. Analysis and prediction of water quality using LSTM deep neural networks in IoT environment. *Sustainability*, *11*(7), 2058.

Ly, H.-B., Le, L. M., Phi, L. V., Phan, V.-H., Tran, V. Q., Pham, B. T., . . . Derrible, S. 2019. Development of an AI model to measure traffic air pollution from multisensor and weather data. *Sensors*, *19*(22), 4941.

Madhumita Murgia, N. T. 2017. DeepMind and national grid in AI talks to balance energy supply. [Energy Sector]. *Financial Times*. 12 March. https://www.ft.com/content/27c8aea0-06a9-11e7-97d1-5e720a26771b

Malek Mohammadi, M., Najafi, M., Salehabadi, N., Serajiantehrani, R., & Kaushal, V. 2020. *Predicting condition of sanitary sewer pipes with gradient boosting tree 'pipelines 2020'* (pp. 80–89). American Society of Civil Engineers.

Malik, P., Pathania, M., & Rathaur, V. K. 2019. Overview of artificial intelligence in medicine. *Journal of Family Medicine and Primary Care*, *8*(7), 2328.

Manne, R. 2021. Machine learning techniques in drug discovery and development. *International Journal of Applied Research*, *7*(4), 21–28.

Manne, R., & Kantheti, S. C. 2021. Application of artificial intelligence in healthcare: Chances and challenges. *Current Journal of Applied Science and Technology*, *40*(6), 78–89.

Marr, B. 2017. First FDA approval for clinical cloud-based deep learning in healthcare. *Forbes*. Forbes Publishing Company.

Masood, A., & Ahmad, K. 2021. A review on emerging artificial intelligence (AI) techniques for air pollution forecasting: Fundamentals, application and performance. *Journal of Cleaner Production*, *322*, 129072.

Midani, F. S., Weil, A. A., Chowdhury, F., Begum, Y. A., Khan, A. I., Debela, M. D., . . . Silverman, J. D. 2018. Human gut microbiota predicts susceptibility to Vibrio cholerae infection. *The Journal of Infectious Diseases*, *218*(4), 645–653.

Mishra, D., & Goyal, P. 2015. Development of artificial intelligence based NO2 forecasting models at Taj Mahal, Agra. *Atmospheric Pollution Research*, *6*(1), 99–106.

Mittal, A., & Bhardwaj, R. 2011. Prediction of daily air pollution using wavelet decomposition and adaptive network-based fuzzy inference system. *International Journal of Environmental Sciences*, *2*(1), 174–184.

Monteleoni, C., Schmidt, G. A., & McQuade, S. 2013. Climate informatics: Accelerating discovering in climate science with machine learning. *Computing in Science & Engineering*, *15*(5), 32–40.

Monteleoni, C., Schmidt, G. A., Saroha, S., & Asplund, E. 2011. Tracking climate models. *Statistical Analysis and Data Mining: The ASA Data Science Journal*, *4*(4), 372–392.

Mudunuru, M. K., Ahmmed, B., Frash, L., & Frijhoff, R. 2023. *Deep learning for modeling enhanced geothermal systems*. Paper presented at the Proceedings of 48th Workshop on Geothermal Reservoir Engineering, Stanford University.

Nations, U. 2021. *Water scarcity*. UN-Water.

Nicholas, F. 2019. *How AI is helping solve climate change, 2023*. www.smashingmagazine. com/search/?q=How%20AI%20Is%20Helping%20Solve%20Climate%20Change

Nost, E., & Colven, E. 2022. Earth for AI: A political ecology of data-driven climate initiatives. *Geoforum, 130*, 23–34.

Ong, B. T., Sugiura, K., & Zettsu, K. 2016. Dynamically pre-trained deep recurrent neural networks using environmental monitoring data for predicting PM 2.5. *Neural Computing and Applications, 27*, 1553–1566.

Oprea, M., Popescu, M., Mihalache, S. F., & Dragomir, E. G. 2017. *Data mining and ANFIS application to particulate matter air pollutant prediction: A comparative study*. Paper presented at the ICAART (2).

Parisa, Z., & Nova, M. 2020. This AI can see the forest and the trees. *IEEE Spectrum, 57*(8), 32–37.

Pearce, F. 2018. *Energy hogs: Can world's huge data centers be made more efficient?* https://e360.yale.edu/features/energy-hogs-can-huge-data-centers-be-made-more-efficient%C2%A0#:~:text=by%20Matt%20Rota-,Energy%20Hogs%3A%20 Can%20World's%20Huge%20Data%20Centers%20Be%20Made%20More,and%20 dramatically%20improve%20energy%20efficiency

Polk, H. 2019. *State of global air 2019: A special report on global exposure to air pollution and its disease burden*. Health Effects Institute.

Polson, N. G., & Sokolov, V. O. 2017. Deep learning for short-term traffic flow prediction. *Transportation Research Part C: Emerging Technologies, 79*, 1–17.

Preeti, H., & Gadgay, B. 2018. Pond cleaning robot. *International Research Journal of Engineering and Technology, 5*(10), 1136–1139.

Qader, M. R., Khan, S., Kamal, M., Usman, M., & Haseeb, M. 2021. Forecasting carbon emissions due to electricity power generation in Bahrain. *Environmental Science and Pollution Research*, 1–12.

Ramkumar, P. N., Kunze, K. N., Haeberle, H. S., Karnuta, J. M., Luu, B. C., Nwachukwu, B. U., & Williams, R. J. 2021. Clinical and research medical applications of artificial intelligence. *Arthroscopy: The Journal of Arthroscopic & Related Surgery, 37*(5), 1694–1697.

Readfearn, G. 2018. WHO launches health review after microplastics found in 90% of bottled water. *Plastics, The Guardian*. 15 March.

Revels, M. 2023. *Predicting physics parameters in nuclear reactors*. Texas A&M University College of Engineering. 21 June. https://engineering.tamu.edu/news/2023/06/ predicting-physics-parameters-in-nuclear-reactors.html.

Rolnick, D., Donti, P. L., Kaack, L. H., Kochanski, K., Lacoste, A., Sankaran, K., … Waldman-Brown, A. 2022. Tackling climate change with machine learning. *ACM Computing Surveys (CSUR), 55*(2), 1–96.

Sarc, R., Curtis, A., Kandlbauer, L., Khodier, K., Lorber, K. E., & Pomberger, R. 2019. Digitalisation and intelligent robotics in value chain of circular economy oriented waste management: A review. *Waste Management, 95*, 476–492.

Saustania. 2019. AI improves efficiency of renewable energy storage. *In Energy*. https://goex plorer.org/ai-improves-efficiency-of-renewable-energy-storage/

Shaheen, M. Y. 2021. *Applications of artificial intelligence (AI) in healthcare: A review*. ScienceOpen Preprints. https://www.scienceopen.com/hosted-document?doi=10.14293/ S2199-1006.1.SOR-.PPVRY8K.v1

Shaw, J., Rudzicz, F., Jamieson, T., & Goldfarb, A. 2019. Artificial intelligence and the implementation challenge. *Journal of Medical Internet Research, 21*(7), e13659.

Shi, C., Liu, D., Cui, Y., Xie, J., Roberts, N. J., & Jiang, G. 2020. Amur tiger stripes: Individual identification based on deep convolutional neural network. *Integrative Zoology, 15*(6), 461–470.

Shivaprakash, K. N., Swami, N., Mysorekar, S., Arora, R., Gangadharan, A., Vohra, K., . . . Kiesecker, J. M. 2022. Potential for artificial intelligence (AI) and machine learning (ML) applications in biodiversity conservation, managing forests, and related services in India. *Sustainability, 14*(12), 7154.

Silva, J., Lucas, P., Araújo, F., Silva, C., Gil, P., Cardoso, A., . . . Salgueiro, P. 2019. *An online platform for real-time air quality monitoring.* Paper presented at the 2019 5th Experiment International Conference (Exp. at '19).

Star, E. 2023. An overview of portfolio manager. 8 September. https://www.energystar.gov/sites/default/files/tools/An%20Overview%20of%20Portfolio%20Manager_April%202023_FINAL_508.pdf

Stein, A. L. 2020. Artificial intelligence and climate change. *Yale Journal on Regulation, 37,* 890.

Su, T.-C., & Yang, M.-D. 2014. Application of morphological segmentation to leaking defect detection in sewer pipelines. *Sensors, 14*(5), 8686–8704.

Subramaniam, S., Raju, N., Ganesan, A., Rajavel, N., Chenniappan, M., Prakash, C., . . . Dixit, S. 2022. Artificial intelligence technologies for forecasting air pollution and human health: A narrative review. *Sustainability, 14*(16), 9951.

Thollander, P., Karlsson, M., Rohdin, P., Wollin, J., & Rosenqvist, J. 2020. *Energy management* (pp. 239–257). Elsevier eBooks. https://doi.org/10.1016/b978-0-12-817247-6.00013-4

Thiyagarajan, K., Kodagoda, S., & Van Nguyen, L. 2017. *Predictive analytics for detecting sensor failure using autoregressive integrated moving average model.* Paper presented at the 2017, 12th IEEE Conference on Industrial Electronics and Applications (ICIEA).

Trends, M. 2022. From buildings to streets, Singapore emerges as the largest digital twin country. *Analytics Insight.* 5 September. https://www.analyticsinsight.net/from-buildings-to-streets-singapore-emerges-as-the-largest-digital-twin-country/

Tyagi, N. 2020. 5 ways ML helps in Uber services optimization. *Analytic Steps.* 12 June. https://www.analyticssteps.com/blogs/5-ways-ml-helps-uber-services-optimization

United Nations Children's Fund (UNICEF) and the World Health Organization (WHO). 2020. State of the world's sanitation: An urgent call to transform sanitation for better health, environments, economies and societies. Summary Report. UNICEF and WHO.

University Energy Management Overview. Energy and Sustainability, Facilities and Campus Services. 2021. https://fcs.cornell.edu/departments/energy-sustainability/energy-managementoverview

Vasisht, D., Kapetanovic, Z., Won, J., Jin, X., Chandra, R., Sinha, S. N., . . . Stratman, S. 2017. *Farmbeats: An IoT platform for data-driven agriculture.* Paper presented at the NSDI.

Vesselinov, V. V., Ahmmed, B., Frash, L., & Mudunuru, M. K. 2022. *GeoThermalCloud: Machine learning for discovery, exploration, and development of hidden geothermal resources.* Technical Report, Proceedings, 47th Workshop on Geothermal Reservoir Engineering.

Vesselinov, V. V., Mudunuru, M. K., Ahmmed, B., Karra, S., & Middleton, R. S. 2020. *Discovering signatures of hidden geothermal resources based on unsupervised learning.* Stanford University Press.

Victor, D. G. 2019. *How artificial intelligence will affect the future of energy and climate.* The Brookings Institution. www.brookings.edu/research/how-artificial-intelligence-willaffect-the-future-of-energy-and-climate/ (accessed 27 October 2021).

Vollrath, M. 2020. How we could supercharge battery development for electric vehicles. *Artificial Intelligence, Energy, Environment.* https://engineering.stanford.edu/magazine/article/how-we-could-supercharge-battery-development-electric-vehicles

Wang, C., Qin, J., Qu, C., Ran, X., Liu, C., & Chen, B. 2021. A smart municipal waste management system based on deep-learning and Internet of Things. *Waste Management, 135,* 20–29.

Wang, P., Liu, Y., Qin, Z., & Zhang, G. 2015. A novel hybrid forecasting model for PM10 and SO2 daily concentrations. *Science of the Total Environment, 505*, 1202–1212.

Wang, Z., Shao, D., Yang, H., & Yang, S. 2015. Prediction of water quality in South to North water transfer project of China based on GA-optimized general regression neural network. *Water Science and Technology: Water Supply, 15*(1), 150–157.

Wei, J., Sanborn, S., & Slaughter, A. 2019. Digital innovation creating the utility of the future. *Deloitte Insights.* https://www2.deloitte.com/us/en/insights/industry/power-and-utilities/digital-transformation-utility-of-the-future.html

Yang, D., Wang, W., Gueymard, C. A., Hong, T., Kleissl, J., Huang, J., . . . Xia, X. A. 2022. A review of solar forecasting, its dependence on atmospheric sciences and implications for grid integration: Towards carbon neutrality. *Renewable and Sustainable Energy Reviews, 161*, 112348.

Ye, Z., Yang, J., Zhong, N., Tu, X., Jia, J., & Wang, J. 2020. Tackling environmental challenges in pollution controls using artificial intelligence: A review. *Science of the Total Environment, 699*, 134279.

Zheng, G., Luo, F., & Chen, W. 2010. Quality prediction of waste water treatment based on immune particle swarm neural networks. *Microprocessors, 31*, 75–77.

Zhou, J., Welling, C. M., Vasquez, M. M., Grego, S., & Chakrabarty, K. 2020. Sensor-array optimization based on time-series data analytics for sanitation-related malodor detection. *IEEE Transactions on Biomedical Circuits and Systems, 14*(4), 705–714.

Zimmerman, N., Presto, A. A., Kumar, S. P., Gu, J., Hauryliuk, A., Robinson, E. S., . . . Subramanian, R. 2018. A machine learning calibration model using random forests to improve sensor performance for lower-cost air quality monitoring. *Atmospheric Measurement Techniques, 11*(1), 291–313.

10 Cerebral Palsy Detection Using Vision Impairment and Machine Learning

S. Beatrice, J. Stella, and John Rose, S.J.

10.1 INTRODUCTION

Cerebral palsy (CP) is a widespread group of disorders that affect a person's posture, mobility, and balance. One in 345 children has been identified with CP, according to estimates from the Centers for Disease Control's Autism and Developmental Disabilities Monitoring (ADDM) Network. CP can be caused by a brain injury or faulty brain development (cerebral problems) or by muscular weakness or dysfunction (Chtourou et al. 2021; Beatrice and Meena 2022). CP affects human movement and posture, but also has other side effects such as seizures, speech difficulties, hearing loss, intellectual disabilities, or even scoliosis and contractures (Upadhyay et al. 2020). Early detection of CP is crucial for maximizing a child's potential and improving their overall well-being. It allows for the timely initiation of interventions that can address motor, communication, and cognitive challenges, ultimately leading to better outcomes and a higher quality of life for individuals with CP.

A number of studies have been used to predict CP. Our work shows that vision can be used as a predictor of CP. A set of standardized tests and measures is used to assess visual acuity, visual field, and eye movements of a child. Using a variety of classical machine learning tools trained on a dataset (Goudiaby et al. 2020), we then predict CP with high precision. Assessment of vision is of course not only important for diagnosis but also an important part of the treatment plan of patients.

In the remainder of this chapter, we first summarize related work. We then define aspects of vision that are useful for diagnosis of CP as well as part of the treatment plan. Next, we describe shortly classic machine learning models that we are using. Finally, we present our results on the efficiency and efficacy of these tools in diagnosing CP.

10.2 RELATED WORK

We can screen for CP by using side-effects, such as kinetic (Kastaniotis et al. 2015; Kang et al. 2017; Kwolek et al. 2019), parametric elliptic Fourier parameters (Yoo and Nixon 2011), skeleton joints (Wang et al. 2016), and human pose (Andriluka et al. 2014). The walking patterns of people with cerebral palsy are recognized via kinetic feature analysis and video stream visualization. For the latter, one can use convolutional neural networks (CNN; Hua et al. 2020) or recurrent neural networks (RNN; Upadhyay et al. 2020).

DOI: 10.1201/9781003405436-12

Human movement and posture (Upadhyay et al. 2020) are known to be impacted by cerebral palsy, but there are additional side effects that can appear, including seizures, speech difficulties, hearing loss, intellectual disabilities, and some that may be much more severe than the ones listed above, such as scoliosis or contractures. To forecast cerebral palsy, the CNN (Hua et al. 2020) algorithm is applied to gait patterns with various hidden layers. The RNN method is also applied to gait data patterns to assess the illness. This could provide an easy-to-use interface and facilities to check for the presence of cerebral palsy in adults as well as infants. Though they do not guarantee the results, it is proven to be effective to detect the presence without the deployment of expensive techniques; for further confirmation a doctor should be consulted. In addition to the traditionally deployed techniques of CP prediction, another major symptom, vision impairment, can also be used to resolve the problem, which is addressed in this chapter.

Rouzbeh and Babaei (2015) described the combination of body limb lengths and skeleton joint angles to recognize the gait. An experiment has been performed on a group of 48 participants, randomly, from different ages. Also considered the cerebral palsy existing in the person without the requirement for a medical consultation based on symptoms. To comprehend the symptoms and offer a conclusion based on them, it employs machine learning algorithms.

In visual examinations, a range of standardized tests and measures are used to assess visual acuity, visual field, and eye movements (Singh et al. 2018). Important information on the severity and characteristics of visual abnormalities linked to cerebral palsy is provided by these assessments. Measurements of visual acuity include Snellen charts and Teller Acuity Cards®. Snellen charts are commonly used to assess visual acuity, which measures how well a person can see objects at a specific distance. This test is often used for adults and older children. Infants and young children who are unable to read letters or speak orally have their visual acuity evaluated with Teller Acuity Cards.

Eye movement is an important component of visual examination. The ability to anticipate movement in space is made possible by the eyes, which also continuously transmit information into the system. However, the compensatory attempts of the visual system may start to obstruct normal maturation; adaptation obstructs learning new things. The youngster who has high postural tone or spasticity frequently uses whole patterns of movement and may commence extension with an upward gaze. Conversely, when the eyes are relaxed, high tonus is lessened, allowing the children's head to drop into gravity. To maximise the likelihood that children with cerebral palsy will progress successfully, it is essential to comprehend the nature of the ongoing interplay between the visual and postural systems. The research we discuss here gives us some understanding of the visual adjustments a child with cerebral palsy makes in order to live.

10.2.1 OBJECTIVES

To determine cerebral palsy by analysing visual impairment using various machine learning models to examine the physical impairment in ambulant children and adolescents and compare all the models that have been used, namely, K-nearest neighbour (KNN), decision tree, and random forest.

STRABISMUS DISORDER TYPES

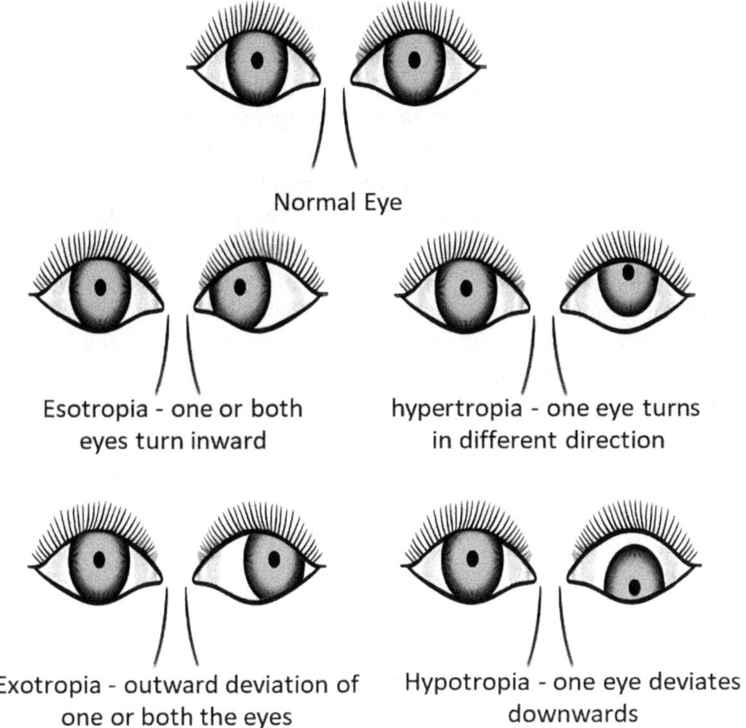

FIGURE 10.1   Squint disorder.

(Source: www.shutterstock.com/image-vector/squint-eye-strabismus-deflection-visual-axes-776025829)

10.3 METHODOLOGY

10.3.1 Vision Impairment

Vision Impairments can be caused by eye conditions like amblyopia ("lazy eye") or strabismus (misaligned or crossed eyes), eye or brain injuries, or birth defects. The term "strabismus," which can alternatively mean "crossed eyes" or "squint," refers to a disorder where the eyes are not correctly aligned (Lee et al. 2014; Wang et al. 2010). The eyes do not always point in the same direction in a person with strabismus. As one eye rotates inward, outward, upward, or downward (esotropia, exotropia, hypertropia, and hypotropia) the other eye keeps a straight line of sight as shown in Figure 10.1.

One eye turning inward and towards the nose is referred to as esotropia, a kind of strabismus. Misalignment can happen often or seldom. Early in childhood, infantile esotropia is a common occurrence, although accommodative esotropia is connected to substantial hyperopia. A multitude of factors might cause acquired esotropia to manifest itself later in life. The brain favours utilizing the opposite eye when one eye is misaligned because of muscular imbalance or refractive problems.

Esotropia can cause amblyopia, disturb binocular vision, and interfere with depth perception. Glasses, vision therapy, patches, and, in rare circumstances, surgery to straighten the eye muscles are all available as forms of treatment. Early intervention is essential to achieve the greatest results.

Exotropia is a kind of strabismus, or eye misalignment, in which one eye is turned outward. It might start later in life or be present from birth, happening sometimes or constantly. Exotropia can affect depth perception and eyesight, which is annoying and painful. Spectacles, eye exercises, and surgery are all possible treatments, depending on the severity. Early identification and intervention are crucial for the treatment of exotropia and the maintenance of visual function. The strabismus condition known as hypertropia, which results in a vertical misalignment, elevates one eye over the other. Mechanical restrictions, muscular weakness, and nerve palsy are just a few of the disorders that can cause it. It can also be hereditary or acquired. Hypertropia may cause double vision, issues with depth perception, and abnormal head tilting as a coping technique. Corrective surgery, prism lenses, eye workouts, and spectacles are among the possible treatments, depending on the underlying reason and the severity of the condition.

A vertical misalignment results from one eye being lower than the other, a condition known as hypotropia. It could be inherited or picked up due to physical restrictions, nerve palsy, or weak muscles. Hypotropia may result in double vision, impaired depth perception, and abnormal head tilting as a coping technique. Corrective surgery, prism lenses, eye workouts, and spectacles are among the possible treatments, depending on the underlying reason and the severity of the condition. Early intervention is crucial for optimum management and visual development.

10.4   DATASET

The real-time electroencephalogram (EEG)-based monitoring system utilizes machine learning to recognize eye movement. The Emotiv Headset is used to capture the users EEG signals wirelessly with the USB receiver. It has 14 channels located at AF3, F7, F3, FC5, T7, P7, O1, O2, P8, T8, FC6, F4, F8, and AF4 electrodes, as shown in Figure 10.2. By considering these channels, the dataset contains 15 independent time series value of their respective electrodes, among that the target variable says whether the person is having cerebral palsy or not.

10.5   MACHINE LEARNING MODELS FOR
PREDICTING CEREBRAL PALSY

The chapter analyses CP detection using various machine learning algorithms such as decision tree, random forest, and KNN with the dataset contains outliers initially and further analysis made without outliers. Machine learning techniques are a powerful tool to analyse CP patients by training and testing the model. It improves the results by identifying the best fit to the detection problem. The performance of the model is analysed by the confusion matrix and by the metrics accuracy, precision and f1_score.

The confusion matrix is an N × N matrix that shows the effectiveness of a classification model, where N is the total number of target classes. In the matrix,

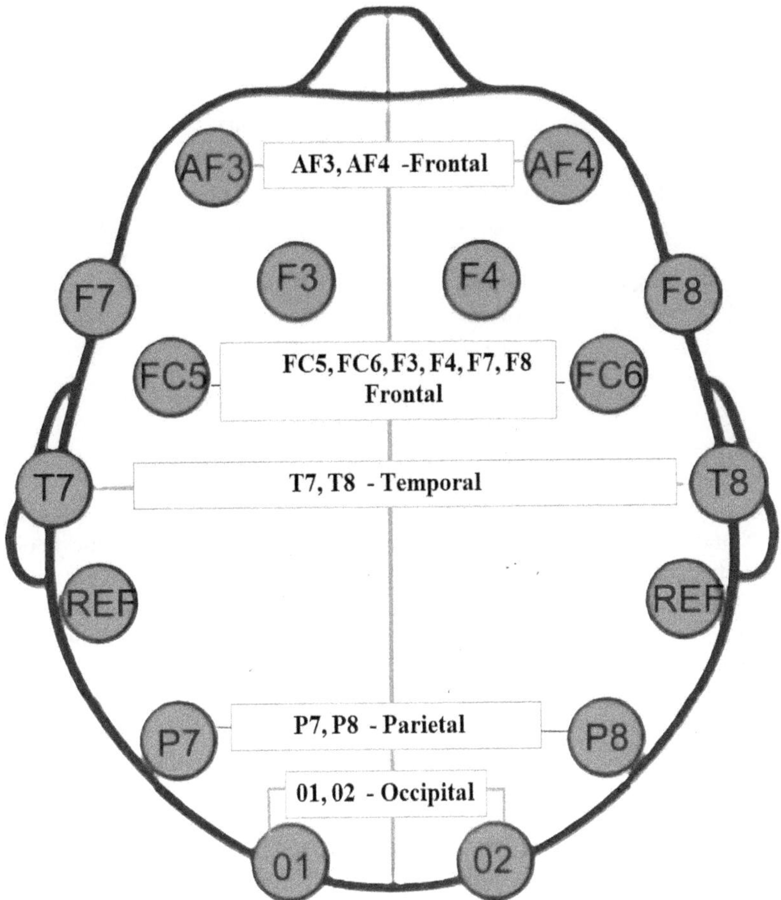

FIGURE 10.2 Location of 14 electrodes of the Emotiv EEG device. (Reproduced with permission from Goudiaby et al. 2020.)

the actual goal values are contrasted with those that the machine learning model anticipated.

The confusion matrix's terminology is as follows:

- True Positives (TP) are when the projected value and the actual value are both positive.
- True negatives (TN) are when the prediction and the actual value are both negative.
- False positives (FP) occur when the forecast is correct, but the result is incorrect. It is recognized as a Type 1 error.
- False negative (FN) occurs when the fact is positive, but the forecast is negative, this is known as a false negative (FN) or Type 2 error.

High TP and TN rates and low FP and FN rates are characteristics of a successful model. A confusion matrix is shown in Table 10.1.

TABLE 10.1
Confusion Matrix

		Actual Values	
		Negative	**Positive**
Predicted Values	**Negative**	**TN** (The predicted value is negative and it is negative)	**FN** (Type 2 error: The predicted value is negative, but it is positive)
	Positive	**FP** (Type 1 error: The predicted value is positive, but it is negative)	**TP** (The predicted value is positive and it is positive)

Outliers:

Outliers are the observations from the sample that lie abnormally in the normal data points. Standard deviation and mean value are calculated to identify the cut-off points for the dataset. If the value exceeds the cut-off points it is calculated as a higher data point; similarly, if it is minimal then it is estimated as a lower data point. After identification of the outliers the data that contains too high values and too small values are removed.

Algorithm for Outlier Removal

- for i in range (values.shape[1]-1)
- μ $\leftarrow$ Mean of values in the column
- σ $\leftarrow$ Standard deviation of values in the column
- lower $\leftarrow$ μ- 4 σ //Lower bound of outlier
- upper $\leftarrow$ μ+ 4 σ //Upper bound of outlier
- for i in range (values.shape[0])
- if (value[j,i] < lower)
- too_small $\leftarrow$ value[j,i]
- if (value[j,i] > upper)
- too_large $\leftarrow$ value[j,i]
- delete the values if values are in too_small or too_large

Implementation of Outlier Removal:

The dataset is loaded into the algorithm to remove the outliers in the dataset and the mean and standard deviation are calculated.

```
#Loading the data
data = eye_data
values = data.values


for i in range(values.shape[1] - 1):
# calculate column mean and standard deviation
  data_mean, data_std = mean(values[:,i]), std(values[:,i])
```

```
eye_data_no_out.describe()
```

	AF3	F7	F3	FC5	T7	P7	O1
count	14304.000000	14304.000000	14304.000000	14304.000000	14304.000000	14304.000000	14304.000000
mean	4298.144530	4007.199968	4261.656384	4120.176654	4339.673482	4618.165981	4070.977384
std	32.384305	27.223294	16.219864	16.905374	11.994652	12.923984	17.708529
min	4198.970000	3913.330000	4197.440000	4067.180000	4308.720000	4566.150000	4026.150000
25%	4280.510000	3990.260000	4250.260000	4107.690000	4331.790000	4611.790000	4057.440000
50%	4293.330000	4004.620000	4262.050000	4120.000000	4338.460000	4617.440000	4069.740000
75%	4308.720000	4020.510000	4268.720000	4128.720000	4346.150000	4625.640000	4082.560000
max	4466.150000	4154.360000	4349.230000	4191.790000	4397.950000	4672.310000	4138.970000

FIGURE 10.3   Dataset with no outliers.

The threshold values are estimated with the result of the standard deviation and then it is used to remove the too large and too small data points.

```
# define outlier bounds
  cut_off = data_std * 4
  lower, upper = data_mean—cut_off, data_mean + cut_off
  # remove too small
  too_small = [j for j in range(values.shape[0])
    if values[j,i] <lower]
  too_large = [j for j in range(values.shape[0])
    if values[j,i] >upper]
```

Figure 10.3 displays the count, mean, standard deviation, minimum, first quartile (Q1), second quartile (Q2), third quartile (Q3), and maximum values of the data gathered from various electrodes.

10.5.1   ANALYSIS OF MACHINE LEARNING MODELS

10.5.1.1   Decision Tree

Decision trees are supervised learning methods used for categorization. To create a model that forecasts the value of a target variable, the goal is to learn simple decision rules generated from the data features. An internal node, a branch node, and a leaf represent the characteristics, rules, and outcomes that create a decision tree's tree-like structure. The DecisionTreeClassifier function executes the machine learning operations, and the fitting model fits the data points by analyzing the Gini index (a measure of impurity in the data) to build a tree. Decision trees are the most straightforward machine learning model that provides explicit criteria for the dataset. The component that results in the highest drop in the Gini index is used to determine the Gini index at the root node. For each subset, this cycle is repeated recursively until a stopping condition is satisfied.

The mathematical representation of the Gini index is

$$\text{Gini impurity} = 1 - \Sigma\big(p(i)\big)^2 \tag{10.1}$$

The confusion matrix is used to assess the performance of the fitted decision tree machine learning model.

Algorithm:

- Select the input dataset and split the data into features (X) and target (Y)
- X ← features, Y ← target
- Split the data into training and testing sets
- Compute with DecisionTreeClassifier function
- dt = DecisionTreeClassifier()
- Fitting the model with Gini index
- dt = dt.fit(X_train, Y_train)
- Evaluating the performance with confusion matrix, accuracy, precision, and F1_score
- Accuracy= accuracy_score(y_true, y_pred)

Algorithm Implementation:

Step 1: The sklearn Python library is used to implement the decision tree algorithm, and it includes a DecisionTreeClassifier function to train and test the model. The dataset is initially divided at random with a test size of 0.33 and the value is supplied within the classifier function.

```
X_train, X_test, y_train, y_test = train_test_split(X, y,
    test_size = 0.33, random_state=20, shuffle=True)
```

Step 2: To estimate the DecisionTreeClassifier() function and train the algorithm, split training data is supplied into the decision tree function. The Gini impurity calculation is used to train the data for the best fit, and it will identify impurities based on the result.

```
clf3 = tree.DecisionTreeClassifier()
clf3 = clf3.fit(X_train,y_train)
```

The model is evaluated using the test dataset based on the best fit, and the train score and test score are computed to examine the model's performance.

```
prediction = clf3.predict(X_test)
y_pred = clf3.predict(X_test)
train_scores.append(clf3.score(X_train,y_train))
test_scores.append(clf3.score(X_test,y_test))
```

Step 3: Plotting the model's accuracy versus the dataset identified for with and without outlier dataset (shown in Table 10.2).

```
accuracy = accuracy_score(y_test, y_pred)
precision = precision_score(y_test, y_pred)
f1_score = f1_score(y_test, y_pred)
```

TABLE 10.2

Comparative Analysis of Machine Learning Models

With Outlier	Accuracy (%)	Precision (%)	F1_Score (%)	Without Outlier	Accuracy (%)	Precision (%)	F1 Score (%)
Decision tree	82.2	79.9	80.3	Decision tree	82.7	79.5	81.0
Random forest	91.7	93.6	90.5	Random forest	93.4	94.7	92.4
K-nearest neighbour	97.1	95.4	93.6	K-nearest neighbour	97.63	96.3	94.4

Step 4: The confusion matrix shown in Figure 10.4 is evaluated, and the algorithm shows

For outlier dataset, TN = 2,156 patients, FN = 438 patients, FP = 392 patients, and TP = 1,735 patients.

Without outlier dataset, TN = 2,258 patients, FN = 483 patients, FP = 390 patients, and TP = 1,813 patients.

With the dataset, the decision tree's findings are probably somewhat precise at recognising cerebral palsy without outliers in the dataset.

10.5.1.2 Random Forest

To provide predictions of cerebral palsy, an ensemble learning approach called random forest combines different decision trees. It functions by constructing a succession of decision trees at each node using bootstrapped samples from the original dataset and randomly chosen feature subsets. During prediction, each tree separately provides its output, and the result is chosen by voting. Random forest does an effective job of handling high-dimensional data and overfitting issues. The collaborative decision-making of multiple trees increases its robustness and generalisation in detecting cerebral palsy. Figure 10.5 shows the random forest model.

Algorithm:

- Select the input dataset and split the data into features (X) and target (Y)
  X ← features, Y ← target
- Split the data into training and testing sets
- Compute with RandomForestClassifier function
  rfc = RandomForestClassifier()
- Fitting the model
  rfc = rfc.fit(X_train, Y_train)
- Evaluating the performance with confusion matrix, accuracy, precision, and F1_score

Algorithm Implementation:

Step 1: The Random forest algorithm is implemented using the sklearn Python library, which has a RandomForestClassifier function to train and test the

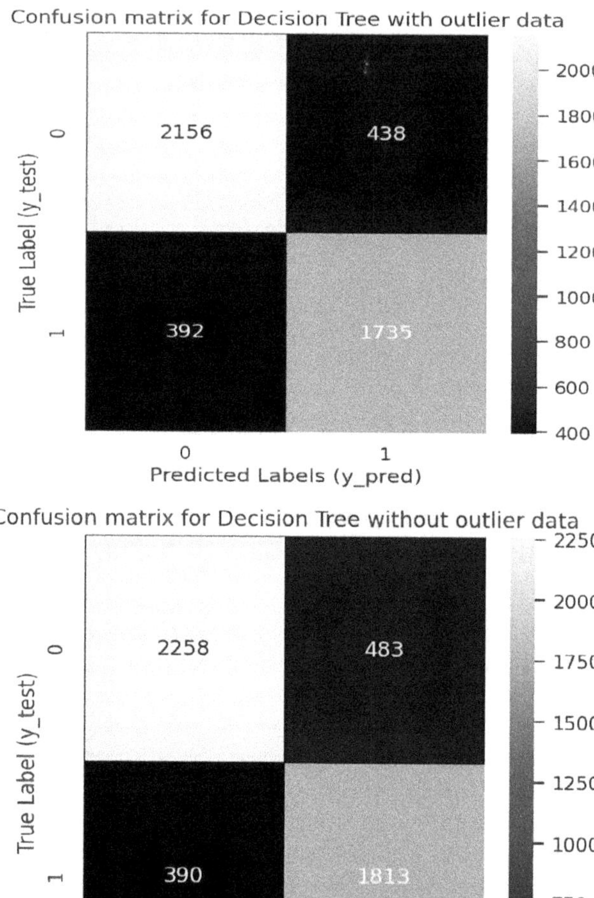

FIGURE 10.4 Confusion matrix for decision tree algorithm with and without outlier data points.

model. Initially the dataset is split randomly with a test size of 0.33 and passed the value inside the classifier function.

```
# importing the library
X_train, X_test, y_train, y_test = train_test_split(X, y,
  test_size = 0.33, random_state=20, shuffle=True) #splitting
  the data
```

Step 2: A metadata estimator that fits a variety of decision tree classifiers to the best model serves as input to the RandomClassifierFunction. The fit

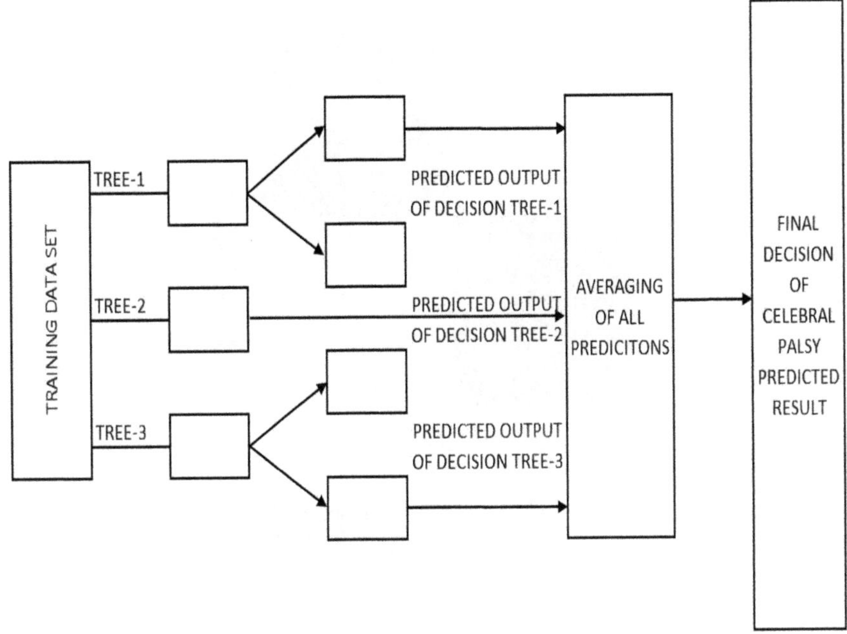

FIGURE 10.5　Random forest model.

(Source: https://en.wikipedia.org/wiki/File:Random_forest_diagram_complete.png)

technique uses the training data to train the algorithm once the model has been initialized.

```
clf4 = RandomForestClassifier()
clf4 = clf4.fit(X_train,y_train)
```

Step 3: After the training procedure is finished, test and train scores are used to assess how well the model performed against the dataset. A score of 0 indicates a total mismatch between the training and model, while a score of 1 indicates a perfect fit.

Step 4: The next phase is the implementation of testing data prediction. A collection of tools is available in the predict() function of the Sklearn library to assess the machine learning model's ability to forecast the testing data.

```
prediction = clf4.predict(X_test)
y_pred = clf4.predict(X_test)
```

Step 5: Plotting the model's accuracy versus the dataset identified for with and without outlier dataset (shown in Table 10.2).

```
accuracy = accuracy_score(y_true, y_pred)
precision = precision_score(y_test, y_pred)
f1_score = f1_score(y_test, y_pred)
```

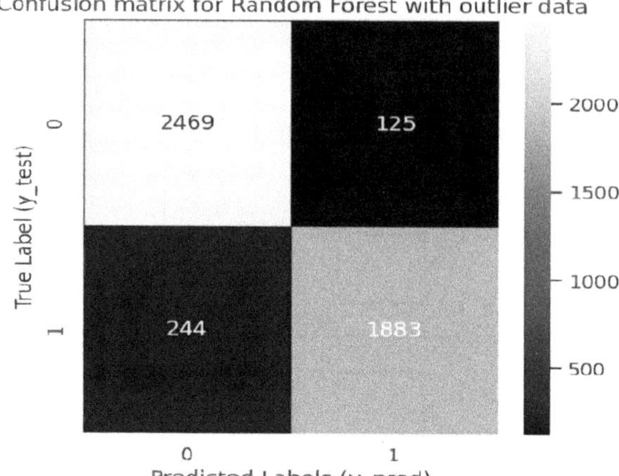

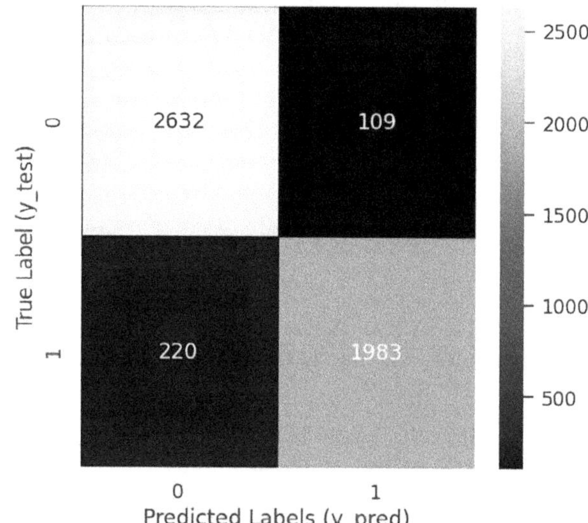

FIGURE 10.6   Confusion matrix for random forest algorithm with and without outlier data.

Step 6: The confusion matrix shown in Figure 10.6 is evaluated, and the algorithm shows

For outlier dataset, TN = 2,469 patients, FN = 125 patients, FP = 244 patients, and TP = 1,883 patients.

Without outlier dataset, TN = 2,632 patients, FN = 109 patients, FP = 220 patients, and TP = 1,983 patients.

10.5.1.3   K-Nearest Neighbour (KNN)

KNN is a supervised algorithmic technique where the testing tuple can be classified based on the majority of K-nearest neighbour category. The purpose of the algorithm

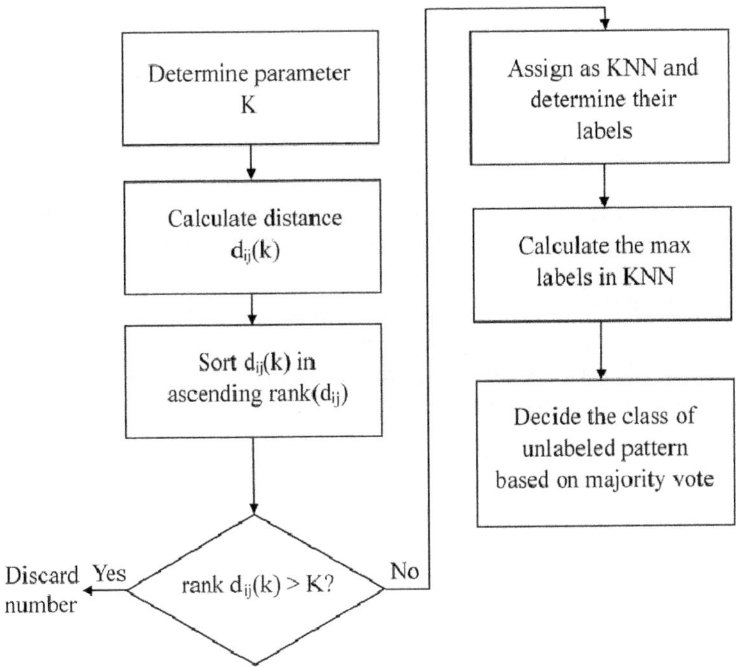

FIGURE 10.7   Flowchart for cerebral palsy prediction using K-nearest neighbour (KNN).

is to classify a new tuple based on the training samples. KNN is a nonparametric technique.

The data will be loaded into the algorithm, and K will be initialized. There will be iterations from the initial value, which is typically 1, up to the total number of training data points to obtain the projected class. Each iteration measures the separation between the test data and each row of the training data. Following the calculation, the distances, in particular the Euclidean distance, will be sorted according to values in ascending order. The algorithm will then retrieve the top k rows from the sorted array along with their most common class, returning the predicted class.

The class designations are combined by a majority vote. The Euclidean distance, designated D(X,Y), is used to calculate the distance. A flowchart for cerebral palsy prediction using KNN is shown in Figure 10.7.

Algorithm

- Select the input dataset and split the data into features (X) and target (Y)
- X ← features, Y← target
- Split the data into training and testing sets
- for each feature(i), compute:
- knn = KNeighbours Classifier (i, weights = "distance")
  Distance $D(X,Y)$ = Euclidean distance between two points $X (x_1,x_2,\ldots\ldots,x_n)$
  and $Y = (y_1,y_2,\ldots,y_n)$ Where $D(X,Y) = \sqrt{\Sigma_{I=1}^{n}(xi - yi)^2}$

- Fitting the model
- knn = knn.fit (X_train, Y_train)
- Evaluating the performance with confusion matrix, accuracy, precision, and F1_score

Algorithm Implementation:

Step 1: In this case, the aim may be 0 for non-CP patients and 1 for CP patients in the datasets of eye movement analysis of EEG signal patterns provided to the KNN algorithm.

Step 2: The KNN method is then calculated after the data points are divided into training and testing groups with a test size of 0.33. The use of the Python sklearn package is required to compute the KNN algorithm. The algorithm iterates the k value from 1 to 15.

```
# Iterating the knn model for k = 1 to 15
for i in range (1,15):
    knn = KNeighborsClassifier(i)
    knn.fit(X_train,y_train)
    prediction = knn.predict(X_test)
```

Step 3: Plotting the model's accuracy versus the dataset identified for with and without outlier dataset (shown in Table 10.2).

Step 4: A confusion matrix may be used to analyse the evaluation for the iterated k values against the data points.

```
#Plotting the Accuracy Chart
import seaborn as sns
f, ax =plt.subplots(figsize = (5,5))
sns.heatmap(cm,annot = True, linewidths= 0.5, linecolor=
    "blue", fmt=".0f", ax=ax)
plt.xlabel("y_pred")
plt.ylabel("y_test")
plt.show()
```

The confusion matrix is shown in Figure 10.8, and the algorithm shows

For outlier dataset, TN = 2,499 patients, FN = 95 patients, FP = 174 patients, and TP = 1,953 patients.

Without outlier dataset, TN = 2,662 patients, FN = 79 patients, FP = 161 patients, and TP = 2,042 patients.

A performance analysis of KNN is shown in Figure 10.9.

Performance analysis of all three models:

Accuracy:

The overall correctness of the detection model is estimated by accuracy as depicted in Equation 10.2.

$$\text{Accuracy} = \frac{TP}{TP + FP + TN + FN} \tag{10.2}$$

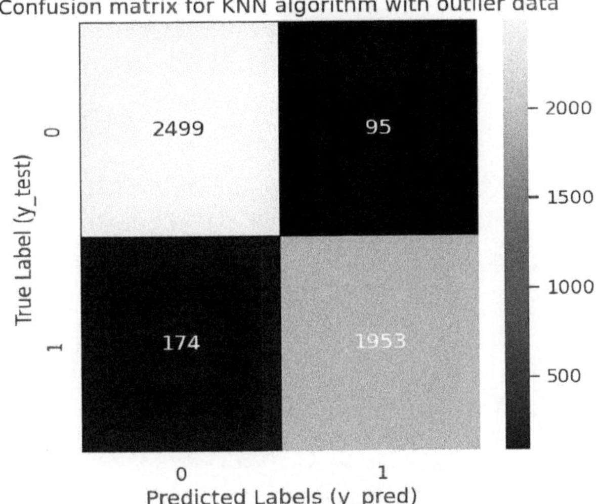

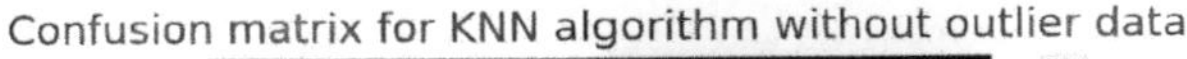

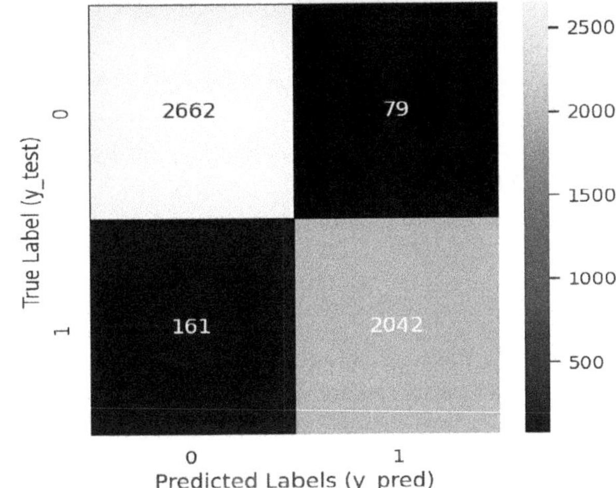

FIGURE 10.8 Confusion matrix for KNN algorithm with and without outlier data.

Precision:

Precision is a performance indicator tool that calculates how many positive predictions made by the model are correct. The formula used to calculate precision is given in Equation 10.3.

$$\text{Precision} = \frac{TP}{TP + FP} \tag{10.3}$$

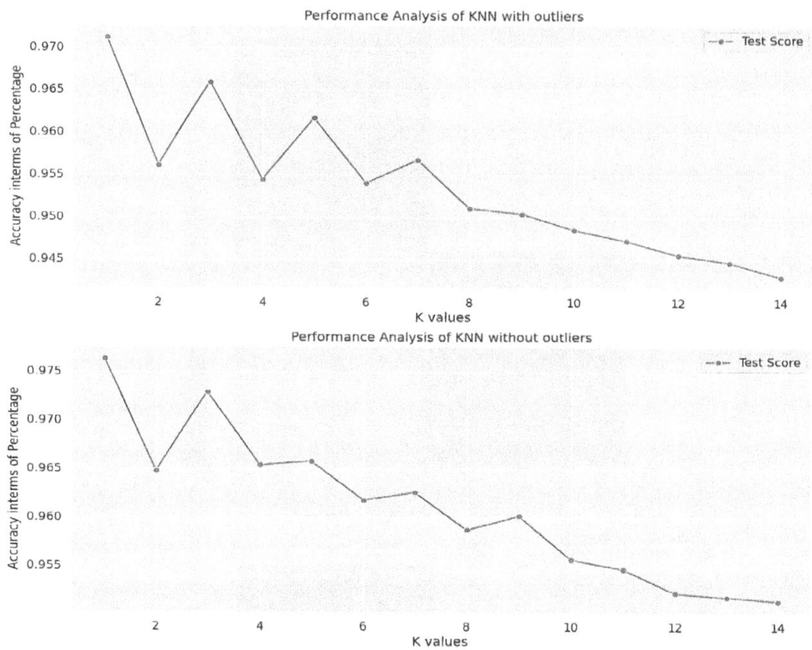

FIGURE 10.9   Accuracy graph for cerebral palsy detection with and without outlier data for K-nearest neighbours (KNN).

F1 Score:

The F1 score balances the trade-off between precision and recall, providing a single metric to evaluate a model's performance. It is depicted in Equation 10.4.

$$\text{F1} - \text{Score} = \frac{2*\left(Precision * Recall\right)}{\left(Precision * Recall\right)} \tag{10.4}$$

```
#Performance analysis of accuracy, precision, and F1 score
from sklearn.metrics import precision_score, recall_score,
  f1_score, accuracy_score
print('Accuracy: %.3f' % accuracy_score(y_test, y_pred))
print('Precision: %.3f' % precision_score(y_test, y_pred))
print('F1 Score: %.3f' % f1_score(y_test, y_pred))
```

A comparison study was made on machine learning models such as KNN, random forest, and decision tree regarding prospective outcomes, and it was shown that the KNN method performs well in comparison to decision trees and random forest. Because of the overfitting issue, the decision tree for CP detection is likely to be very tightly fitted to the training dataset, which will result in less satisfactory accuracy. However, random forest outperforms decision trees, which use a collection of decision trees and a voting method to identify CP. The detection accuracy of KNN appears to be quite near to the genuine value in terms of precision and accuracy, and it precisely identifies the data points. The performance analysis of classical machine learning models is displayed in Table 10.2.

10.6 CONCLUSION

Multimedia information processing can be used in medical applications for cerebral palsy, especially when screening for early detection is risky. Screening needs to be done for every child based on simple measurements by nurse practitioners in the rural village every time. This may consume lots of time and be subject to human error. Using the eye movement data collected from the emotive devices we can quickly identify the results by performing analysis on the data. Classical machine learning models have more explanatory results than deep learning algorithms such as CNN and RNN variants because the neural network can be trained with updated weights and might cluster into different regions for the detection process. This may cause problems in the neural network so the classic machine learning model of KNN outperforms the random forest algorithm in our opinion.

REFERENCES

Andriluka, M., Pishchulin, L., Gehler, P., & Schiele, B. 2014. 2d human pose estimation: New benchmark and state of the art analysis. In *Proceedings of the IEEE Conference on Computer Vision and Pattern Recognition* (pp. 3686–3693), IEEE.

Beatrice, S., & Meena, J. 2022. Overhauled approach to effectuate the Amelioration in EEG analysis. *Intelligent Automation & Soft Computing*, *33*(1).

Chtourou, I., Fendri, E., & Hammami, M. 2021. Person re-identification based on gait via part view transformation model under variable covariate conditions. *Journal of Visual Communication and Image Representation*, *77*, 103093.

Goudiaby, B., Othmani, A., & Nait-ali, A. (2020). EEG Biometrics for Person Verification. In A. Nait-ali (ed.), *Hidden Biometrics. Series in BioEngineering*. Springer. https://doi. org/10.1007/978-981-13-0956-4_3

Hua, C. H., Kim, K., Huynh-The, T., You, J. I., Yu, S. Y., Le-Tien, T., . . . & Lee, S. 2020. Convolutional network with twofold feature augmentation for diabetic retinopathy recognition from multi-modal images. *IEEE Journal of Biomedical and Health Informatics*, *25*(7), 2686–2697.

Kang, Z., Deng, M., & Wang, C. 2017. Frontal-view human gait recognition based on Kinect features and deterministic learning. In *2017 36th Chinese Control Conference (CCC)* (pp. 10834–10839). IEEE.

Kastaniotis, D., Theodorakopoulos, I., Theoharatos, C., Economou, G., & Fotopoulos, S. 2015. A framework for gait-based recognition using Kinect. *Pattern Recognition Letters*, *68*, 327–335.

Kwolek, B., Michalczuk, A., Krzeszowski, T., Switonski, A., Josinski, H., & Wojciechowski, K. 2019. Calibrated and synchronized multi-view video and motion capture dataset for evaluation of gait recognition. *Multimedia Tools and Applications*, *78*, 32437–32465.

Lee, T. K., Belkhatir, M., & Sanei, S. 2014. A comprehensive review of past and present vision-based techniques for gait recognition. *Multimedia Tools and Applications*, *72*, 2833–2869.

Rouzbeh, S. O. H. R. A. B., & Babaei, M. A. H. D. I. 2015. Human gait recognition using body measures and joint angles. *International Journal*, 6(4), 2305–1493.

Singh, J. P., Jain, S., Arora, S., & Singh, U. P. 2018. Vision-based gait recognition: A survey. *IEEE Access*, 6, 70497–70527.

Upadhyay, J., Paranjpe, R., Purohit, H., & Joshi, R. 2020. Biometric identification using gait analysis by deep learning. In *2020 IEEE Pune Section International Conference (Pune-Con)* (pp. 152–156). IEEE.

Wang, J., She, M., Nahavandi, S., & Kouzani, A. 2010. A review of vision-based gait recognition methods for human identification. In *2010 International Conference on Digital Image Computing: Techniques and Applications* (pp. 320–327). IEEE. www.cdc.gov/ncbddd/cp/facts.html

Wang, Y., Sun, J., Li, J., & Zhao, D. 2016. Gait recognition based on 3D skeleton joints captured by Kinect. In *2016 IEEE International Conference on Image Processing (ICIP)* (pp. 3151–3155). IEEE.

Yoo, J. H., & Nixon, M. S. 2011. Automated markerless analysis of human gait motion for recognition and classification. *ETRI Journal*, 33(2), 259–266.

11 ECG-Based Diagnosis of Sudden Cardiac Death Using Machine Learning

Lohith Ashwa, Sivakannan Subramani,
and Xavier Savarimuthu, S.J.

11.1 INTRODUCTION

Sudden cardiac death (SCD), the abrupt, unexpected death of a person signifies a common passing from a cardiovascular reason, demonstrated by unexpected disintegration of awareness inside one hour of the start of extraordinary indications. SCD occurs when there is an unexpected loss of blood flow caused by the failure of the heart to pump adequately. Reasons for SCD or sudden cardiac arrest include heart dysfunction and atherosclerotic illness. Less common causes include lack of oxygen, very low potassium and heart failure. A number of inherited disorders may also increase the risk, including long QT syndrome (Aro et al. 2011; Teodorescu et al. 2011). Sudden cardiac arrest may be due to heart attack. Most sudden deaths are caused by drastic cardiac arrhythmias. The three major types of presenting rhythms are ventricular tachyarrhythmia (either ventricular tachycardia [VT] or ventricular fibrillation [VF]), bradyasystole and pulse-less electrical activity (PEA). While these arrhythmias are mostly caused by an underlying heart condition (up to 8% of patients who undergo SCD have coronary artery disease (CAD; Cheng et al. 2017), they are often the first manifestation of a cardiac problem. Cardiac arrests may be reversed by using a defibrillator. This device delivers an electric shock to restore the normal heart rhythm (Chieng and Paul 2018; Šmíd and Rokyta 2017). However, since most cardiac arrests occur out of hospital in an unmonitored environment and a shock must be administered within minutes after the start of the arrhythmia, the overall survival rate for cardiac arrest is lower than 5% (Cheng et al. 2017; London et al. 2019). It is therefore important to accurately determine which patients are at risk for developing dangerous arrhythmias in order to implement optimal treatment and prevention strategies (Porthan et al. 2018). When an underlying cardiac condition is diagnosed in time, it can often be managed to prevent deterioration. Patient management involves lifestyle interventions; pharmacotherapy and/or device therapy (Chieng and Paul 2018). Lifestyle interventions can aim at preventing deterioration of both disease and comorbidities. In patients with certain conditions, such as hypertrophic cardiomyopathy (Shenasa and Shenasa 2017) or long QT syndrome (LQTS), intense physical activity is known to provoke arrhythmias (Aro et al. 2011). This specific patient group can thus be restricted from endurance training. For CAD

DOI: 10.1201/9781003405436-13

prevention in contrast, a sedentary lifestyle is known to potentially cause deterioration (El-Sherif 2015; Koene et al. 2017) and these patients could thus benefit from additional activity. The goal of pharmacotherapy is to control and improve the general heart condition. It can include discontinuation of known pro-arrhythmic drugs or prescription of anti-arrhythmic drugs such as beta-blockers.

The heart is fundamentally significant as it is one of the most vital muscles in the human body, intricately connected with veins to form the cardiovascular system. It pumps blood to every cell of the body. However, heart muscle dysfunction can occur. Heart failure, a common disorder, gradually develops, resulting from the heart's inability to efficiently pump blood to every cell of the body (Cheng et al. 2017). The heart can be weakened by coronary episodes, characterized by persistent hypertension or a dysfunction of one of the heart valves. Additionally, cardiovascular failure is often not recognized until it progresses to the advanced stage known as congestive heart failure, which leads to fluid accumulation in the lungs, feet and abdominal cavity. Congestive heart failure is one of the conditions that can lead to sudden cardiac death, along with other heart diseases such as coronary artery disease (CAD), high blood pressure (BP), valvular heart disease (VH), diabetes, and alcohol abuse (Shenasa and Shenasa 2017). As per European Heart Network and European Society of Cardiology, more than 4 million individuals die from cardiovascular sicknesses in Europe and 1.9 million in European Union (EU) which is 47% passings in Europe and 40% in EU. Since there is no specific diagnostic test for congestive heart failure, a combination of clinical assessments, such as patient history, ECG (Electrocardiogram), physical examinations, chest echocardiography, or radiography, is essential for diagnosis (Masetic and Subasi 2016). The ECG is non-interfering (invasive) apparatus that note down the electrical action of the human heart and show asymmetry of the pulses. ECG mindfully examines and note down the electrical driving forces of the heart and show conceivable harm to the heart or asymmetry of the pulses. Therefore, Electrocardiogram is a principal gadget for choosing the result and the soundness of the cardiovascular framework. In addition, it is huge to characterize exact and opportune conclusion of doctors to keep away from more harm and to decide legitimate techniques and approaches (Porthan et al. 2018). Still, the issue happens when there is insufficient number of specialists to experience the prerequisite of out patients.

This chapter describes the use of random forest algorithms to investigate the performance of a classifier for ECG signals classification in diagnosing SCD through supervised ECG data processing.

11.1.1 Electrocardiogram

An electrocardiogram (ECG) is a simple and brief method used to assess the heart function. The ECG is a representation of the electrical action of the heart. A particular waveform of ECG reveals the action of every heart component. Anodes are situated on the torso, arms, and legs. By associating the anodes to an electrocardiograph, the electrical action of the heart is determined, deciphered, and printed out for further translation. The wave that is created from the upper chambers of the heart

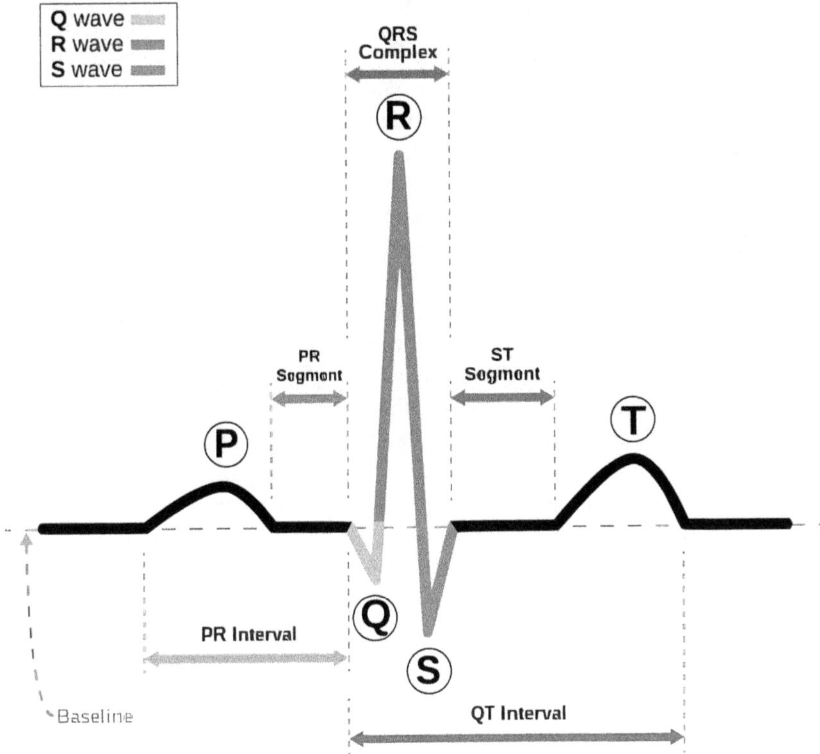

FIGURE 11.1 Schematic representation of a normal electrocardiograph. (Source: www.en.wikipedia.org/wiki/Electrocardiography)

is called the P wave. The P wave signifies that the atria (the two upper chambers of the heart) are contracting to pump out blood. The mix of three graphical redirections seen on a regular electrocardiogram is known as the "QRS complex" (Yücelbaş et al. 2017; Thanuja and Balakrishnan 2013). This shows the ventricles (the two lower chambers of the heart) are contracting to pump out blood. The PR interim equates to the time estimated from the earliest starting point of the P wave to the start of the R (or Q) wave (Yücelbaş et al. 2017). The PR interim shows the aggregate of time the electrical driving forces take to arrive at the ventricles from the sinus node. The next short upward portion is known as the "ST fragment." The ST section means the period and time length from the finish of the constriction of the ventricles to the start of the rest time frame before the ventricles start to contract for the following beat. The following upward bend is known as the "T wave." In ECG the T wave speaks to the repolarization of the ventricles. The "R-R interim" is the interim from the pinnacle of one QRS complex to the pinnacle of the following as appeared on an electrocardiogram which evaluates the ventricular rate (Teodorescu et al. 2011). The wave form of an ordinary ECG is shown in Figure 11.1.

11.1.2 Sudden Cardiac Death and Electrocardiogram Signal Affiliation

Many studies of heart failure focus on diagnosis and classification of SCD. In recent years, many well-known studies have provided information about automated diagnosis and classification of SCD. ECG is regarded as one of the most systematic features to diagnose SCD. Periodic variations in the time span of a heartbeat, also known as RR intervals (time interval from one R wave to the next R wave) of an ECG have been reported to be associated with sleep apnea episodes. These fluctuations within the individual's cardiovascular system are expressed as bradycardia during apnea followed by tachycardia upon its cessation (Wongdhamma et al. 2013). Various research studies have been accomplished to detect SCD by using the features extracted from the RR interval. These studies quantified the RR interval by using median, mean, interquartile range (IQR), and the standard deviation of the change in RR intervals. A new methodology is proposed in this chapter by conflation of RR-interval-based features of the ECG signal. Performance evaluation of this method is done by measuring the classification performance in determining the SCD. Support vector machines (SVMs) and random forest are used as a classification method (also known as supervised learning) to investigate apnea episode detection. Therefore, it is necessary to develop an effective and automated diagnostic system based on ECG recordings, combined with the application of machine learning techniques for classification of heart diseases. These diagnostic systems will aid medical experts in detecting the irregularities in the cardiovascular system (Mietus et al. 2000). The diagnostic system will first process the ECG signals taken from different subjects and hence decomposed into few features performing feature extraction. Extracted ECG signals are used to detect different types of heart failure by using various machine learning techniques (Masetic and Subasi 2016).

11.2 FEATURE EXTRACTION

Heart rate variability (HRV) is the physiological phenomenon of variation in the time interval between heartbeats. It is measured by changing the interval between individual beats. Another term used is RR variability, where R is the point corresponding to the apex of the QRS complex, which is the name given to the combination of three graphic deviations seen on a typical EKG. RR is the interval between consecutive Rs. For HRV analysis, all RR intervals of the ECG signal are calculated by Thomkin's algorithm and time series data are generated. Time series data of RR intervals are resampled at 4 Hz to extract indicators in the frequency domain, which is one of the methods of linear analysis. Linear feature vectors in the time and frequency domains are extracted, and nonlinear HRV feature vectors are also extracted.

11.3 PATTERN RECOGNITION

11.3.1 Random Forest

Random forest is an arrangement technique that combines different tree indicators in that manner that each tree hold tight an estimation of haphazardly picked vector scatter among all trees in woods similarly. In this way, in the arbitrary backwoods calculation, an irregular vector θk is delivered, free from the past irregular vectors and dispersed to all trees, and each tree is developed utilizing preparing set and irregular vector θk which brings about assortment of tree-organized classifiers $\{h(x, \theta_k), k = 1, \ldots\}$ at input vector x. In arbitrary backwoods calculation, inspiration mistake is given by

$$PE^* = P_{X,Y}(mg(X,Y)<0) \tag{11.1}$$

where subscripts X and Y are irregular vectors that signify the likelihood is over the X,Y space and mg is the edge task which processes the degree to which the normal number of votes at irregular vectors for the correct yield rise above the normal decision in favor of some other yield. Edge work is characterized as

$$mg(X,Y) = av_k I(h_k (x) = Y)\text{-}max_{j \neq y}max\ av_k I (h_k (X) = j) \tag{11.2}$$

where I (.) is the indicator function.

Two parameters that estimate the precision of individual classifier and reliance between classifiers are quality and relationship, separately. An irregular wood with arbitrary arrangement is found by choosing a little gathering of info factors on each hub capriciously. In this chapter, irregular woodland comprised of 20 trees, each built while thinking about 6 arbitrary highlights. Superb qualities for trees and irregular classification are procured by allotting divergent estimations of trees and arbitrary arrangement, utilizing the order exactness as embeddings work. Along these lines, 20 as the quantity of trees gave the brilliant precision results. In a simple way it is referred as an ensemble method that takes a subset of observations and a subset of variables to create a tree for decision. It builds several such decision trees and merges them to obtain a prediction that is more accurate and stable. It is a direct result of the fact that we get the final decision higher than the best judge by getting the maximum vote from a jury of impartial judges.

11.3.2 Support Vector Machine

The subject of bolster vector machines was first presented by Boser et al. (1992). They presented a preparation calculation that expanded the edge between preparing examples and choice limits. The support vector machine (SVM) is a hyper plane that isolates positive and negative arrangements of models, with maximal edge. SVM is associated to the highlights of part techniques, utilized in high-dimensional component space for processing a dab item. Piece strategies tackle the issues of quadratic increment in memory while putting away the highlights and time required ascertaining the classifier's discriminant work by abstaining from mapping the information into the high-dimensional element space. While considering the dataset referenced above, comprising sets $(x_1,y_1), (x_2,y_2) \ldots \ldots (x_n,y_n)$ where the hyper plane is defined by

$$\{x: f (x) = x^T \beta + \beta_0 = 0\} \tag{11.3}$$

where β is a unit vector $\|\beta\| = 1$.

A categorization rule created by $f(x)$ is

$$G(x) = \text{sign}\,[X^T \beta + \beta_0] \qquad (11.4)$$

Now, when classes are separable, a function f(x) with $y_i\,f(x_i)$ for all i values and hyper plane that produces the biggest margin between the training points for classes −1 and 1 can be determined.

The optimization problem for this concept is presented by the following formula subject to

$$Y_i\,(\chi\beta + \beta_0) \geq M,\ I = 1, \ldots .,N, \qquad (11.5)$$

Thus, edge is made and it is M unit beside hyper plane from different sides. In this chapter SVM classifier is used. Direct, nonlinear and RBF pieces were also tried, yet computational time expanded in nonlinear RBF portions and exactness were fundamentally decreased. Subsequently, we used portion for SVM classifier which gives best precision among different parts with $\omega = 1$ and $\sigma = 1$. These qualities give ideal grouping precision with sensible computational time. Be that as it may, we tried various estimations of ω and σ, for example, 0.5 and 2, 5 and 0.5, individually.

11.4 CASE STUDY: PREDICTION USING SUPPORT VECTOR MACHINE AND RANDOM FOREST

Thirteen features were selected for predicting SCD. The initial phase was data collection. The 13 features are some common parameters that provide necessary data in predicting SCD. One of the important features is ECG data. After the datasets are collected they are read and processed. Thereafter, have to fill the null values present, and this is done by taking the mean values and filling up in the particular row where the values are not present and this process is called standardization.

The data is split into a training set and a test set. This will help to identify the features and labels. Labels are nothing but the data that specifies whether SCD is present or not: 1 represents that the patient has SCD and 0 represents that the patient does not. The percentage for training set is 70% and for test set it is 30% of data which is taken into consideration. A classifier is created and we call it as predicted label because it is used to predict SCD based on the attributes specified. Two algorithms have been used in this study, the SVM classifier and the random forest algorithm, to get better prediction by comparing both algorithms for accurate results. These classifiers are used mainly for the purpose of evaluation, and the random forest algorithm produced better results when compared to the SVM classifier.

Soon after the evaluation, this is obtained by allotting 13 columns for attributes and one column for label which identifies SCD. The data is visualized by various forms such as graphs, tables, waveforms, parallel coordinates, clusters, flowcharts, etc.,

11.4.1 PERFORMANCE AND EVALUATION

The system's performance is measured in terms of precision (P), recall (R) and accuracy (A). Precision can be seen as a measure of exactness or quality, whereas recall

TABLE 11.1

Comparison between Random Forest Classifier and Support Vector Machine

Classifiers	Accuracy	Precision	Recall
Random forest classifier	81.19	87.06	77.09
Support vector machine	76.06	84.42	72.05

is a measure of completeness or quantity. In simple terms, high precision means that an algorithm returned substantially more relevant results than irrelevant, high recall means that an algorithm returned most of the relevant results. The F-measure is a measure that combines precision and recall and is the harmonic mean of precision and recall.

P = Number of correct answers produced/Total number of answers produced

R = Number of correct answers produced/Total number of possible correct answers

F-Measure = 2PR/(P + R), the traditional F-measure or balanced F-score.

11.4.2 Results

The results obtained for SCD prediction on considering 13 sets of features, namely serum cholesterol, blood sugar, ECG, heart rate, exercise-induced asthma (EIA), segment division (ST), physical standard test (PST), non-invasive mechanical ventilation (NMV), thallium stress test, age, sex, chest pain type and blood pressure, are adequate. Initially normalization and standardization procedures are carried out to get the values in the range from 0 and 1 and the standard deviation and mean are found to make the classification easier. We use two classifiers to evaluate the results, that is, SVM classifier and random forest classifier. The results for SVM classifier had accuracy = 76.06, precision = 84.42 and recall = 72.05 and for random forest accuracy = 82.9, precision = 87.9 and recall = 79.68 (Table 11.1). The random forest classifier provided better results in all measures.

11.5 SUMMARY

Sudden cardiac death (SCD) is one of the hardest diseases to predict as death may occur with no prior indication. This chapter discusses an efficient system capable of predicting SCD using 13 preselected features evaluated using an SVM classifier and random forest algorithm. This system can be very useful in saving the lives of patients having cardiac problems, who are likely to experience SCD in the future. In the future, with more datasets and multiple features, we will be able to predict SCD with more precision, and which could also lead to the development of a hardware device for preventing SCD.

REFERENCES

Aro, A. L., Huikuri, H. V., Tikkanen, J. T., Junttila, M. J., Rissanen, H. A., Reunanen, A., & Anttonen, O. 2011. QRS-T angle as a predictor of sudden cardiac death in a middle-aged general population, *Europace*, 14(6), 872–876. doi:10.1093/europace/eur393

Boser, B. E., Guyon, I. M., & Vapnik, V. N. 1992. A training algorithm for optimal margin classifiers, Proceedings of the 5th Annual Workshop on Computational Learning Theory (COLT'92), Pittsburgh, 27–29 July 1992, 144–152.

Cheng, Y.-J., Li, Z.-Y., Yao, F.-J., Xu, X.-J., Ji, C.-C., Chen, X.-M., & Wu, S.-H. 2017. Early repolarization is associated with a significantly increased risk of ventricular arrhythmias and sudden cardiac death in patients with structural heart diseases, *Heart Rhythm*, 14(8), 1157–1164. doi:10.1016/j.hrthm.2017.04.022

Chieng, D., & Paul, V. 2018. Current device therapies for sudden cardiac death prevention—the ICD, subcutaneous ICD and wearable ICD, *Heart, Lung and Circulation*. doi:10.1016/j. hlc.2018.09.011

El-Sherif, N. 2015. Sudden cardiac death in ischemic heart disease: Pathophysiology and risk stratification, *The Egyptian Journal of Critical Care Medicine*, 3(1), 17–22. doi:10.1016/j. ejccm.2015.02.001

Koene, R. J., Adkisson, W. O., & Benditt, D. G. 2017. Syncope and the risk of sudden cardiac death: Evaluation, management, and prevention, *Journal of Arrhythmia*, 33(6), 533–544. doi:10.1016/j.joa.2017.07.005

London, K. S., Hartwell, C., Cesar, S., Sarquella-Brugada, G., & White, J. L. 2019. Can sudden cardiac death risk in the young be identified in the emergency department?, *Journal of Emergency Nursing*. doi:10.1016/j.jen.2019.09.009

Masetic, Z., & Subasi, A. 2016. Congestive heart failure detection using random forest classifier, *Computer Methods and Programs in Biomedicine*, 130, 54–64. doi:10.1016/j. cmpb.2016.03.020

Mietus, J. E., Peng, C. K., Ivanov, P. C., & Goldberger, A. L. 2000. Detection of obstructive sleep apnea from cardiac interbeat interval time series, *Computers in Cardiology 2000*. Vol. 27 (Cat. 00CH37163). doi:10.1109/cic.2000.898634

Porthan, K., Kenttä, T., Niiranen, T. J., Nieminen, M. S., Oikarinen, L., Viitasalo, M., . . . Tikkanen, J. T. 2018. ECG left ventricular hypertrophy as a risk predictor of sudden cardiac death, *International Journal of Cardiology*. doi:10.1016/j.ijcard.2018.09.104

Shenasa, M., & Shenasa, H. 2017. Hypertension, left ventricular hypertrophy, and sudden cardiac death, *International Journal of Cardiology*, 237, 60–63. doi:10.1016/j. ijcard.2017.03.002

Šmíd, J., & Rokyta, R. 2017. Atrial fibrillation and its relation to cardiac diseases and sudden cardiac death, *Cor et Vasa*, 59(4), e325–e331. doi:10.1016/j.crvasa.2017.06.005

Teodorescu, C., Reinier, K., Uy-Evanado, A., Navarro, J., Mariani, R., Gunson, K., & Chugh, S. S. 2011. Prolonged QRS duration on the resting ECG is associated with sudden death risk in coronary disease, independent of prolonged ventricular repolarization, *Heart Rhythm*, 8(10), 1562–1567. doi:10.1016/j.hrthm.2011.06.011

Thanuja, R., & Balakrishnan, R. 2013. Real time sleep apnea monitor using ECG, 2013 IEEE Conference on Information and Communication Technologies. doi:10.1109/ cict.2013.6558237

Wongdhamma, W., Le, T. Q., & Bukkapatnam, S. T. S. 2013. Wireless wearable multi-sensory system for monitoring of sleep apnea and other cardiorespiratory disorders, 2013 IEEE International Conference on Automation Science and Engineering (CASE). doi:10.1109/ coase.2013.6654074

Yücelbaş, Ş., Yücelbaş, C., Tezel, G., Özşen, S., Küççüktürk, S., & Yosunkaya, Ş. 2017. Predetermination of OSA degree using morphological features of the ECG signal, *Expert Systems with Applications*, 81, 79–87. doi:10.1016/j.eswa.2017.03.049

Section III

AI and Its Applications in Environmental Science

12 AI in Environmental Applications

Prasenjeet Acharjee

12.1 INTRODUCTION

12.1.1 OVERVIEW OF ENVIRONMENTAL CHALLENGES

The world is currently facing a range of environmental challenges that are posing significant threats to the health and well-being of people and the planet. Some of the most pressing environmental challenges include:

- Climate change. Climate change is one of the biggest environmental challenges the world is facing today. The rise in global temperatures due to human activities is causing widespread impacts such as sea level rise, more frequent and severe weather events, and changes in ecosystems and biodiversity.
- Biodiversity loss. The loss of biodiversity is occurring at an unprecedented rate, with around one million species at risk of extinction due to human activities such as deforestation, pollution, and climate change.
- Water scarcity. As the world's population continues to grow, water scarcity is becoming a major issue in many regions of the world, leading to conflicts over access to water resources.
- Air pollution. Air pollution is a major health hazard, causing millions of premature deaths each year, and contributing to climate change.
- Plastic pollution. The growing problem of plastic pollution is having devastating impacts on marine life and ecosystems, as well as on human health.
- Land use change. Changes in land use, such as deforestation and urbanization, are driving biodiversity loss, increasing greenhouse gas emissions, and reducing the capacity of ecosystems to provide essential services such as clean water and air.

Addressing these environmental challenges requires urgent action at both the global and local levels, including the adoption of sustainable practices and policies, increased investment in renewable energy and technology, and increased public awareness and education.

12.1.2 ROLE OF AI IN ADDRESSING ENVIRONMENTAL ISSUES

Artificial intelligence (AI) is increasingly being recognized as a powerful tool in addressing many of the environmental challenges that we face. Here are some of the

DOI: 10.1201/9781003405436-15

ways in which AI can play a role in addressing environmental issues, as outlined in this chapter:

- Environmental monitoring. AI can be used to monitor the environment by analyzing data from various sources, such as satellites, sensors, and drones. This enables real-time monitoring of environmental parameters, such as air and water quality, land use, and biodiversity. AI algorithms can analyze this data to identify patterns and trends, enabling scientists and policymakers to make informed decisions about managing natural resources and mitigating environmental risks.
- Climate change mitigation and adaptation. AI can help in reducing greenhouse gas emissions and optimizing the use of renewable energy. For instance, AI can be used to optimize energy systems, predicting and optimizing energy demand, and reducing energy consumption. AI algorithms can also assist in developing climate models that can help decision-makers develop effective strategies to adapt to climate change impacts.
- Sustainable agriculture. AI can optimize crop yields while minimizing inputs like water and fertilizer. AI can enable precision farming, where crops can be monitored and managed on an individual plant basis, thereby optimizing inputs and minimizing waste. This not only leads to better resource management, but also increases food security while reducing the environmental impact of agriculture.
- Wildlife conservation. AI can be used to monitor and track wildlife populations and their habitats. AI algorithms can be used to identify species, track movements, and understand behaviour. This information can be used to develop effective conservation strategies and to mitigate human-wildlife conflicts.
- Sustainable urban development. AI can optimize urban systems, such as transportation and energy use, reducing environmental impacts such as air pollution and greenhouse gas emissions. AI algorithms can also be used for urban planning and design, optimizing land use and reducing the environmental impact of urbanization.

Overall, AI has a significant role to play in addressing many of the environmental challenges that we face today. However, the ethical considerations and challenges associated with AI use should not be ignored. Responsible and sustainable use of AI is necessary, and this requires transparency, accountability, and stakeholder engagement. If used responsibly, AI has the potential to provide valuable insights into environmental data and enable effective decision-making for sustainable management and protection of our planet.

12.2 AI TECHNIQUES FOR ENVIRONMENTAL MONITORING

AI techniques are increasingly employed in environmental monitoring to identify and address issues. A key application involves analyzing remote sensing data, using machine learning to extract insights from satellite imagery, tracking land cover changes, deforestation, and habitat loss. AI also aids in species identification, utilizing images of animals, birds, and plants. In climate modelling, deep learning and neural networks analyze extensive climate data to predict future patterns. Additionally, AI enhances air quality monitoring, predicting pollution levels and sources, and forecasting air quality. Water resource management benefits from AI, predicting water availability and depletion. Waste management and circular economy efforts benefit from AI, identifying and sorting recyclable materials.

There are a few ethical and social implications to consider. Responsible AI use is crucial to mitigate potential negative impacts. AI holds significant promise for environmental management, requiring ethical and conscientious integration to ensure benefits while minimizing drawbacks.

12.2.1 Remote Sensing and Image Processing for Land Use and Land Cover Mapping

Remote sensing and image processing are powerful tools for land use and land cover mapping, enabling the creation of detailed and accurate maps of the earth's surface. In this chapter, we will explore the role of remote sensing and image processing in land use and land cover mapping.

Remote sensing involves the use of sensors on satellites or aircraft to collect data about the earth's surface. These sensors capture data in various parts of the electromagnetic spectrum, including visible light, infrared, and microwave radiation. Image processing techniques are then used to analyze the data and create maps of different land use and land cover types.

One of the primary applications of remote sensing and image processing is in the creation of land use and land cover maps. Land use refers to the human activities that take place on the land, such as agriculture, urbanization, and forestry. Land cover refers to the physical characteristics of the land, such as vegetation, water bodies, and bare soil. Land use and land cover maps are essential for understanding patterns of land use change, identifying areas of environmental concern, and supporting land management decisions.

There are various techniques used for land use and land cover mapping, including supervised and unsupervised classification. In supervised classification, the analyst uses training data to teach the algorithm how to classify the different land use and land cover types. In unsupervised classification, the algorithm identifies clusters of pixels with similar spectral characteristics and assigns them to different land use and land cover classes.

Recent advances in remote sensing technology, including high-resolution satellite imagery and advanced image processing techniques, have improved the accuracy and resolution of land use and land cover maps. For example, object-based image analysis is a technique that uses geographic objects, such as buildings or trees, as the basis for classification, resulting in more accurate and detailed maps.

Convolutional neural networks (CNNs) are a popular deep learning method for processing remote sensing imagery. They consist of layers of convolutional and pooling operations that can automatically learn and extract relevant features from the input data, leading to improved classification accuracy (Zhang et al. 2016).

Here are some real-life use cases of remote sensing and image processing for land use and land cover mapping:

- One real-life use case is the classification of land cover types using the publicly available Sentinel-2 satellite imagery dataset, which offers multispectral data with high spatial resolution (10 m–60 m). This dataset can be accessed through the Copernicus Open Access Hub (https://scihub.copernicus.eu/).

Here is an example of Python code using the Keras library to build and train a CNN model for land cover classification:

```python
1.  import numpy as np
2.  import keras
3.  from keras.models import Sequential
4.  from keras.layers import Dense, Dropout, Flatten
5.  from keras.layers import Conv2D, MaxPooling2D
6.
7.  # Load the preprocessed Sentinel-2 data (X) and
    corresponding land cover labels (y)
8.  X = np.load('sentinel2_data.npy')
9.  y = np.load('land_cover_labels.npy')
10.
11. # Split the data into training and testing sets
12. from sklearn.model_selection import train_test_split
13. X_train, X_test, y_train, y_test = train_test_split(X, y,
    test_size=0.2, random_state=42)
14.
15. # Normalize the data
16. X_train = X_train.astype('float32')
17. X_test = X_test.astype('float32')
18. X_train /= 255
19. X_test /= 255
20.
21. # Convert labels to one-hot encoding
22. num_classes = len(np.unique(y))
23. y_train = keras.utils.to_categorical(y_train, num_classes)
24. y_test = keras.utils.to_categorical(y_test, num_classes)
25.
26. # Define the CNN architecture
27. model = Sequential()
28. model.add(Conv2D(32, kernel_size=(3, 3),
    activation='relu', input_shape=(64, 64, 3)))
29. model.add(MaxPooling2D(pool_size=(2, 2)))
30. model.add(Conv2D(64, kernel_size=(3, 3),
    activation='relu'))
31. model.add(MaxPooling2D(pool_size=(2, 2)))
32. model.add(Flatten())
33. model.add(Dense(128, activation='relu'))
34. model.add(Dropout(0.5))
35. model.add(Dense(num_classes, activation='softmax'))
```

```
36.
37. # Compile the model
38.              model.compile(loss=keras.losses.categorical_
    crossentropy, optimizer=keras.optimizers.Adam(),
    metrics=['accuracy'])
39.
40. # Train the model
41.     model.fit(X_train,    y_train,    batch_size=128,
    epochs=20,    verbose=1,    validation_data=(X_test, y_test))
42.
```

- The BigEarthNet dataset (https://bigearth.net/) is another publicly available dataset containing Sentinel-2 satellite images, specifically designed for land cover classification tasks. It consists of 590,326 image patches with annotations for the 43 land cover classes defined by the CORINE Land Cover taxonomy.
- Monitoring deforestation in the Amazon rainforest. One real-life use case is monitoring deforestation in the Amazon rainforest using Landsat satellite imagery. The Amazon rainforest is a vital ecosystem for global climate regulation and biodiversity. Monitoring deforestation in the region is crucial for understanding its environmental impact and informing conservation efforts. For a comprehensive tutorial on using the Google Earth Engine Python API, refer to the official documentation: https://developers.google.com/earth-engine/guides/python_install.
- Monitoring urbanization using night-time light data. Another real-life use case is monitoring urbanization and urban growth using night-time light data from the Visible Infrared Imaging Radiometer Suite (VIIRS) dataset. This dataset captures nighttime light intensity, which can be used to assess urban growth and its impacts on the environment. For more information about the VIIRS Nighttime Day/Night Band dataset, refer to the official website of the National Oceanic and Atmospheric Administration: www.ngdc.noaa.gov/eog/viirs.

Real-life use cases of remote sensing and image processing techniques in land use and land cover mapping, such as monitoring deforestation in the Amazon rainforest and urban growth using night-time light data, demonstrate the value of these technologies in understanding environmental changes and their implications. Leveraging publicly available datasets such as Landsat and VIIRS, and platforms such as Google Earth Engine, researchers and policymakers can gain critical insights into the dynamics of ecosystems and urban areas, informing more effective and sustainable decision-making for environmental management and conservation.

In addition to the use cases discussed, there are many other potential applications of remote sensing and image processing techniques in land use and land cover mapping, including but not limited to:

- Monitoring the impact of natural disasters (e.g., floods, wildfires, and droughts)
- Assessing agricultural productivity and monitoring crop health

- Identifying and monitoring illegal land use activities, such as logging and mining
- Assessing the impact of climate change on ecosystems and land cover dynamics

12.2.1.1 Validation Methods for Remote Sensing and Image Processing in Land Use and Land Cover Mapping

To validate the results obtained from remote sensing and image processing, several validation methods can be employed. Ground truth data collection involves field surveys and sample points to verify the accuracy of the mapped areas. Comparative analysis with existing land cover databases and historical data can further validate the mapping outcomes. Accuracy assessment metrics such as confusion matrices, producer's accuracy, and user's accuracy provide quantitative measures to evaluate the correctness of the classification results. Additionally, cross-validation techniques, such as k-fold validation, can ensure the model's robustness and generalization to new data. Integrating validation methods into the remote sensing and image processing workflow enhances the reliability of land use and land cover mapping, ultimately supporting effective environmental management strategies.

As the availability of satellite data and advancements in AI-based image processing techniques continue to grow, we can expect even more novel applications and insights to emerge. These developments will help researchers, policymakers, and stakeholders better understand and address pressing environmental challenges and contribute to more effective and sustainable management of our planet's resources.

12.2.2 Machine Learning for Species Identification and Classification

Machine learning has become an essential tool in the field of species identification and classification. By automating the process of recognizing and cataloguing plant and animal species, machine learning algorithms can significantly improve the speed and accuracy of species identification, contributing to biodiversity conservation, ecological monitoring, and biological research.

One popular machine learning approach for species identification and classification is using deep learning techniques, such as convolutional neural networks (CNNs). These models have shown great success in various image classification tasks, including species recognition (Wäldchen & Mäder 2018a).

A real-life use case is the classification of plant species using the publicly available Flavia dataset (http://flavia.sourceforge.net/), which contains 1,907 leaf images from 32 different species. Here's an example of Python code using the Keras library to build and train a CNN model for plant species classification:

```
1. import numpy as np
2. import keras
3. from keras.models import Sequential
4. from keras.layers import Dense, Dropout, Flatten
5. from keras.layers import Conv2D, MaxPooling2D
6. from keras.preprocessing.image import ImageDataGenerator
```

```
 7.
 8. # Load the preprocessed Flavia data (X) and corresponding
    plant species labels (y)
 9. X = np.load('flavia_data.npy')
10. y = np.load('flavia_labels.npy')
11.
12. # Split the data into training and testing sets
13. from sklearn.model_selection import train_test_split
14. X_train, X_test, y_train, y_test = train_test_split(X, y,
    test_size=0.2, random_state=42)
15.
16. # Normalize the data
17. X_train = X_train.astype('float32')
18. X_test = X_test.astype('float32')
19. X_train /= 255
20. X_test /= 255
21.
22. # Convert labels to one-hot encoding
23. num_classes = len(np.unique(y))
24. y_train = keras.utils.to_categorical(y_train, num_classes)
25. y_test = keras.utils.to_categorical(y_test, num_classes)
26.
27. # Define the CNN architecture
28. model = Sequential()
29. model.add(Conv2D(32, kernel_size=(3, 3),
    activation='relu', input_shape=(128, 128, 3)))
30. model.add(MaxPooling2D(pool_size=(2, 2)))
31. model.add(Conv2D(64, kernel_size=(3, 3),
    activation='relu'))
32. model.add(MaxPooling2D(pool_size=(2, 2)))
33. model.add(Flatten())
34. model.add(Dense(128, activation='relu'))
35. model.add(Dropout(0.5))
36. model.add(Dense(num_classes, activation='softmax'))
37.
38. # Compile the model
39.                           model.compile(loss=keras.losses.
    categorical_crossentropy, optimizer=keras.optimizers.
    Adam(), metrics=['accuracy'])
40.
41. # Apply data augmentation
42.    datagen   =   ImageDataGenerator(rotation_range=20,
    zoom_range=0.15, width_shift_range=0.2, height_shift_
    range=0.2, shear_range=0.15, horizontal_flip=True, fill_
    mode="nearest")
43. datagen.fit(X_train)
44.
45. # Train the model
46. model.fit_generator(datagen.flow(X_train, y_train,
    batch_size=32), epochs=100, verbose=1, validation_data=
    (X_test, y_test))
47.
```

TABLE 12.1

Recent Studies Using Deep Learning for Animal Identification

Taxa	#Taxa	#Images	Architecture	Accuracy*
Mammals	26	32,240	ResNet-101	69.0%
Mammals	48	3,200,000	ResNet-101	93.8%
Ant genera	57	150,088	AlexNet	83.0%
Aquatic macro-invertebrate	29	11,832	AlexNet	85.6%
Fishes	23	27,370	–individual–	98.6%
Fishes	15	29,000	–individual–	94.0%
Insects	10	550	ResNet-101	98.7%

* Accuracy is reported for the best-performing model per paper without manually preprocessing.
Source: Wäldchen & Mäder 2018b, p. 7.

Apart from the Flavia dataset, there are other publicly available datasets that can be used to apply machine learning techniques for species identification and classification, such as:

- iNaturalist dataset (www.inaturalist.org/pages/developers). A large-scale dataset containing millions of images of plants, animals, and fungi, annotated by citizen scientists and experts. This dataset can be used for training deep learning models to identify a wide range of species.
- BirdCLEF dataset (www.imageclef.org/lifeclef/2021/bird). A dataset containing audio recordings of bird vocalizations, which can be used for bird species identification based on their songs and calls. Machine learning models such as CNNs and recurrent neural networks (RNNs) can be employed to process and classify these audio recordings.

Table 12.1 compiles recent research employing deep learning methodologies for the identification of animal species through image analysis. Despite ecologists amassing substantial quantities of superior data, image datasets for diverse animal categories remain infrequent, constituting a significant impediment for the implementation of advanced machine learning approaches (Wäldchen & Mäder 2018b).

These are just a few examples of the many applications of machine learning for species identification and classification. By leveraging the power of machine learning and the growing availability of species-related data, researchers can more effectively monitor and study biodiversity, and policymakers can make more informed decisions for the conservation and management of ecosystems.

12.2.2.1 Validation Methods for Machine Learning in Species Identification and Classification

Machine learning techniques are instrumental in identifying and classifying various species based on image data. Ensuring the accuracy and reliability of these models is crucial for informed conservation decisions. Validation methods can aid in evaluating the performance of these machine learning algorithms. Cross-validation techniques, such as k-fold cross-validation, help assess the model's ability to generalize to new, unseen data. Receiver operating characteristic (ROC) curves and area nnder the curve (AUC) scores measure the model's discriminatory power and performance. Precision, recall, and F1 score provide insights into the model's effectiveness in correctly identifying species. Utilizing separate validation datasets, distinct from the training dataset, helps assess the model's performance on unseen data. Additionally, incorporating domain expertise and crowd-sourced data validation can further enhance the accuracy of species identification models. By systematically validating machine learning-based species identification models, environmental researchers can ensure reliable results for biodiversity conservation efforts.

12.2.3 Neural Networks for Climate Modelling

Neural networks have emerged as a powerful tool for climate modelling, enabling more accurate predictions of weather patterns and the impacts of climate change. By training neural networks on historical climate data, researchers can identify complex patterns and relationships that are difficult to capture using traditional climate modelling techniques (Rasp et al. 2018).

One common use of neural networks in climate modelling is to predict future temperature trends using historical temperature data. Neural networks have shown great promise in the field of climate modelling, enabling more accurate and efficient predictions of weather patterns and the impacts of climate change. By training advanced neural network architectures such as long short-term memory on publicly available climate datasets, researchers can identify complex relationships and patterns that are difficult to capture with traditional climate modelling techniques. This can lead to more reliable predictions and better decision-making for climate change mitigation and adaptation efforts.

In addition to temperature prediction, neural networks can be employed for various other climate modelling tasks, such as:

- Precipitation prediction. Neural networks can be used to predict precipitation patterns, which is crucial for agriculture, water resource management, and flood prediction.
- Extreme weather event prediction. Neural networks can help forecast extreme weather events such as hurricanes, tornadoes, and droughts, enabling more effective disaster preparedness and response.
- Climate change impact assessment. Neural networks can be used to model the complex interactions between climate variables, ecosystems, and human activities, facilitating better assessment of the potential impacts of climate change on various sectors, including agriculture, water resources, and public health.

There are numerous publicly available datasets that can be used for climate modelling using neural networks. Some of these datasets include:

- ERA5 Reanalysis Data (https://cds.climate.copernicus.eu/cdsapp#!/dataset/reanalysis-era5-single-levels?tab=form). ERA5 is a global climate reanalysis dataset provided by the European Centre for Medium-Range Weather

Forecasts (ECMWF). It contains hourly data on various climate variables, including temperature, precipitation, wind speed, and pressure, with a spatial resolution of 0.25 degrees.

- NOAA Climate Data (www.ncei.noaa.gov/access/index.html). The National Oceanic and Atmospheric Administration (NOAA) provides a wide range of climate and weather datasets, including temperature, precipitation, and oceanic data. Some popular datasets are the Global Historical Climatology Network (GHCN) and the Climate Prediction Center (CPC) datasets.
- NASA EarthData (https://earthdata.nasa.gov/). NASA EarthData provides access to various satellite and model-based datasets related to climate, atmosphere, ocean, and land surface. Some popular datasets include the Global Land Data Assimilation System (GLDAS) and the Modern-Era Retrospective Analysis for Research and Applications (MERRA).
- WorldClim (www.worldclim.org/). WorldClim is a set of high-resolution global climate data, including data on monthly average temperature and precipitation for the period 1970 to 2000. The data can be used for climate modelling and other environmental applications.
- Global Precipitation Climatology Project (GPCP) (https://gpm.nasa.gov/mission/gpcp). The GPCP is a long-term satellite-based dataset that provides monthly and daily precipitation data from 1979 to the present. The dataset is created by combining information from multiple satellites and in situ measurements.
- Berkeley Earth Surface Temperature (BEST) dataset (http://berkeleyearth.org/data/). The BEST dataset provides monthly temperature and uncertainty data for land and ocean. The dataset covers the period from 1750 to the present, with a spatial resolution of 1 degree.
- PRISM Climate Group (www.prism.oregonstate.edu/). The PRISM dataset provides high-resolution climate data for the United States, including temperature and precipitation data on a monthly, daily, and hourly basis.
- CMIP6 (Coupled Model Intercomparison Project Phase 6) (https://esgf-node.llnl.gov/projects/cmip6/). CMIP6 is a collection of climate model output data from various global climate models. The data covers historical climate simulations as well as future projections under different greenhouse gas emission scenarios.

In this example, I'll demonstrate how to use the ERA5 Reanalysis Data for predicting future temperature using a simple feedforward neural network with Keras and TensorFlow. This is a basic example for educational purposes and not intended to be a fully fledged climate model.

- First, download the ERA5 dataset from the Climate Data Store (CDS): https://cds.climate.copernicus.eu/cdsapp#!/dataset/reanalysis-era5-single-levels?tab=form. For this example, we will use the "2m_temperature" variable with a monthly frequency and a specific geographical region (latitude: 30–40, longitude: −90–80) for a period from 2000 to 2020.
- Save the downloaded dataset as a NetCDF file (e.g., 'ERA5_temp.nc').

- Install the required Python packages:

```
1. pip install netCDF4 pandas tensorflow
```

- Now, let's create a Python script to preprocess the data and train a feedforward neural network:

```
 1. import numpy as np
 2. import pandas as pd
 3. import netCDF4
 4. from tensorflow.keras.models import Sequential
 5. from tensorflow.keras.layers import Dense
 6. from sklearn.preprocessing import MinMaxScaler
 7. from sklearn.model_selection import train_test_split
 8.
 9. # Load ERA5 data
10. filename = 'ERA5_temp.nc'
11. dataset = netCDF4.Dataset(filename)
12.
13. # Extract temperature data
14. temp = dataset.variables['t2m'][:]
15. time = dataset.variables['time']
16. dates = netCDF4.num2date(time[:], time.units)
17.
18. # Calculate monthly mean temperature
19. temp_df = pd.DataFrame(temp.mean(axis=(1, 2)),
    columns=['Temperature'], index=pd.to_datetime(dates))
20. temp_df = temp_df.resample('M').mean()
21.
22. # Scale the temperature data
23. scaler = MinMaxScaler()
24. data_scaled = scaler.fit_transform(temp_df.values)
25.
26. # Create input/output pairs for the neural network
    model
27. X, y = [], []
28. window_size = 12
29. for i in range(len(data_scaled)-window_size):
30.     X.append(data_scaled[i:i+window_size].flatten())
31.     y.append(data_scaled[i+window_size])
32.
33. X, y = np.array(X), np.array(y)
34.
35. # Split the data into training and testing sets
36. X_train, X_test, y_train, y_test = train_test_split(X,
    y, test_size=0.2, random_state=42, shuffle=False)
37.
38. # Build the feedforward neural network model
39. model = Sequential()
40. model.add(Dense(64, activation='relu', input_shape=(X_
    train.shape[1],)))
```

```
41. model.add(Dense(32, activation='relu'))
42. model.add(Dense(1))
43. model.compile(loss='mse', optimizer='adam')
44.
45. # Train the model
46. model.fit(X_train, y_train, epochs=100, batch_size=32,
    verbose=1, shuffle=False)
47.
48. # Evaluate the model
49. predictions = model.predict(X_test)
50. predictions = scaler.inverse_transform(predictions)
51. y_test = scaler.inverse_transform(y_test)
52.
53. # Compute the prediction error
54. error = np.sqrt(np.mean((predictions-y_test)**2))
55. print("Prediction error:", error)
56.
```

In this script, we preprocess the ERA5 dataset, create input/output pairs for the neural network model, and train a feedforward neural network using Keras and TensorFlow.

Note that this is a simplified example of using neural networks for climate modelling. For more advanced and accurate models, you would need to consider additional climate variables, use more sophisticated neural network architectures, and potentially integrate other sources of data.

12.2.3.1 Validation Methods for Neural Networks in Climate Modelling

Validation methods can encompass multiple aspects of climate modelling. Historical data validation involves comparing model outputs with past climate records to gauge the model's ability to replicate past conditions accurately. Cross-validation techniques, such as temporal cross-validation, assess the model's performance across different time periods. Metrics like mean absolute error, root mean squared error, and correlation coefficients quantify the model's prediction accuracy. Sensitivity analysis examines the model's response to varying input parameters to identify the most influential variables. Additionally, ensemble techniques involving multiple neural networks can enhance model reliability by accounting for uncertainty. Incorporating validation methods into neural network–based climate modelling enhances confidence in the model's predictions, contributing to more robust climate assessments and adaptation strategies.

When working with climate data, integrating climate variables like precipitation, humidity, solar radiation, and wind speed into neural networks enhances model accuracy. These variables can be integrated as input features or through multimodal neural networks. Addressing uncertainties is essential; Bayesian deep learning, Monte Carlo dropout, and ensemble methods provide uncertainty estimates alongside predictions for robust decision-making. Incorporating domain knowledge, such as physical laws, into neural networks using physics-informed neural networks or physically based loss functions during training improves generalization. In conclusion, while the example uses a simple feedforward neural network for temperature prediction, opportunities abound for advanced models. Incorporating variables, advanced architectures, domain knowledge, and uncertainty estimation enhances a climate model's accuracy and contributes to addressing climate change impacts.

12.2.4 Reinforcement Learning for Smart Energy Management

Reinforcement learning (RL) has gained significant attention in the field of smart energy management, primarily due to its ability to optimize decision-making processes in dynamic and uncertain environments. Smart energy management aims to optimize energy generation, distribution, and consumption, considering factors such as renewable energy sources, grid stability, and energy demand. RL can be used in various applications within smart energy management, such as demand response, energy storage optimization, and microgrid control.

- Demand response: RL can be applied to optimize the scheduling and control of flexible loads, such as electric vehicles or heating, ventilation, and air conditioning (HVAC) systems. By learning the optimal control policies for these flexible loads, RL algorithms can help reduce energy consumption and peak demand, while maintaining user comfort.
- Energy storage optimization: Energy storage systems, such as batteries or thermal storage, play a crucial role in integrating renewable energy sources into the grid. RL can be used to optimize the charging and discharging of energy storage systems, considering factors like energy prices, grid constraints, and renewable energy availability.
- Microgrid control: RL can be employed for the optimal control and operation of microgrids, which are small-scale power systems that can operate independently or interconnected with the main grid. RL algorithms can learn the optimal control policies for microgrid components, such as generators, energy storage systems, and flexible loads, to optimize energy efficiency, grid stability, and the integration of renewable energy sources.

Here are some real-world datasets of energy loads and prices that can be used for your reinforcement learning application:

- PJM Interconnection Load and Price Data. PJM Interconnection is a regional transmission organization in the United States that operates the largest electricity market in the world. They provide historical load and price data for their region. You can access the data through their website:
  - Load data: https://dataminer2.pjm.com/feed/hrl_load_metered/definition
  - Price data: https://dataminer2.pjm.com/feed/rt_lmp_final/definition
- New York Independent System Operator (NYISO) Data. NYISO operates the electricity market in the state of New York. They provide historical load and price data through their website:
  - Load data: http://mis.nyiso.com/public/P-58Blist.htm
  - Price data: http://mis.nyiso.com/public/P-221list.htm

- Australian Energy Market Operator (AEMO) Data. AEMO operates the electricity market in Australia. They provide historical demand and price data on their website:
  - Demand data: www.aemo.com.au/energy-systems/electricity/national-electricity-market-nem/data-nem/aggregated-data
  - Price data: www.aemo.com.au/energy-systems/electricity/national-electricity-market-nem/data-nem/settlements-and-pricing
- UK National Grid ESO Data. The UK National Grid ESO operates the electricity market in the United Kingdom. They provide historical demand and price data through their portal:
  - Demand data: https://data.nationalgrideso.com/demand
  - Price data: https://data.nationalgrideso.com/pricing
- Open Power System Data (OPSD). OPSD is a platform that provides open-source data for power systems around the world, including electricity load and price data for several countries in Europe. You can access the data through their website:
  - Load data: https://data.open-power-system-data.org/time_series/
  - Price data: https://data.open-power-system-data.org/day_ahead_prices/

Please note that the format and structure of the data may differ between sources. You will need to preprocess and format the data to fit the reinforcement learning framework in your example code.

Nakabi & Toivanen (2021) propose a model of a microgrid system containing a wind turbine, energy storage, thermostatically controlled loads, price-responsive loads, and a connection to the main grid. An energy management system coordinates these different elements using both direct control of the thermostatically controlled loads and batteries and indirect control of the price-responsive loads by setting electricity prices. The problem is formulated mathematically as a Markov decision process. Several deep reinforcement learning algorithms, including DQN, DDPG and A3C, are implemented and tested on this model. The learning curves show that the DQN algorithms have better stability compared to policy-based methods like A3C (Nakabi & Toivanen 2021, p. 10).

12.2.4.1 Validation Methods for Reinforcement Learning in Smart Energy Management

Simulation-based validation involves emulating real-world scenarios within controlled environments to assess the reinforcement learning algorithm's performance. Comparative analysis with baseline strategies, such as rule-based systems, validates the algorithm's superiority in energy optimization. Tracking performance metrics like energy consumption reduction, cost savings, and carbon emissions reduction quantifies the impact of reinforcement learning–based energy management. Field trials and pilot implementations validate algorithm effectiveness in real-world settings. Additionally, sensitivity analysis evaluates the algorithm's performance under varying conditions and parameters. Integrating validation methods into reinforcement learning–driven energy management ensures reliable results, fostering sustainable energy consumption patterns and resource conservation.

12.3 AI APPLICATIONS IN ENVIRONMENTAL MANAGEMENT

AI is a transformative force for addressing environmental challenges by employing advanced algorithms and data analysis techniques. This chapter explores diverse AI applications in environmental management, emphasizing real-life use cases and their impact on environmental understanding and preservation.

AI's pivotal role is evident in air quality monitoring and management. Machine learning algorithms, combined with remote sensing data, predict pollution levels and pinpoint sources (Castelluccio et al. 2015; Singh & Chauhan 2020). IBM's Green Horizons initiative uses AI to forecast air pollution in Beijing, offering policy recommendations (Liu et al. 2019).

Water resource management benefits from AI, enhancing flood prediction, water distribution, and leak detection. Google's partnership with the Central Valley Water Board in California employs AI and satellite imagery to track groundwater usage (Rodell et al. 2018).

AI-driven robots like AMP Robotics optimize waste sorting, enabling efficient recycling (AMP Robotics). AI also optimizes waste collection logistics and smart bins for efficient waste management. Biodiversity conservation, climate change mitigation, and reforestation benefit from AI. Machine learning identifies species, monitors habitats, and guides conservation efforts (Cohen et al. 2010). Platforms such as Wildbook use computer vision for wildlife population monitoring (Berger-Wolf et al. 2017).

AI enhances climate models, predicts deforestation using satellite imagery, and optimizes renewable energy generation (Mohamad & Teh 2018). In summary, AI transforms environmental management, from air quality to biodiversity conservation, contributing to sustainable development and planet preservation.

12.3.1 AIR QUALITY MONITORING AND MANAGEMENT

Air quality monitoring and management are crucial for maintaining public health and mitigating the adverse effects of air pollution on the environment. AI has emerged as a powerful tool for tackling this issue by enabling the development of advanced monitoring and prediction systems that offer timely insights and solutions.

One of the main challenges in air quality monitoring is the accurate prediction of air pollutant concentrations, which can vary spatially and temporally. Machine learning techniques, such as regression models, have been widely used for predicting air pollution levels based on historical data and weather conditions (Singh & Chauhan 2020). For example, the Gaussian process regression (GPR) model is a popular technique for predicting air quality in urban areas (Saiohai et al. 2023).

A publicly available dataset that can be used for air quality prediction is the Air Quality in Madrid dataset available on the Kaggle platform. The dataset consists of hourly measurements of various air pollutants and meteorological variables collected from multiple monitoring stations in Madrid, Spain, from 2001 to 2018. You can download the dataset from the following link:

<u>Madrid Air Quality Dataset:</u> www.kaggle.com/decide-soluciones/air-quality-madrid

To build a GPR model using this dataset, you can use the following Python code, assuming you have the necessary libraries installed:

```python
1.  import pandas as pd
2.  import numpy as np
3.  from sklearn.model_selection import train_test_split
4.  from sklearn.gaussian_process import
    GaussianProcessRegressor
5.  from sklearn.gaussian_process.kernels import RBF,
    WhiteKernel
6.  from sklearn.metrics import mean_squared_error, r2_score
7.
8.  # Load the dataset
9.  data = pd.read_csv("madrid_air_quality.csv")
10.
11. # Preprocess the dataset
12. # (1) Drop missing values
13. data = data.dropna()
14.
15. # (2) Extract relevant features (e.g., meteorological
    variables) and target (e.g., NO2 concentration)
16. X = data[['temperature', 'pressure', 'humidity']]
17. y = data['NO2']
18.
19. # Split the data into train and test sets
20. X_train, X_test, y_train, y_test = train_test_split(X, y,
    test_size=0.2, random_state=42)
21.
22. # Define the GPR model
23. kernel = RBF(length_scale=1.0) + WhiteKernel(noise_
    level=1.0)
24. gpr = GaussianProcessRegressor(kernel=kernel)
25.
26. # Train the model
27. gpr.fit(X_train, y_train)
28.
29. # Predict NO2 concentration
30. y_pred = gpr.predict(X_test)
31.
32. # Evaluate the model
33. mse = mean_squared_error(y_test, y_pred)
34. r2 = r2_score(y_test, y_pred)
35. print(f"MSE: {mse}")
36. print(f"R2 Score: {r2}")
37.
```

In this code snippet, we preprocess the dataset by removing missing values, extracting relevant features, and splitting it into training and testing sets. We then define a GPR model with a combination of a radial basis function (RBF) kernel and a WhiteKernel, train it on the training set, and make predictions on the test set. Finally, we evaluate the model's performance using the mean squared error (MSE) and the coefficient of determination (R^2 score).

Incorporating additional features such as past pollutant levels or spatial dependencies between monitoring stations could further improve the model's performance.

Beyond predicting pollutant concentrations, AI can also be applied in air quality management to identify the sources of pollution and suggest effective emission control strategies. For instance, the IBM Green Horizons initiative used machine learning to predict air pollution levels in Beijing and provided policy recommendations for controlling emissions (Liu et al. 2019). Another application is the use of AI in developing low-cost air quality sensors that can be deployed in large networks to provide real-time, high-resolution data on air pollution.

In conclusion, AI has the potential to revolutionize air quality monitoring and management by offering accurate predictions of air pollutant concentrations and actionable insights for pollution mitigation. By leveraging AI techniques, we can better understand and address the challenges of air pollution, ultimately contributing to improved public health and environmental sustainability.

12.3.2 Water Resource Management and Conservation

Water resource management and conservation are essential for ensuring the availability of clean water for current and future generations. AI techniques, such as machine learning and deep learning, have been increasingly employed to improve the efficiency and effectiveness of water resource management, addressing challenges related to water quality, supply, and demand.

One of the key applications of AI in water resource management is the prediction of water demand, which allows for better planning and management of water distribution systems. Machine learning models, such as time series forecasting models, can be used to predict water demand based on historical data and relevant features, such as weather conditions and demographic information (Wu et al. 2020).

A publicly available dataset suitable for water demand prediction is the Hourly Water Supply and Demand dataset available on the OpenML platform. This dataset contains hourly water supply and demand data from January 1, 2013, to December 31, 2016, for a city in Portugal. The dataset can be accessed using the following link: Hourly Water Supply and Demand Dataset: www.openml.org/d/40957

In addition to water demand prediction, AI can be applied in several other aspects of water resource management, such as monitoring water quality, predicting hydrological events such as floods and droughts, and optimizing water distribution networks. Machine learning algorithms can analyze large volumes of water quality data to identify contamination sources, predict pollutant concentrations, and guide mitigation efforts (Dai et al. 2021).

Here are some publicly available data sources related to water quality monitoring, predicting hydrological events, and optimizing water distribution networks:

- Water Quality Monitoring
  - USGS Water Quality Data. The United States Geological Survey (USGS) provides water quality data collected from various sources,

including rivers, lakes, and wells. The dataset includes information about various water quality parameters, such as dissolved oxygen, pH, and water temperature. USGS Water Quality Data (https://waterdata.usgs.gov/nwis/qw)

- UCI Machine Learning Repository—Water Quality Dataset. The dataset contains water quality data from various sources, including rivers, lakes, and wells. It covers several water quality parameters such as pH, electrical conductivity (EC), total dissolved solids (TDS), etc. UCI Water Quality Dataset (https://archive.ics.uci.edu/ml/datasets/Water+Quality)

- Predicting Hydrological Events (Floods and Droughts)
  - NOAA National Centers for Environmental Information. NOAA provides a wide range of datasets related to weather and climate events, including precipitation data, temperature data, and streamflow data, which can be used for predicting floods and droughts. NOAA National Centers for Environmental Information (www.ncei.noaa.gov/access/)
  - Global Runoff Data Centre (GRDC). GRDC provides global river discharge data collected from over 9,000 stations in 160 countries. The data can be used for hydrological modelling, flood prediction, and water resource assessment. Global Runoff Data Centre (www.bafg.de/GRDC/EN/Home/homepage_node.html)

- Optimizing Water Distribution Networks
  - EPANET. EPANET is a software developed by the US Environmental Protection Agency (EPA) to model water distribution networks. The software includes a dataset of a hypothetical water distribution network with information about pipes, junctions, reservoirs, and tanks. Researchers can use this dataset to test and develop optimization algorithms for water distribution networks. EPANET Example Dataset (https://github.com/OpenWaterAnalytics/epanet-example-networks)
  - C-Town Water Distribution Network Dataset. This dataset, provided by the University of Exeter, represents a medium-sized town's water distribution network. The dataset includes information about pipes, nodes, pumps, and valves, which can be used to test and develop optimization algorithms for water distribution networks. C-Town Dataset (https://emps.exeter.ac.uk/engineering/research/cws/resources/benchmarks/design-resilience/)

These data sources can be valuable resources for researchers and practitioners working on water resource management projects that involve AI techniques.

Using the publicly available datasets mentioned earlier, researchers and practitioners can develop AI models for various water resource management tasks. These models can help predict water quality, anticipate hydrological events, and optimize water distribution networks, contributing to more efficient and sustainable water management.

For example, researchers can use the USGS Water Quality Data or the UCI Water Quality Dataset to develop machine learning models that predict water quality parameters such as dissolved oxygen, pH, and water temperature. By understanding the relationships between various water quality parameters and potential pollution sources, these models can help identify contamination sources and guide remediation efforts.

Similarly, datasets from NOAA and GRDC can be used to create AI models that predict hydrological events like floods and droughts. These models can leverage precipitation data, temperature data, and streamflow data to provide early warnings of potential flooding or drought conditions. Early warning systems can help communities better prepare for and respond to hydrological events, reducing the adverse impacts on people, property, and the environment.

Last, water distribution network datasets such as the EPANET Example Dataset or the C-Town Water Distribution Network Dataset can be used to develop AI-based optimization algorithms for water distribution systems. These algorithms can help water utilities manage their distribution networks more efficiently by optimizing parameters such as water pressure, flow rates, and energy consumption. In turn, this can lead to reduced water losses, lower energy costs, and improved service quality for consumers.

In summary, AI applications in water resource management offer numerous opportunities to enhance the efficiency, sustainability, and resilience of water systems. By leveraging publicly available datasets and AI techniques, researchers and practitioners can develop innovative solutions to the complex challenges faced in managing and conserving our vital water resources.

12.3.3 Waste Management and Circular Economy

Waste management and the circular economy are closely related, as both aim to minimize waste, reduce resource consumption, and maximize the value of products and materials throughout their lifecycle. AI techniques have been increasingly employed in waste management, facilitating the transition to a circular economy by optimizing waste collection, recycling, and resource recovery processes.

One application of AI in waste management is the optimization of waste collection routes. Machine learning algorithms can analyze historical waste generation data, along with geographic and demographic information, to design more efficient waste collection routes that minimize fuel consumption, vehicle wear, and greenhouse gas emissions.

A publicly available dataset suitable for waste collection route optimization is the Waste Collection Benchmark Problem (WCBP) dataset. The dataset contains information about waste collection points, waste generation rates, and collection time windows in a Brazilian city. The dataset can be accessed using the following link: Waste Collection Benchmark Problem Dataset, www.sciencedirect.com/science/article/pii/S2351978916306424?via%3Dihub

Although obtaining publicly available datasets for recycling and material recovery applications can be challenging, researchers and practitioners can create their own datasets using images of waste materials or by collaborating with recycling facilities. Such datasets can then be used to train AI models for waste identification and classification tasks.

In summary, AI techniques can significantly contribute to waste management and the circular economy by optimizing waste collection routes, enhancing recycling processes, and guiding policy interventions for waste reduction. By leveraging custom datasets or publicly available data sources, researchers and practitioners can develop innovative solutions to address the complex challenges faced in waste management and promote a more sustainable future.

12.4 ETHICAL AND SOCIAL IMPLICATIONS OF AI IN ENVIRONMENTAL APPLICATIONS

The integration of AI in environmental applications has ushered in a new era of technological advancement with profound ethical and social implications. While AI holds the potential to revolutionize environmental monitoring, resource management, and conservation efforts, its deployment raises pertinent ethical concerns. One primary concern involves the bias and fairness of AI algorithms utilized in environmental decision-making processes. Biases in training data or algorithmic design can perpetuate environmental injustices, disproportionately affecting marginalized communities (Barocas et al. 2019). Furthermore, the opacity of certain AI systems, such as deep neural networks, can hinder the interpretability of decision-making processes, impeding accountability and hindering public trust in automated environmental management (Hamon et al. n.d.). Socially, the increased reliance on AI may lead to job displacement in sectors traditionally involved in data collection and analysis, necessitating proactive measures for workforce reskilling and transition (Ernst et al. 2018). Balancing the potential benefits of AI in environmental applications with these ethical and social considerations requires a comprehensive and collaborative approach involving policymakers, technologists, and ethicists to ensure equitable and sustainable AI integration.

12.5 FUTURE DIRECTIONS AND CHALLENGES OF AI IN ENVIRONMENTAL APPLICATIONS

As AI continues to evolve and its applications in the environmental domain expand, future research and development efforts must address a range of challenges and opportunities. The following are some key areas to consider in terms of future directions and challenges of AI in environmental applications.

Integration of AI techniques with Earth System Models. Earth System Models (ESMs) are complex computational tools that simulate the interactions between the atmosphere, oceans, land, and biosphere. Integrating AI techniques with ESMs can improve their accuracy and computational efficiency, leading to better predictions of climate change and related impacts (Huntingford et al. 2019).
  - **Real-life use case**: AI-based parameterizations can be employed to improve the representation of cloud processes in ESMs, which is a significant source of uncertainty in climate projections (Schneider et al. 2017).

Interdisciplinary collaboration. Future AI applications in the environmental field will benefit from interdisciplinary collaboration involving experts from fields such as ecology, hydrology, atmospheric science, and social science. This can lead to the development of more comprehensive and accurate models, better understanding of system dynamics, and more effective environmental management strategies (Vinuesa et al. 2020).

- **Real-life use case**: The Microsoft AI for Earth program fosters collaboration between AI researchers and environmental scientists to develop innovative solutions for pressing environmental issues, such as biodiversity conservation, climate change, and resource management (Microsoft 2017).

Explainability and transparency. As AI models become more complex, ensuring their explainability and transparency will be critical in gaining trust from decision-makers and stakeholders. This will involve developing techniques that can effectively communicate the underlying mechanisms and reasoning behind AI predictions, while also addressing concerns related to fairness, bias, and accountability (Barredo Arrieta et al. 2020).

- **Real-life use case**: Explainable AI (XAI) techniques can be applied to make predictions from AI-based flood forecasting models more interpretable for decision-makers, facilitating better-informed and timely responses to flood events (Tiddi & Schlobach 2022).

Data availability and quality. AI model effectiveness in environmental applications relies on data quality and availability. Future work should target data limitations, utilizing satellite imagery, remote sensing, crowdsourcing, and data fusion to address coverage and resolution gaps.

- **Real-life use case**: Using crowdsourced data from mobile applications, such as eBird, can help improve AI-based species distribution models by providing additional data points for model training (Beery et al. 2021).

Scalability and generalizability. AI models often perform well on specific tasks and datasets but may struggle to generalize to new contexts or regions. Future research should focus on developing methods to improve the scalability and generalizability of AI models in environmental applications, such as transfer learning, domain adaptation, and data augmentation techniques.

- **Real-life use case**: Transfer learning techniques can be employed to adapt AI models trained on air quality data from one region to another, improving their predictive performance and generalizability across different geographic areas (Iskandaryan et al. 2020).

Policy and regulation. The development and deployment of AI in environmental applications will need to be supported by appropriate policy and regulatory frameworks that promote responsible innovation, ensure data privacy and security, and address potential negative consequences. These frameworks should be informed by ongoing interdisciplinary dialogue and stakeholder engagement to ensure that AI-driven solutions align with societal values and environmental objectives (Cath et al. 2017).

- **Real-life use case**: The European Union's AI policy framework aims to provide guidance and regulation for AI applications in various sectors, including the environment, to ensure responsible development and use (European Commission 2021).

In conclusion, the future directions and challenges of AI in environmental applications present both opportunities and potential pitfalls. By addressing these challenges and embracing collaboration, transparency, and inclusivity, AI can play a transformative role in addressing environmental issues, promoting sustainability, and improving our understanding of the earth's systems.

12.6 CONCLUSION

The integration of artificial intelligence (AI) in environmental applications holds immense potential to address global challenges and foster sustainable development. AI contributes significantly to environmental monitoring, resource management, and policymaking by enhancing our understanding of complex systems and providing actionable insights for decision-makers.

AI's applications in environmental monitoring, such as remote sensing for land use mapping, machine learning for species identification, and neural networks for climate modeling, have progressed remarkably. These techniques process extensive environmental data efficiently, enabling accurate assessments of critical phenomena. Furthermore, AI optimizes air quality, water conservation, and waste management, supporting sustainable resource utilization. Despite its promise, AI deployment in the environment raises ethical and social concerns. Addressing transparency, privacy, and accountability is crucial for responsible AI innovation (Barredo Arrieta et al. 2020). The future of AI in the environment involves navigating technical and ethical challenges while harnessing technology for a resilient world. By investing in interdisciplinary research, promoting responsible AI, and fostering collaboration, AI can play a crucial role in environmental management and global well-being (Cath et al. 2017; Schneider et al. 2017).

REFERENCES

Barocas, S., Hardt, M., & Narayanan, A. 2019. *Fairness and Machine Learning: Limitations and Opportunities*. fairmlbook.org.

Barredo Arrieta, A., Díaz-Rodríguez, N., Del Ser, J., Bennetot, A., Tabik, S., Barbado, A., Garcia, S., Gil-Lopez, S., Molina, D., Benjamins, R., Chatila, R., & Herrera, F. 2020. Explainable Artificial Intelligence (XAI): Concepts, Taxonomies, Opportunities and Challenges Toward Responsible AI. *Information Fusion*, *58*, 82–115. https://doi. org/10.1016/j.inffus.2019.12.012

Beery, S., Cole, E., Parker, J., Perona, P., & Winner, K. 2021. Species Distribution Modeling for Machine Learning Practitioners: A Review. *ACM SIGCAS Conference on Computing and Sustainable Societies (COMPASS)*, 329–348. https://doi.org/10.1145/3460112.3471966

Berger-Wolf, T. Y., Rubenstein, D. I., Stewart, C. V., Holmberg, J. A., Parham, J., Menon, S., Crall, J., Van Oast, J., Kiciman, E., & Joppa, L. 2017. *Wildbook: Crowdsourcing, Computer Vision, and Data Science for Conservation*. http://arxiv.org/abs/1710.08880

Castelluccio, M., Poggi, G., Sansone, C., & Verdoliva, L. 2015. *Land Use Classification in Remote Sensing Images by Convolutional Neural Networks*. http://arxiv.org/abs/1508.00092

Cath, C., Wachter, S., Mittelstadt, B., Taddeo, M., & Floridi, L. 2017. Artificial Intelligence and the 'Good Society': The US, EU, and UK Approach. *Science and Engineering Ethics*. https://doi.org/10.1007/s11948-017-9901-7

Cohen, W. B., Yang, Z., & Kennedy, R. 2010. Detecting Trends in Forest Disturbance and Recovery Using Yearly Landsat Time Series: 2. TimeSync—Tools for Calibration and Validation. *Remote Sensing of Environment, 114*(12), 2911–2924. https://doi.org/10.1016/j.rse.2010.07.010

Dai, H., Huang, G., Wang, J., Zeng, H., & Zhou, F. 2021. Prediction of Air Pollutant Concentration Based on One-Dimensional Multi-Scale CNN-LSTM Considering Spatial-Temporal Characteristics: A Case Study of Xi'an, China. *Atmosphere, 12*(12), 1626. https://doi.org/10.3390/atmos12121626

Ernst, E., Merola, R., & Samaan, D. 2018. *International Labour Organization.* www.ilo.org/publns.

EUROPEAN COMMISSION. 2021, April 21. *Laying Down Harmonised Rules on Artificial Intelligence (Artificial Intelligence Act) and Amending Certain Union Legislative Acts.* https://Eur-Lex.Europa.Eu/Legal-Content/EN/TXT/HTML/?Uri=CELEX:52021PC0206&from=EN.

Hamon, R., Hamon, R., Junklewitz, H., Sanchez, I., & European Commission. Joint Research Centre. n.d. *Robustness and Explainability of Artificial Intelligence: From Technical to Policy Solutions.* https://doi.org/10.2760/57493

Huntingford, C., Jeffers, E. S., Bonsall, M. B., Christensen, H. M., Lees, T., & Yang, H. 2019. Machine Learning and Artificial Intelligence to Aid Climate Change Research and Preparedness. *Environmental Research Letters, 14*(12), 124007. https://doi.org/10.1088/1748-9326/ab4e55

Iskandaryan, D., Ramos, F., & Trilles, S. 2020. Air Quality Prediction in Smart Cities Using Machine Learning Technologies based on Sensor Data: A Review. *Applied Sciences, 10*(7), 2401. https://doi.org/10.3390/app10072401

Liu, J., Kiesewetter, G., Klimont, Z., Cofala, J., Heyes, C., Schöpp, W., Zhu, T., Cao, G., Gomez Sanabria, A., Sander, R., Guo, F., Zhang, Q., Nguyen, B., Bertok, I., Rafaj, P., & Amann, M. 2019. Mitigation Pathways of Air Pollution from Residential Emissions in the Beijing-Tianjin-Hebei region in China. *Environment International, 125*, 236–244. https://doi.org/10.1016/j.envint.2018.09.059

Microsoft. 2017. *AI for Earth.* Https://Www.Microsoft.Com/En-Us/Ai/Ai-for-Earth.

Mohamad, F., & Teh, J. 2018. Impacts of Energy Storage System on Power System Reliability: A Systematic Review. *Energies, 11*(7), 1749. https://doi.org/10.3390/en11071749

Nakabi, T. A., & Toivanen, P. 2021. Deep Reinforcement Learning for Energy Management in a Microgrid with Flexible Demand. *Sustainable Energy, Grids and Networks, 25*, 100413. https://doi.org/10.1016/j.segan.2020.100413

Rasp, S., Pritchard, M. S., & Gentine, P. 2018. *Deep Learning to Represent Subgrid Processes in Climate Models.* https://doi.org/10.5281/zenodo.1402384

Rodell, M., Famiglietti, J. S., Wiese, D. N., Reager, J. T., Beaudoing, H. K., Landerer, F. W., & Lo, M. H. 2018. Emerging Trends in Global Freshwater Availability. *Nature, 557*(7707), 651–659. https://doi.org/10.1038/s41586-018-0123-1

Saiohai, J., Bualert, S., Thongyen, T., Duangmal, K., Choomanee, P., & Szymanski, W. W. 2023. Statistical PM2.5 Prediction in an Urban Area Using Vertical Meteorological Factors. *Atmosphere, 14*(3), 589. https://doi.org/10.3390/atmos14030589

Schneider, T., Teixeira, J., Bretherton, C. S., Brient, F., Pressel, K. G., Schär, C., & Siebesma, A. P. 2017. Climate Goals and Computing the Future of Clouds. *Nature Climate Change, 7*(1), 3–5. https://doi.org/10.1038/nclimate3190

Singh, R. P., & Chauhan, A. 2020. Impact of Lockdown on Air Quality in India During COVID-19 Pandemic. *Air Quality, Atmosphere & Health, 13*(8), 921–928. https://doi.org/10.1007/s11869-020-00863-1

Tiddi, I., & Schlobach, S. 2022. Knowledge Graphs as Tools for Explainable Machine Learning: A Survey. *Artificial Intelligence, 302*, 103627. https://doi.org/10.1016/j.artint.2021.103627

Vinuesa, R., Azizpour, H., Leite, I., Balaam, M., Dignum, V., Domisch, S., Felländer, A., Langhans, S. D., Tegmark, M., & Fuso Nerini, F. 2020. The Role of Artificial Intelligence in Achieving the Sustainable Development Goals. *Nature Communications*, *11*(1), 233. https://doi.org/10.1038/s41467-019-14108-y

Wäldchen, J., & Mäder, P. 2018a. Machine Learning for Image Based Species Identification. In *Methods in Ecology and Evolution* (Vol. 9, Issue 11, pp. 2216–2225). British Ecological Society. https://doi.org/10.1111/2041-210X.13075

Wäldchen, J., & Mäder, P. 2018b. Machine Learning for Image Based Species Identification. *Methods in Ecology and Evolution*, *9*(11), 2216–2225. https://doi.org/10.1111/2041-210X.13075

Wu, Y., Jing, W., Liu, J., Ma, Q., Yuan, J., Wang, Y., Du, M., & Liu, M. 2020. Effects of Temperature and Humidity on the Daily New Cases and New Deaths of COVID-19 in 166 Countries. *Science of The Total Environment*, *729*, 139051. https://doi.org/10.1016/j.scitotenv.2020.139051

Zhang, L., Zhang, L., & Du, B. 2016. Deep Learning for Remote Sensing Data: A Technical Tutorial on the State of the Art. *IEEE Geoscience and Remote Sensing Magazine*, *4*(2), 22–40. https://doi.org/10.1109/MGRS.2016.2540798

13 Automation of Adsorption Processes Using AI
Recent Trends and Prospects

Saizel Pathania, S. Daksa, Sanjanavhas Srinivasan,
and Xavier Savarimuthu, S.J.

13.1 INTRODUCTION

The earth is referred to as a blue planet due to the abundance of water, particularly the vast expanses of oceans and seas that cover about 71% of its surface. Water plays a crucial role in supporting diverse life forms and their ecosystems. It is often known for its unique properties contributing to utility potential. Water is odourless and colourless signifying prominent usage in other liquid compositions. Over the years there has been a rapid decline in the availability of freshwater due to diversification and multiple addition as a utility value. Scarcity is one of the issues that has increased across the globe due to population expansion and parallel resource demand. Some of the reasons are industrial effluent release, inadequate infrastructure, and lack of water management systems (Gleick and Cooley 2021).

Furthermore, the water footprint is essential to understand the direct and indirect water consumption patterns in different industrial sectors (Hogeboom 2020). Every sector's composition of wastewater is different and significant. It includes organic, inorganic matter, metals, and microbes from industrial effluent. There are various treatment methods based on the raw materials present in the water, which include physical, chemical, biological, and hybrid methods (Quach-Cu et al. 2018). Significant advancements have been achieved over the last few decades in creating innovative, effective, and economical methods for eliminating different contaminants from wastewater (Dhote et al. 2012). In recent years, there has been development of green and sustainable wastewater technologies, enabling better sustainable practices (Mojiri and Bashir 2022). There is a need for a monitoring solution focussing on the adsorption process, which generally requires greater input of raw materials and monitoring to reduce membrane fouling and reduced efficiency (Mojiri and Bashir 2022).

In recent years, the use of artificial intelligence (AI) has increased exponentially because of its ability to be used in diverse domains; integration with environmental sciences being one among the many. AI refers to the field of computer science and technology that focuses on creating intelligent machines that can perform tasks that typically require human intelligence. The incorporation of AI in water management

DOI: 10.1201/9781003405436-16

has shown considerable success, exemplified by key projects such as water quality monitoring (Sahoo and Kim 2013), optimizing irrigation practices (Singh et al. 2012) water treatment process optimization (Amaral et al. 2019), and so on. AI techniques are being employed in water treatment and desalination to optimise the process and provide practical solutions to water pollution and scarcity. The application of AI has contributed to modest implementation, flexibility, design simplicity, and generalisation (Suraj Kumar Bhagat et al. 2023).

This chapter discusses applications of AI in the treatment of water using different models, such as artificial neural network (ANN), random forest (RF), radial basis function (RBF) kernel, and adaptive neuro-fuzzy inference systems (ANFIS), that help to analyse the data and predict the best adsorbent that can be used for the treatment of wastewater. The chapter also highlights the adsorption process associated with the removal of pollutants such as heavy metals, dyes, chemicals, etc., while focusing on the merits and demerits of applications associated with machine learning (ML) and AI. Furthermore, limitations to be addressed, such as poor data availability, modelling, and analysis in certain methodologies associated with the adsorption process are also discussed.

13.2 DATA COLLECTION

Data collection is an integral part of any analysis or prediction. It is the process of gathering data for use in business decision-making, strategic planning, research, and other purposes. Data in wastewater management can be collected from various sources, including sensors, manual measurements, laboratory analyses, and historical records. Here are some common methods of data collection in wastewater management.

1. Sensors: Sensors can be used to collect real-time data on various parameters, including pH, temperature, flow rate, dissolved oxygen, and others. Sensor data can be used to monitor process performance, detect anomalies, and optimise treatment processes.
2. Manual measurements: Manual measurements can be taken by plant operators or technicians to collect data on various parameters, including pH, temperature, and dissolved oxygen. These measurements can be used to calibrate sensors or verify sensor readings.
3. Laboratory analyses: Samples of wastewater can be collected and analysed in a laboratory to determine the concentrations of various contaminants, including nutrients, metals, and organic compounds. Laboratory data can be used to verify sensor readings and ensure compliance with regulatory requirements.
4. Historical records: Historical records can be used to track trends in process performance, identify patterns, and forecast future performance. Historical data can include information on flow rates, treatment efficiency, energy consumption, and other parameters.

Hence, these data sources can be used to monitor process performance, detect anomalies, optimise treatment processes, and ensure compliance with regulatory requirements.

13.3 AI TECHNIQUES

AI has seen exponential growth in almost every domain over the years. Its application in environment sciences offers a wide scope but it is unexploited yet. Different techniques of AI can be used for various purposes, such as wastewater treatment operations, water reuse, water saving, and cost reduction through prediction, diagnosis, assessment, and simulation. In this chapter, the focus is placed on cost reduction by employing different algorithms to predict the best adsorbent that can be used for optimally treating water. Figure 13.1, shows the subcategories of AI techniques that are explored in detail in this chapter.

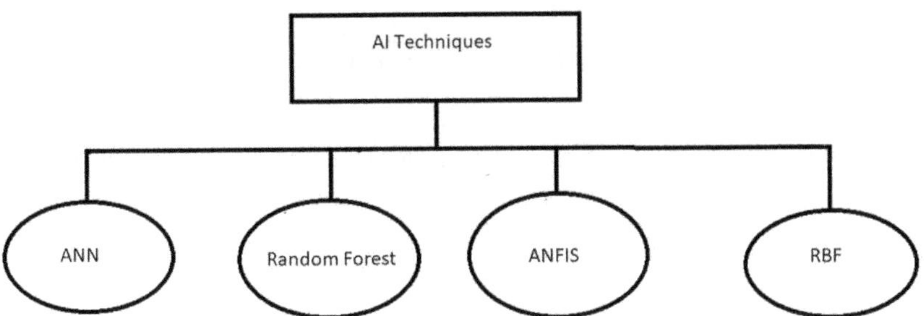

FIGURE 13.1 AI techniques. ANFIS = adaptive neuro-fuzzy inference systems; ANN = artificial neural network; RBF = radial basis function.

13.3.1 RANDOM FOREST

Random forest (RF) is a supervised machine learning algorithm that is widely used for classification and regression tasks. Figure 13.2 shows the architecture of an RF. It uses concepts such as entropy, information gain, root nodes, and leaf nodes to build multiple decision trees and combine their predictions to produce a final prediction. Each decision tree is trained on a random subset of the data, and the final prediction is made by combining the predictions of all the trees.

The Gini index and entropy functions are generally used to measure the quality of a split. A higher Gini index means there is higher impurity in the data. The choice between the two can be made by the user. Both entropy and Gini index produce similar results but the Gini index is computationally faster than entropy as there is no use of logarithms as in entropy. It is also to be noted that entropy obtains slightly better results. Since there is just a slight difference in their results, the Gini index is generally preferred.

Mathematically, the Gini index and entropy can be written as:

$$\text{Gini index} = 1 - \Sigma^{n}_{i=1}(P_i)^2 \tag{13.1}$$

$$= 1 - [(P_{\text{Positive class}})^2 + (P_{\text{Negative class}})^2] \tag{13.2}$$

$$\text{Entropy}(S) = - \Sigma\, p(x)*\log_2(p(x)) \tag{13.3}$$

$$E(S) = - [P_{\text{(Positive class)}}*\log_2 P_{\text{(Positive class)}} + P_{\text{(Negative class)}}*\log_2 P_{\text{(Negative class)}}] \tag{13.4}$$

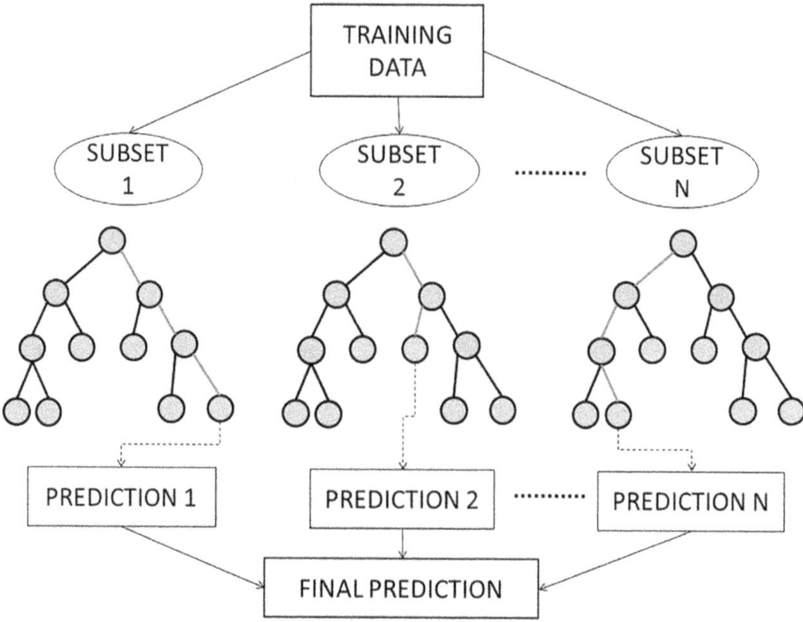

FIGURE 13.2 Architecture of a random forest (https://learnopencv.com/generative-and-discriminative-models/).

The working of a RF operation is given here:

1. After the data has been collected by different methods, the RF operation takes in a set of data that has already been labelled with the correct outcomes. This data is used to train the algorithm to make accurate predictions.
2. The algorithm randomly selects a subset of the data and a subset of the features to use for each decision tree.
3. A decision tree is built using the selected data and features. The tree is constructed by selecting the best feature to split the data based on the information gained. The process is repeated until a leaf node is reached, which represents a final prediction.
4. The process of building decision trees is repeated many times to create a forest of trees.
5. To make a prediction, each decision tree in the forest is given the input data, and it makes its own prediction.
6. The final prediction is then made by combining the predictions of all the decision trees in the forest, either by taking a vote or averaging the results.

Hence, the RF algorithm can be used in wastewater treatment to model the adsorption of contaminants such as dyes, heavy metals, and organic compounds by various adsorbents. The RF algorithm is a machine learning algorithm that can handle complex and nonlinear relationships between input variables and output variables. In wastewater treatment, RF can be used to identify the important parameters that affect the adsorption process, such as contact time, initial concentration, temperature, pH, and adsorbent dosage. The algorithm can also be used to predict the adsorption capacity or removal efficiency of the adsorbent. RF can provide accurate predictions even when the data has noise or missing values. Moreover, RF can handle high-dimensional data and can perform variable selection, which makes it a useful tool for wastewater treatment applications.

13.3.2 Artificial Neural Network

One of the many applications of artificial neural networks (ANNs) in wastewater management is in the detection of anomalies and faults in the treatment process. An architecture of ANN is shown in Figure 13.3. An ANN can be trained to recognize patterns in the data and to identify any deviations from normal behaviour, such as changes in the flow rate, the pH level, or the concentration of pollutants. This can help to detect equipment malfunctions, leaks, and other issues that may affect the efficiency of the treatment process.

An ANN is a type of machine learning algorithm that is modelled after the structure and function of the human brain. It consists of a network of interconnected nodes, called neurons, that work together to process and analyse data.

The process of building an ANN involves feeding it with data and allowing the network to learn patterns and relationships within the data. This is done by adjusting the weights and biases of the neurons through a process called training, using a training dataset. Once trained, the ANN can be used to make predictions on new data, by feeding the data into the network and obtaining the output from the final layer of neurons. The output can then be compared to the actual output to evaluate the accuracy of the model.

For further understanding, the net input can be calculated as follows:

$$Y_a = x_1.w_1 + x_2.w_2 + x_3.w_3 \ldots x_m.w_m \tag{13.5}$$
$$i.e.,\ net\ input\ is = \Sigma_i^m x_i.w_n \tag{13.6}$$

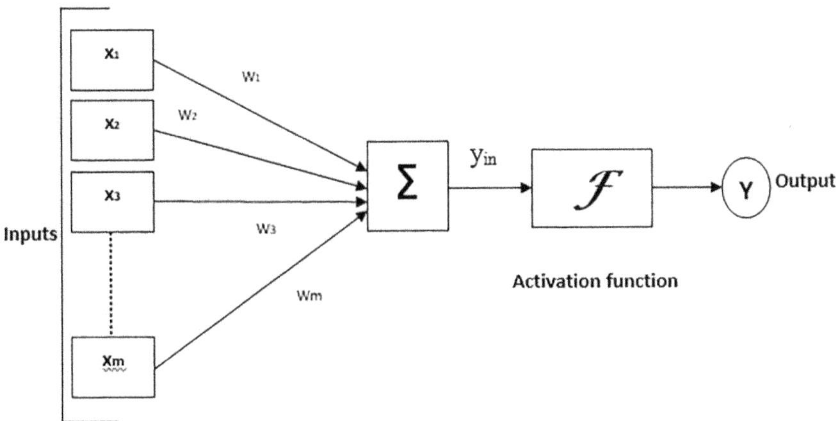

FIGURE 13.3   The architecture of an artificial neural network.

where,

y_a = net input x_i = input w_n = weights

To calculate the output, we apply activation function over the net input as:

$$Y = F(y_a) \tag{13.7}$$

Hence, the output is calculated.

13.3.3 Radial Basis Function Network and Multilayer Perceptron Network

A radial basis function network (RBFN) can also be utilised for optimising the operating conditions of a wastewater treatment process. By using an optimisation algorithm such as a genetic algorithm or particle swarm optimization, the RBFN model can identify the optimal values of the process parameters such as flow rate, pH, and contact time that result in the maximum removal efficiency of the pollutant.

Moreover, RBFN can be applied for real-time monitoring of the wastewater treatment process parameters. By using input variables such as flow rate, pH, and turbidity, the RBFN model can predict the quality of the effluent in real time and alert the operators if any deviation from the optimal conditions occurs. This can help prevent the discharge of untreated or poorly treated wastewater into the environment.

RBFN is a type of neural network that uses radial basis functions as activation functions. It is a three-layer feed-forward neural network, with an input layer, a hidden layer, and an output layer, as shown in Figure 13.4.

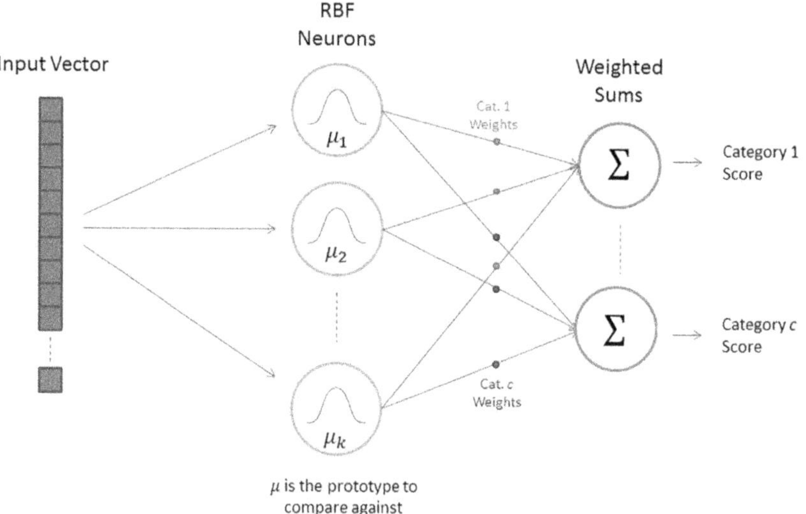

FIGURE 13.4 Architecture of a radial basis function (RBF) network (https://mccormickml.com/2013/08/15/radial-basis-function-network-rbfn-tutorial/).

In an RBFN, the hidden layer neurons are activated by radial basis functions, which are functions that depend only on the distance from the neuron to a centre point. The centres of the radial basis functions are typically chosen based on the input data, either by clustering the data points or by using some other method. The weights between the hidden layer and output layer are then calculated using a linear regression method.

Multilayer preception (MLP) is also a type of neural network, but it uses a different type of activation function called a sigmoid function. It is also a three-layer feed-forward neural network, with an input layer, a hidden layer, and an output layer. The hidden layer neurons are activated by the sigmoid function, which is a smooth curve that maps any input value to a value between 0 and 1.

The main difference between RBFN and MLP is in the type of activation function used in the hidden layer. RBFNs use radial basis functions, which are based on distance from the centre point, while MLPs use sigmoid functions, which are based on a smooth curve. The choice between these two types of neural networks depends on the nature of the problem being solved and the characteristics of the input data.

13.3.4 ADAPTIVE NEURO-FUZZY INFERENCE SYSTEM

The adaptive neuro-fuzzy inference system (ANFIS) is a hybrid computational model that combines the principles of fuzzy logic and neural networks, shown in Figure 13.5. It is a machine learning algorithm that is capable of learning from data and making decisions based on the learned knowledge.

ANFIS is a type of neural network adapted from the Takagi-Sugeno fuzzy interference system. ANFIS uses a set of rules that are expressed in terms of fuzzy "if-then" statements to model complex nonlinear systems. These are:

Rule 1: If x is A_1 and y is B_1, then $f_1 = p_1{}^*x + q_1{}^*y + r_1$
Rule 2: If x is A_2 and y is B_2, then $f_2 = p_2{}^*x + q_2{}^*y + r_2$

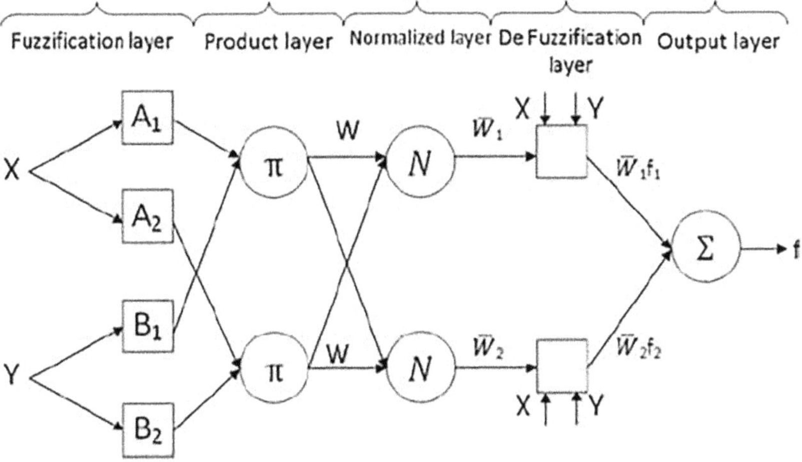

FIGURE 13.5 Architecture of an adaptive neuro-fizzy inference system.

where x and y are two inputs and one output z.

Yılmaz (Şişman) (2003) describes the ANFIS neural network structure as containing five layers excluding the input layer (Layer 0):

Layer 0 is the input layer with n nodes where n is the number of inputs to the system.

Layer 1 is the fuzzification layer where the nodes are responsible for determining the membership value of an input to a particular linguistic term. To achieve this, a Gaussian function is employed, with the mean of the function representing a specific point within the input range. By using this Gaussian function, the fuzzification layer is able to quantify the degree of membership of an input to a linguistic term. The Gaussian function with the mean is:

$$\mu Ai(x) = \frac{1}{1 + \left[\dfrac{x - ci}{ai}\right]^{2bi}} \ ; \ \text{where } ai, bi, ci \text{ are parameters for the function.} \quad (13.8)$$

Layer 2 is the layer in which each node provides the strength of the rule by means of a multiplication operator. The membership values represented by $\mu Ai\,(x_0)$ and $\mu Bi\,(x_1)$ are multiplied to find the firing strength of a rule where the variable x_0 has a linguistic value of Ai, and xi has a linguistic value of Bi, in the antecedent part of Rule i.

$$\mathrm{W}i = \mu Ai(x_0) * \mu Bi(x_1) \quad (13.9)$$

Layer 3 is the normalization layer, which normalizes the strength of all rules according to the following equations.

$$\overline{W}i = \frac{Wi}{\sum R_{j=1} W_j} \ ; \ \text{where } Wi \text{ is the firing strength of the } i\text{th rule that was}$$

computed in layer 2. $\quad (13.10)$

Layer 4 is a layer of adaptive nodes. Every node is responsible for the computation of a linear function where the function coefficients are adapted by using the error function of the multilayer feed-forward neural network.

$$\overline{W}ifi = \overline{W}i\left(p_0 * x_0 + p_1 * x_1 + p_2\right) \quad (13.11)$$

where $i = n + 1$ and n is the number of inputs to the system and $\overline{W}i$ is the output of layer 3.

Layer 5 is the output layer where the function is the summation of net outputs of the nodes in layer 4. The output is calculated as:

$$\sum_i \overline{w}ifi = \frac{\sum_i wifi}{\sum_i wi} \quad (13.12)$$

where $\bar{w}_i f_i$ is the output of nodes I obtained in layer 4. The overall output is the summation of the rule consequents.

The algorithm is capable of adjusting the membership functions and rule parameters based on input data, which makes it a powerful tool for modelling complex systems with uncertainty and imprecision. ANFIS has found application in a wide range of fields, including control systems, prediction, and optimization.

Hence, we see the importance of AI in wastewater management is very crucial. It helps not just by giving optimal and practical solutions but also by reducing the cost of treating water.

13.4 NEED FOR AI

13.4.1 WasteWater Treatment

The process involves treatment of wastewater discharged from both commercial and artificial sources for reducing the impact on the direct usage of water. The various types of treatments include biologically, physiochemical, technologically enabled methods to treat the effluents (Mojiri and Bashir 2022). The treatment method is selected based on the characteristics of the wastewater. Each process has its unique character, which involves not only the cost, but also includes feasibility, environmental impact, sludge production, operation difficulty, and pretreatment requirements. Even though there are many processes, only a few have been practised in the industrial sectors, which have its own advantages and disadvantages. The pollutants are removed by means of physiological and/or biological processes. Greenhouse gas emissions (GHG) are released from the wastewater treatment plant during the treatment process; this has been considered a critical source of GHG emissions. Also, energy consumption at the plant causes the release of CO_2 (Siril Singh et al. 2023).

13.4.2 Adsorption

This process involves the accumulation of the solid particles onto the surface of a solid porous medium in a given contact period (Popoola 2019). The pore size and high internal surface area of the adsorbent are essential for efficient removal of contaminants and, further, the pore sizes are designed for the process accordingly (Yousef et al. 2020). There are natural and synthetic adsorbents depending on the source (Quach-Cu et al. 2018). Synthetic adsorbents are designed based on the raw materials used for the adsorption process. These include use of chemicals and polymer-based materials such as polycyclic aromatic hydrocarbons (PAHs), coal fly-ash and nanomaterials such as alumina and silica (Alam et al. 2022).

There are also crucial factors such as water quality and flow rate which are to be addressed in parallel with the capital and operational costs (Albatrni and Qiblawey 2020). Another major issue associated with the process is membrane fouling caused by microorganisms, organic, and colloidal waste substances present in the feed water (Ashfaq et al. 2022). Reusability of the contaminant also becomes an additional cost and the process is of relatively less utility if the efficiency of the product is lower compared to initial properties since it has undergone the treatment process. Over time the contaminants could increase in the waste generation and disposal could be an issue associated but it could be resolved to a certain extent using regeneration (Albatrni and Qiblawey 2020).

Hence, AI can play a crucial role in wastewater management by improving the efficiency and effectiveness of several processes involved in the treatment and reuse of wastewater. Some ways in which AI can be helpful are discussed here.

1. Predictive maintenance: AI can analyse real-time data from sensors and predict when equipment may fail or require maintenance. This can help prevent unplanned downtime while reducing the need for emergency repairs.
2. Optimising treatment processes: AI can optimise wastewater treatment processes by analysing data from sensors, predicting changes in wastewater composition, and adjusting treatment parameters accordingly. This can lead to better treatment efficiency, reduced energy consumption, and lower operating costs.
3. Decision support: AI can provide decision support for plant operators and engineers by analysing data from multiple sources, including historical data, sensor data, and weather forecasts. This can help optimise treatment processes, reduce costs, and improve overall system performance.
4. Real-time monitoring and control: AI can provide real-time monitoring and control of wastewater treatment processes by analysing data from sensors and adjusting treatment parameters in real time. This can lead to better process control and improved treatment efficiency.

Hence, in wastewater treatment, AI brings out new technology over traditional technology.

13.5 APPLICATIONS OF AI

An adsorbent is a material or substance that has the ability to adsorb or collect molecules or particles on its surface. AI techniques were proved to be effective in studying the relationships between variables in water treatment. Some studies have also predicted the removal of multiple pollutants from the water with the help of AI. There are different applications for AI.

13.5.1 REMOVAL OF DYES

AI models are used to predict the adsorption performance of various absorbents used for the removal of either a single dye or multiple dyes. Data is collected to treat water from different sources, such as tough simulated wastewater or real textile wastewater to evaluate the performance of the adsorbent and the model used.

A notable use case of an RF algorithm for the removal of dyes has been proposed by A. R. Bagheri et al. (2015), which involves achieving a high R^2 value of 0.96 and a low mean square error (MSE) of 0.0021 on CG Dye adsorption using copper sulphide nanoparticles loaded on activated carbon (AC) as the adsorbent. The input variables for the model included "initial dye concentration," "adsorbent amount," and "sonication time," while the dependent variable to be predicted was "adsorption capacity."

Another example of the potential application of ANNs in the removal of dyes, studied by Çelekli et al. (2016), is demonstrated by a study involving basic red dye and walnut husk as the adsorbent, where an R^2 value of 0.9901 with a sum of squared errors (SSE) of 0.2303 was obtained. The ANNs were trained using input variables such as temperature, contact time, initial dye concentration, adsorbent particle size, and pH, while the output variable was the "removal efficiency" of the dye.

13.5.2 Removal of Heavy Metals

Applications of AI can also be exploited by evaluating the removal of heavy metals using different adsorbents. There are different studies focused on the simultaneous adsorption of multiple metals from aqueous phase or single metal adsorption.

AI application in wastewater treatment has been studied by Singh et al. (2013), such as in the case of chlorophenol dye removal using coconut shell carbon as an adsorbent. Input variables for the experiment included contact time, chlorophenol concentration, temperature, and pH, with "removal efficiency" serving as the output variable. Two machine learning algorithms, namely RBFN and the MLP network, were utilized for this experiment. The RBFN model achieved an R^2 value of 0.96 with an MSE of 6.03.

Another notable application of AI studied by Foroutan et al. (2020) is in the adsorption of heavy metals from the aqueous phase, with Cr(VI) as the target metal and clay-based adsorbents. Input variables in this experiment included contact time, temperature, and metal concentration, pH, and adsorbent dose, and "removal efficiency" served as the output variable. The model of choice for this study was the ANFIS, with R^2 value of 0.9997 and MSE of $1.288E^{-06}$.

Hence, we see the importance of AI in wastewater management is very crucial. It helps not just by giving optimal and practical solutions but also by reducing the cost of treating water.

13.6 DISCUSSION

In recent years, the field of wastewater treatment has witnessed significant advancements with the integration of AI algorithms offering innovative approaches to optimize the treatment processes and enhance overall efficiency. With the use of AI, researchers and practitioners have been able to tackle complex challenges associated with wastewater treatment, such as pollutant removal, resource recovery, and process optimization.

The four main AI algorithms discussed in this chapter are usually used in addressing these challenges (Table 13.1). These include neural networks, genetic algorithms, and fuzzy logic. Where neural networks, which are inspired by the human brain, excel in complex pattern recognition tasks and have proven effective in predicting treatment performance and optimizing process parameters, genetic algorithms mimic the process of natural evolution to find optimal solutions for complex optimization problems. Fuzzy logic, in contrast, deals with uncertainty and imprecision in data, making it suitable for handling the inherent variability and complexity of water treatment systems.

TABLE 13.1

Comparison of AI Algorithms

Serial Number	Dye/Metal	Adsorbent	AI Algorithm	Model Performance	Studied By
1.	Basic red	Walnut husk	ANN	$R^2 = 0.9991$, SSE = 0.2303	Çelekli et al. (2016)
2.	CG	Copper sulphide nanoparticles loaded on activated carbon	RF	$R^2 = 0.9657$, MSE = 0.0021	Bagheri et al. (2015)
3.	Cr (VI)	Clay-based adsorbents	ANFIS	$R^2 = 0.9997$, MSE = 1.288E-06	Foroutan et al. (2020)
4.	Chlorophenol	Coconut shell carbon	RBFN and MLPN	$R^2 = 0.96$, MSE = 6.03	K. P. Singh et al. (2013)

ANFIS = adaptive neuro-fuzzy inference system; ANN = artificial neural network; MLPN = multilayer perceptron network; MSE = mean square error; RBFN = radial basis function network; SSE = sum of squared errors.

R^2 represents how well the independent variable(s) can predict or explain the variability in the dependent variable. It ranges between 0 and 1, where 0 indicates that the independent variable(s) have no explanatory power, and 1 indicates a perfect fit where the independent variable(s) can explain all the variability in the dependent variable. Since the models have R^2 values of more than 0.9, it suggests a stronger relationship between the variables and a better fit of the regression model. Thus, we can conclude that the ANN and ANFIS work the best among the other algorithms.

While AI algorithms perform wonders, there are a few limitations that can be observed regarding treating wastewater. For example, the availability of the data is poor. Moreover, the selection of the data is concerning as well in addition to poor reproducibility and less evidence of applications in real water treatment. These challenges can be tackled with more experiments and employment of better data sampling techniques, such as random sampling, stratified sampling, and so on.

13.7 CONCLUSION

AI hybrid techniques can be considered as more efficient approaches that are being employed extensively not just for predicting the removal of various pollutants but that can also be employed for the purpose of water quality analysis, process optimisation, and autonomous decision-making. As discussed, algorithms such as ANN, ANFIS, random forests, and RBFN are most commonly used; ANN is known to perform slightly better to predict the best adsorbent to treat water in optimal cost and time. Other than these algorithms, variations of genetic algorithms, such as ANN-GA, RSM-GA, or feed-forward backpropagations with ANN can be used for the detection and optimal solution for the treatment of water. Hence, traditional wastewater management techniques with the help of AI are beneficial in delivering better outcomes.

REFERENCES

Alam, G., Ihsanullah, I., Naushad, M., & Sillanpää, M. 2022. Applications of artificial intelligence in water treatment for optimization and automation of adsorption processes: Recent advances and prospects, *J. Chem. Eng.* 427, 130011. https://doi.org/10.1016/j.cej.2021.130011

Albatrni, H., & Qiblawey, H. 2020. Comparison study between adsorption and membrane technologies for the removal of mercury, *Sep. Purif. Technol.* 257, 117833. https://doi.org/10.1016/j.seppur.2020.117833.

Amaral, A., Gillot, S., Garrido-Baserba, M., Filali, A., Karpinska, A. M., Plósz, B. G., . . . & Rosso, D. 2019. Modelling gas–liquid mass transfer in wastewater treatment: When current knowledge needs to encounter engineering practice and vice versa, *Water Sci. Technol.* 80(4), 607–619.

Ashfaq, M., & Al-Ghouti, M. 2022. Reverse osmosis membrane fouling and its physical, chemical, and biological characterization. https://doi.org/10.1016/B978-0-323-89977-2.00008-7.

Bagheri, A. R., Ghaedi, M., Hajati, S., Ghaedi, A. M., Goudarzi, A., & Asfaram, A. 2015. Random forest model for the ultrasonic-assisted removal of chrysoidine G by copper sulfide nanoparticles loaded on activated carbon; response surface methodology approach, *RSC Adv.* 5, 59335–59343. https://doi.org/10.1039/C5RA08399K.

Çelekli, A., Bozkurt, H., & Geyik, F. 2016. Artificial neural network and genetic algorithms for modeling of removal of an azo dye on walnut husk, *Desalin. Water Treat.* 57, 15580–15591. https://doi.org/10.1080/19443994.2015.1070759.

Dhote, J., Ingole, S., & Chavhan, A. 2012. Review on wastewater treatment technologies, *Int. J. Eng. Res. Technol.* 1(5), 1–10.

Foroutan, R., Peighambardoust, S. J., Mohammadi, R., Omidvar, M., Sorial, G. A., & Ramavandi, B. 2020. Influence of chitosan and magnetic iron nanoparticles on chromium adsorption behavior of natural clay: Adaptive neuro-fuzzy inference modeling, *Int. J. Biol. Macromol.* 151, 355–365. https://doi.org/10.1016/j.ijbiomac.2020.02.202.

Gleick, Peter H., & Cooley, H. 2021. Freshwater scarcity, *Annu. Rev. Environ. Resour.* 46, 319–348, SSRN: https://ssrn.com/abstract=3953089 or http://dx.doi.org/10.1146/annurev-environ-012220-101319

Hogeboom, R. J. 2020. The water footprint concept and water's grand environmental challenges, *One Earth* 2(3), 218–222, ISSN 2590–3322. https://doi.org/10.1016/j.oneear.2020.02.010.

Mojiri, A., & Bashir, M. 2022. Wastewater treatment: Current and future techniques, *Water* 14, 448. https://doi.org/10.3390/w14030448

Popoola, L. 2019. Nano-magnetic walnut shell-rice husk for Cd(II) sorption: Design and optimization using artificial intelligence and design expert, *Heliyon* 5, e02381. https://doi.org/10.1016/j.heliyon.2019.e02381.

Quach-Cu, J., Herrera-Lynch, B., Marciniak, C., Adams, S., Simmerman, A., & Reinke, R. A. 2018. The effect of primary, secondary, and tertiary wastewater treatment processes on antibiotic resistance gene (ARG) concentrations in solid and dissolved wastewater fractions, *Water* 10(1), 37. https://doi.org/10.3390/w10010037.

Sahoo, P. K., & Kim, K. 2013. A review of the arsenic concentration in paddy rice from the perspective of geoscience, *J. Geosci.* 17, 107–122.

Singh, K. P., Gupta, S., Ojha, P., & Rai, P. 2013. Predicting adsorptive removal of chlorophenol from aqueous solution using artificial intelligence based modeling approaches, *Environ. Sci. Pollut. Res.* 20, 2271–2287. https://doi.org/10.1007/s11356-012-1102-y.

Singh, S. N., Srivastava, G., & Bhatt, A. 2012. Physicochemical determination of pollutants in wastewater in Dheradun, *Curr. World Environ.* 7, 133–138.

Singh, S., Yadav, R., & Singh, A. N. 2023. Applications of waste-to-economy practices in the urban wastewater sector: Implications for ecosystem, human health and environment,

Waste Management and Resource Recycling in the Developing World, Elsevier eBooks, Pages 625–646. https://doi.org/10.1016/b978-0-323-90463-6.00010-5

Suraj Kumar Bhagat, Karl Ezra Pilario, Olusola Emmanuel Babalola, Tiyasha Tiyasha, Muhammad Yaqub, Chijioke Elijah Onu, Konstantina Pyrgaki, Mayadah W. Falah, Ali H. Jawad, Dina Ali Yaseen, Noureddine Barka, Zaher Mundher Yaseen, 2023. Comprehensive review on machine learning methodologies for modeling dye removal processes in wastewater, Journal of Cleaner Production,Volume 385,135522,ISSN 0959-6526,https://doi.org/10.1016/j.jclepro.2022.135522.

Yılmaz (Şişman), N. A. 2003. A temporal neuro-fuzzy approach for time-series analysis (PhD—Doctoral Program). Middle East Technical University.

Yousef, R., Qiblawey, H., & El-Naas, M. H. 2020. Adsorption as a process for produced water treatment: A review, *Processes* 8(12), 1657. https://doi.org/10.3390/pr8121657

14 Application of Machine Learning Algorithms in the Field of Bioacoustics

Fiona Ghosh, Niranjana Nair, Pranav Elayadam, and Xavier Savarimuthu, S.J.

14.1 INTRODUCTION

The environment has come under immense pressure due to various anthropogenic activities. Our actions, including land degradation, deforestation, and pollution, have had an irreversible, deleterious effect on our environment. According to research published in January 2016 on the climatic, biological, and geochemical traces of human activity found in sediments and ice cores, it was recommended that the period from the middle of the 20th century should be recognized as a separate geological epoch from the Holocene, coined unofficially as the Anthropocene (Waters et al. 2016). This highlights the extent of influence on our planet in the short span of our existence. One of the crucial components that have been adversely affected is biodiversity. According to the Convention on Biological Diversity (CBD), the term "biological diversity" refers to the variety of living things from all sources, such as terrestrial, marine, and other aquatic habitats, as well as the ecological communities to which they belong. This encompasses diversity within and between species, as well as a diversity of ecosystems. In short, biodiversity is life and what makes life possible on our planet. Biodiversity loss is a major environmental issue affecting the planet. Worldwide, populations of mammals, birds, fish, reptiles, and amphibians have decreased by 68% since 1970, according to World Wildlife Fund (WWF) research (Almond et al. 2020). The main drivers of this decline are habitat loss, overexploitation, climate change, and pollution. In addition, a study published in the journal *Nature* in 2019 revealed that insect populations are declining at an alarming rate of 2.5% a year, which could have devastating consequences for ecosystems that rely on insects for pollination, pest control, and decomposition (Wagner et al. 2021). The decline of biodiversity has effects on society and the economy, in addition to the natural environment. For example, crop production might decrease due to the loss of pollinators, and human livelihoods could be impacted by the loss of natural resources like water and timber. Biodiversity monitoring is critical for understanding the health of ecosystems and informing conservation efforts. By tracking changes in species populations and distributions, scientists can identify areas of concern and develop targeted strategies to protect biodiversity. Studies have shown that effective monitoring programs can improve conservation outcomes. A study published

DOI: 10.1201/9781003405436-17

in *Nature Communications* demonstrated how monitoring can help identify areas of high conservation value and guide habitat restoration efforts (Venter et al. 2016). Monitoring can also provide valuable data on the impact of environmental stressors such as climate change and pollution on biodiversity (O'Connor et al. 2020). This information is essential for developing effective mitigation and adaptation strategies. In short, biodiversity monitoring is essential for ensuring the long-term survival of species and ecosystems.

A method of biodiversity monitoring is through bioacoustics. Bioacoustics is an interdisciplinary field that focuses on the study of sound production, propagation, and reception in animals, particularly in the context of their natural environments. It combines elements of biology, ecology, acoustics, and data analysis to understand the acoustic behaviours of organisms and their ecological implications. Bioacoustics has gained significant importance in recent years as a powerful tool for biodiversity monitoring, species conservation, and ecological research. Traditionally, bioacoustics has involved manual data collection and analysis, which can be time-consuming, expensive, and prone to errors. However, advancements in artificial intelligence (AI) offer a promising solution to this problem. AI can help automate bioacoustics studies such as species identification, habitat mapping, and population tracking. Large datasets may be analyzed by machine learning algorithms to find patterns that would be challenging for people to see. As a result, better conservation policies and the eventual protection of our planet's natural resources might result from an increase in the precision and effectiveness of biodiversity monitoring. To prevent unforeseen outcomes, it is crucial to make sure that these technologies are utilized morally and responsibly.

In this chapter, we review the different algorithms used in various research works on bioacoustics conducted using different datasets, methods, and objectives. We compared research articles and papers where mosquito, frog, and cicada sounds were analyzed using various algorithms, including convolutional neural networks (CNNs), linear discriminant analysis, decision trees, and support vector machines (SVMs). The accuracy rates provided by each algorithm were compared and analyzed to narrow down the most suitable algorithm for use in wildlife conservation measures. Advancements in AI offer great potential for improving biodiversity monitoring and conservation efforts. By automating data collection and analysis processes, AI can greatly improve the efficiency and accuracy of biodiversity monitoring, allowing for more targeted conservation strategies to protect our planet's natural resources.

14.2 DATA COLLECTION

The first step in a bioacoustics study is to collect a large number of sound recordings from the specific animal species of interest. Researchers use special devices like hydrophones or microphones, either in enclosed spaces or outdoors, to capture these sounds. This requires careful planning and preparation, selecting the right tools and settings, processing the recordings to reduce background noise, and storing the data securely.

Gathering the data may involve working in challenging or remote areas, and researchers must follow laws and ethical guidelines when studying wildlife.

Sometimes, additional information needs to be added to the recorded data, either manually or using machine learning techniques.

In a study, flight tones from both laboratory-grown and wild-caught mosquitoes were recorded using the microphone Telinga EM-23 (Noda et al. 2016). Another study used a database called birdclef2021 (Murakami et al. 2021) to compare CNN and decision tree (DT) algorithms. Different recording standards were followed in another research work, where an automated digital recording system was used, consisting of a Nomad Jukebox 3 digital mp3 player and recorder, along with a Sony ECM-MS907 microphone. The recorder collected one minute of sound every 30 minutes for five consecutive days at each site (Acevedo et al. 2009).

14.3   DATA CLEANING

In bioacoustics, data cleaning is a vital process where researchers remove any incorrect or irrelevant signals from the collected sound data to ensure it is accurate and of high quality. This involves getting rid of background noise, like wind or other animal sounds, and identifying and categorizing the sounds of interest, such as vocalizations from a specific species. The goal is to improve the accuracy of the data.

After cleaning the data, researchers check its reliability, standardize the format, and carefully examine it for errors or inconsistencies. Data cleaning is essential because it ensures that the subsequent data analysis is meaningful and can effectively address the research question or problem at hand.

A method for data cleaning involves isolating animal sounds from other noises, and a research study used the software tool called Raven Pro 1.2.1 to accomplish this (Acevedo et al. 2009).

14.4   DATA EXPLORATION

Data exploration methodology refers to the process of studying the recorded sound data to understand its properties and features better. This involves presenting the data in different ways, like graphs or charts, to spot patterns or trends. It also involves identifying which acoustic features, such as pitch or loudness, are essential for addressing the research question or problem. The primary goal of data exploration is to gain insights into the dependable data and determine the most suitable data analysis methods for achieving the research objectives. In this study the data has been represented with the help of diagrams, charts, and tables.

14.5   DATA MODELING

The method of using machine learning and deep learning algorithms to create models that can predict or categorize sounds based on their acoustic properties is known as data modeling methodology in the field of bioacoustics. Selecting the appropriate machine learning algorithm is necessary for this. CNN, DT, SVM, linear discriminant analysis (LDA), are a few examples of machine learning techniques that have been used in many papers discussed here. The research subject or query, as well as the qualities of the sound data, influence the algorithm selection. Separating the sound data into training and testing sets, choosing and extracting relevant acoustic features, training and evaluating the model, optimizing it to increase performance, and verifying the final model are further steps involved in data modeling. The goal of data modeling is to create models that can correctly forecast or categorize noises, which can provide information about the behaviour and ecology of various species and contribute to conservation efforts.

14.5.1 CONVOLUTIONAL NEURAL NETWORK

Architecture of convolutional neural network is shown in Figure 14.1. CNN is a deep learning method that can be used in bioacoustics research to analyze animal vocalizations. CNNs have been used in a variety of bioacoustics investigations, including for the classification of bird songs, the identification of frog calls, and the analysis of whale vocalizations. Because they can learn hierarchical representations of the acoustic features that are crucial for differentiating between various call types, CNNs are particularly advantageous for analyzing bioacoustics data. CNNs can recognize intricate patterns and structures in the auditory data and use these features to produce precise classifications by applying a sequence of convolutional and pooling layers.

For example, if researchers are studying the vocalizations of different bird species, they might use CNNs to identify the combination of frequency and temporal features that best distinguishes between different species. The CNN model would then be used to classify new vocalizations based on their acoustic features, allowing researchers to accurately identify the species of the calling bird.

The capacity of CNNs to learn features directly from the unprocessed audio input, without the need for considerable feature engineering, is a benefit in bioacoustics research. Additionally, by adjusting the model's architecture and hyperparameters, CNNs can be applied to a variety of bioacoustic classification issues. A CNN model's

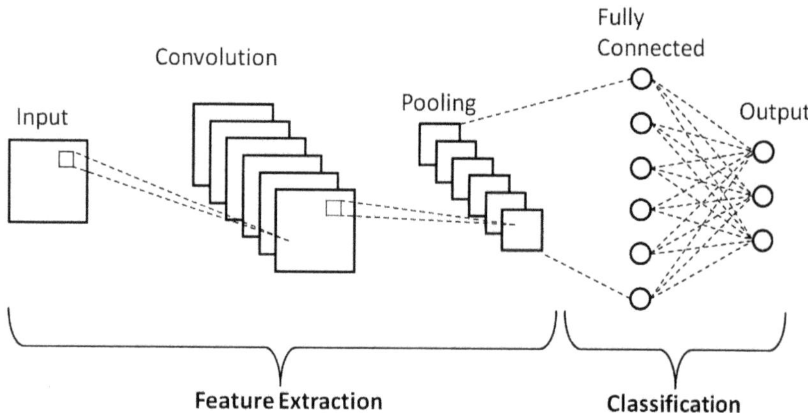

FIGURE 14.1 Architecture of a convolutional neural network.

(Source: https://www.upgrad.com/blog/basic-cnn-architecture/)

training and optimization, however, can be computationally expensive and necessitate a significant amount of labeled data. Additionally, the interpretability of CNNs may be constrained because the learned features might not always match the vocalizations' useful acoustic characteristics.

In one study, particularly, a region-based (R)-CNN employed a variety of large capacity CNNs and extracted the characteristics from numerous fully connected layers, a voting process that classifies calls as a whole by first classifying each syllable of a call, then choosing the species that received the most votes to classify the call as a whole. We obtained a max classification of 88.45% using this method and the features derived from layer fc-rcnn, with a mean and median of 76.61% and 76.84%, respectively (Strout et al. 2017).

14.5.2 Support Vector Machine

Support vector machines (SVMs), a well-liked machine learning method, are utilized in bioacoustics investigations to analyze animal vocalizations based on their acoustic characteristics is shown in Figure 14.2. SVMs operate by locating a hyperplane that maximizes the distance between various vocalization classes. Selecting a set of acoustic features that are pertinent to the classification issue, such as frequency, duration, or spectral shape, is often the first step taken by researchers using SVMs in bioacoustics research. After that, they would train an SVM model with these attributes to identify various call kinds based on their acoustic properties.

For example, if researchers were studying the vocalizations of different marine mammal species, they might use SVM to identify the combination of spectral and temporal features that best distinguishes between different species. The SVM model

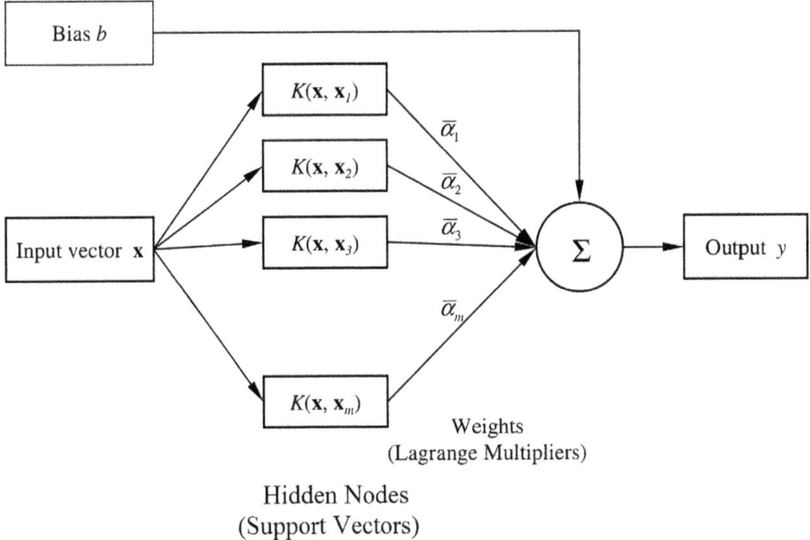

FIGURE 14.2   Architecture of a support vector machine.

would then be used to classify new vocalizations based on their acoustic features, allowing researchers to accurately identify the species of the calling animal. SVM has the advantage of handling nonlinear correlations between the acoustic features and the class labels, which enables it to precisely categorize intricate patterns in animal vocalizations. SVMs can be modified to handle unbalanced datasets, which are typical in bioacoustics research, and are also reasonably resistant to overfitting.

SVMs do, however, have significant drawbacks, including sensitivity to the kernel function selection and the possibility for overfitting if the number of features is very high in comparison to the number of observations. SVMs in bioacoustics research depend on careful feature selection, suitable model tuning, and validation on different datasets, much like any machine learning technique.

14.5.3   Linear Discriminant Analysis

LDA is a statistical method used in bioacoustics investigations to categorize and examine animal vocalizations based on their acoustic characteristics. LDA finds a linear combination of features that maximizes the distance between various vocalization groups. Selecting a set of acoustic features that are important to the classification problem, such as frequency, duration, or spectral shape, is often the first step taken by researchers using LDA in bioacoustics research. After that, they would train an LDA model with these features to identify various call kinds based on their acoustic properties.

The combination of frequency and duration parameters that best distinguishes between various species may be found using LDA, for example, if researchers were examining the vocalizations of several bird species. The LDA model would then be used to categorize fresh vocalizations according to their acoustic characteristics, enabling researchers to precisely identify the calling bird's species. One benefit of LDA is that it has a long history of successful use in a range of species and circumstances. It is also a well-known and commonly utilized technique in bioacoustics research. LDA is an effective tool for researchers who are new to the subject of bioacoustics because it is also fairly straightforward to use and interpret. However, LDA has some drawbacks, including the linearity assumption, which might not always be true, and the risk of overfitting if the number of features is excessively high in comparison to the number of observations.

14.5.4   Decision Tree

Figure 14.3 shows the architecture of decision tree. In bioacoustics studies, decision trees are a type of machine learning algorithm that can be used to classify and analyze animal vocalizations. Each decision is represented by a node in the decision tree, and each outcome is represented by a branch. A decision tree is a graphical representation of a series of decisions and their related outcomes. Researchers may begin by specifying a set of features that are pertinent to the classification problem, such as frequency, duration, or spectral structure, to construct a decision tree for analyzing animal vocalizations. The decision tree algorithm would then be trained using these features to distinguish between various call kinds based on their acoustic properties.

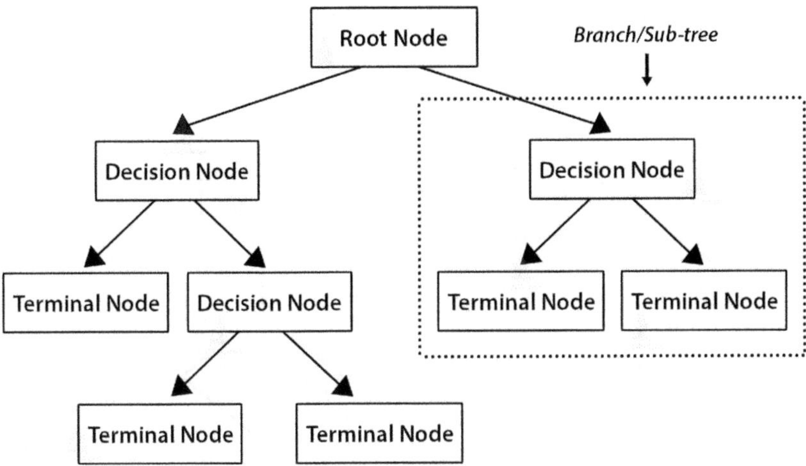

FIGURE 14.3   The architecture of a decision tree.

(Source: https://aiplanet.com/blog/introduction-to-decision-tree-algorithm/)

A decision tree for identifying bird songs, for instance, might comprise nodes for judgments based on the frequency, length, and presence or absence of specific harmonic or modulation properties. The algorithm would choose a different branch in the tree depending on how each decision turned out, eventually classifying the call as belonging to a certain bird species or call type. Determining which features are most crucial for differentiating between various call types and understanding the underlying structure of the classification problem are both made possible by decision trees' advantage of being simple to interpret and visualize.

Decision trees are another popular machine learning algorithm that can be visualized to understand the decision-making process. By visualizing the tree structure and the feature importance, we can understand which features are most informative for the classification task.

In summary, data visualization plays a crucial role in understanding the behaviour of different machine learning algorithms, including SVM, CNN, BNN, LDA, and decision trees. By visualizing the internal workings of these algorithms, researchers can gain insights into how they work, and make more informed decisions about their implementation and optimization.

14.6   FILTERS TO REMOVE NOISE

14.6.1   CONVOLUTIONAL NEURAL NETWORK

Studies have proved that in many situations the CNN model has been used for voice detection in animals. In one study, approximately 897 classification records were considered. To effectively record and to analyze the sound from the cattle, sound monitoring devices and systems were configured. These monitors are competent to record sounds through microphones and microcontrollers. The model consisted of

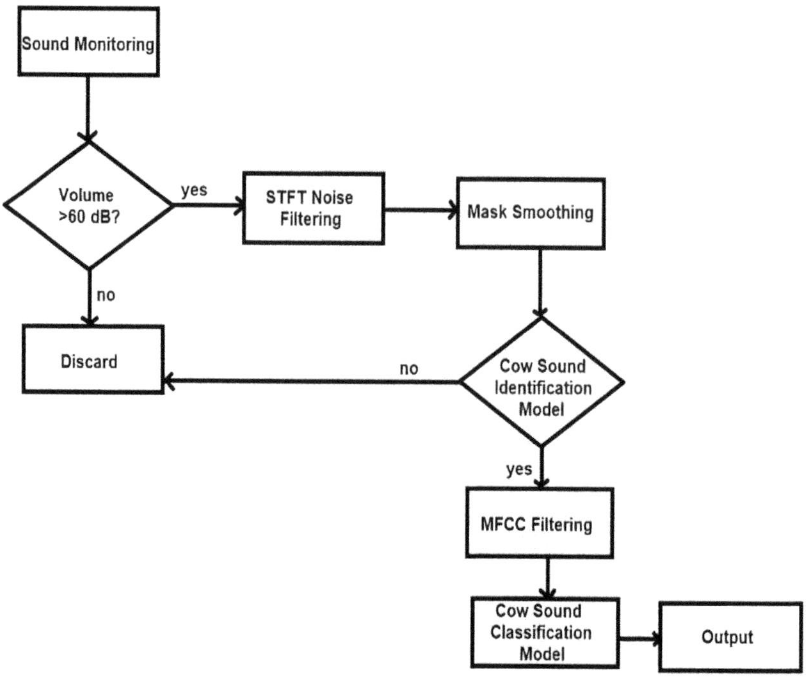

FIGURE 14.4 Flow chart showing the collection of audio data and noise filtering in the on-site monitoring system.

two activation functions: Softmax and a rectified linear aviation unit (Relu). The equation used by the CNN to train the model with weights is:

$$qt+1 = qt + Dqt \tag{14.1}$$

To remove noise from the data, a short-term Fourier transform (STFT) filter was used. This filter helps in removing and eliminating noises from the sound signal and improves its quality. The computation of STFTs is done by splitting the time signal into equal lengths of smaller groups and computing the Fourier transform separately on each of them and so the Fourier spectrum is revealed for each smaller part. In the recording system only sounds that are higher than 60 dB are stored, to segregate other voices from the cattle's sounds. There were 2,048 audio frames that were used between the STFT columns. The hop size used was 256 and the length of the window was assigned to be 2,048 (Jung et al. 2021).

In Figure 14.4, flowchart shows the steps that are involved to remove unwanted noise from the data and all the criteria based on which the filtering is done, and Table 14.1 shows the accuracy of CNN model for cattle vocalization classification without noise filter (Jung et al. 2021).

TABLE 14.1

Accuracy of Convolutional Neural Network Model for Cattle Vocalization Classification without Noise Filter

	Without Noise Filtering	With Noise Filtering
True positive	140	141
False positive	12	11
True negative	189	199
False egative	19	10
True recognition rate (%)	92.10	92.76
False recognition rate (%)	90.86	95.21
Accuracy (%)	91.38	94.18

14.6.2 Decision Tree

In a study conducted it was concluded that decision trees are one of the models that can be used in bioacoustics. Studies have proved that decision trees have been useful in the field of bioacoustics. In this study reference has been made to the voice of a vast number of types of nocturnal animals. For this study there were around 408 sound samples taken into consideration of 18 species. To verify the accuracy rate, 4,477 feature segmentations were made use of. To identify different kinds of nocturnal animal species on the basis of audio signals that were recorded, a system to automatically identify mixed nocturnal animal vocalizations was put forward.

The sound signal that was recorded was saved in a 16-bit mono format and it was resampled at 44.1 kHz. For a good flow in further processing, each sound signal's amplitude was normalized within the range [-1,1]. To filter the noise during the signal analysis a de-noise filter has been applied. After this is done the de-noise parameters need to be retrieved. To retrieve the de-noise parameters, Fourier analysis and spectral noise gating are applied. This step also helps to "purify" the data which is present in the noisy environment. To perform the signal preprocessing step, the block thresholding method was applied. An STFT was calculated by the time-frequency de-noising procedures of the noisy signals. In order to weaken the noise signals, these resulting coefficients were operated in Table 14.2 (Chen et al. 2015), and the Figure 14.5 shows that the identification system architecture for nocturnal animal vocalization (Chen et al. 2015).

14.6.3 Linear Discriminant Analysis

According to previous studies conducted it has been observed that in bioacoustics, a linear discriminant model is one such technique that can be used. The signals are taken as inputs for the prediction. These inputs are broken down into a set of syllables. For the purpose of recognition, the syllables serve as the basic acoustic units.

TABLE 14.2

Recognition Results of Nocturnal Animals

Species	True Positive Rate	False Positive Rate	Precision	Recall	F-Measure
Hylachinesis	1	0.001	0.952	1	0.976
Rhacophorus taipeianus	0.882	0.003	0.928	0.886	0.901
Microhylafissipes	0.876	0.018	0.872	0.878	0.869
Buergeria japonica	0.798	0.017	0.835	0.798	0.812
Fejervaryalimnocharis	0.919	0.011	0.922	0.918	0.928
Rhacophorus prasinatus	0.861	0.009	0.875	0.859	0.862
Kaloula pulchra	0.864	0.001	0.912	0.861	0.886
Hoplobatrachusrugulosus	1	0.001	0.92	1	0.958
Babina okinavana	0.781	0.023	0.819	0.762	0.799
Hylaranaguentheri	0.922	0.008	0.832	0.924	0.875
Kurixalusidiotocus	0.942	0.023	0.892	0.941	0.925
Caprimulgus affinis	1	0	1	1	1
Apus affinis	0.667	0	1	0.667	0.8
Tannasozanensis	1	0	1	1	1
Cryptotympanatakasagona	1	0	1	1	1
Otus spilocephalus	0.842	0	1	0.842	0.914
Gryllusbimaculatus	1	0	1	1	1
Gorsachiusmelanolophus	1	0	1	1	1

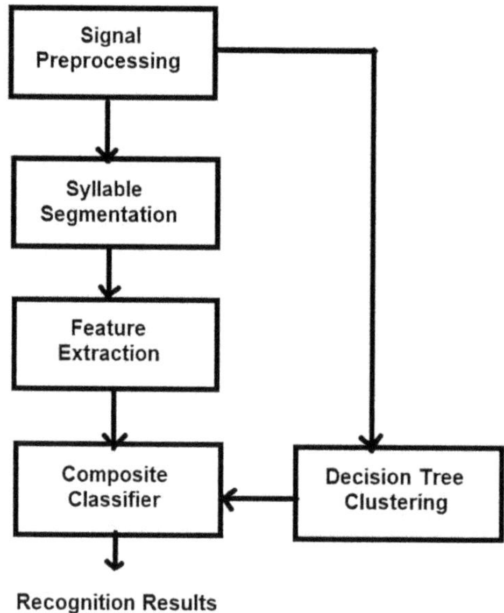

FIGURE 14.5 Identification system architecture for nocturnal animal vocalization.

Step 1: STFT is used to calculate the spectrogram of the inputs, which are bioacoustic signals.

Step 2: Begin with n = 0.

Step 3: Identify f_n and t_n such that $|S(f_n,t_n)| \geq |S(f, t)|$, for all pairs of (f, t). Assign the position of the nth syllable as (f_n,t_n).

Step 4: Calculate the amplitude A_n (0) in decibels using the formula 20 log10 |S $(f_n,t_n)|$ and set the frequency parameter W_n (0) as f_n. If A_n (0) < A_0 (0) — 20 dB, stop the segmentation process as the amplitude of the nth syllable is too small, and there is no need to extract more syllables.

Step 5: Starting from (f_n,t_n), track the highest peak of |S(f, t)| for t <, t_n until A_n (t) < A_n (0)—β dB, where β is the stopping criterion (default value: 20). Similarly, track the highest peak of |S(f, t)| for t < t_n until A_n (t) < A_n (0)—β dB. This step defines the starting time $(t_n - t_s)$ and ending time $(t_n + t_e)$ of the nth syllable around t_n.

Step 6: Store the amplitude trajectory of the nth syllable as a function A_n (s), where s varies from $(t_n - t_s)$ to $(t_n + t_e)$.

Step 7: Remove the region of the nth syllable by setting $S(f,[(t_n - t_s),t_n + t_e])$ to 0. Increment n by 1 and return to Step 3 to find the next syllable (Lee et al. 2006).

14.6.4 SUPPORT VECTOR MACHINE

SVM is another model that can be used in the field of bioacoustics as proven by early research is shown in Figure 14.6. This study described how SVM has been utilized in bioacoustics. The sound collection InsectSingers dataset has been used to get the

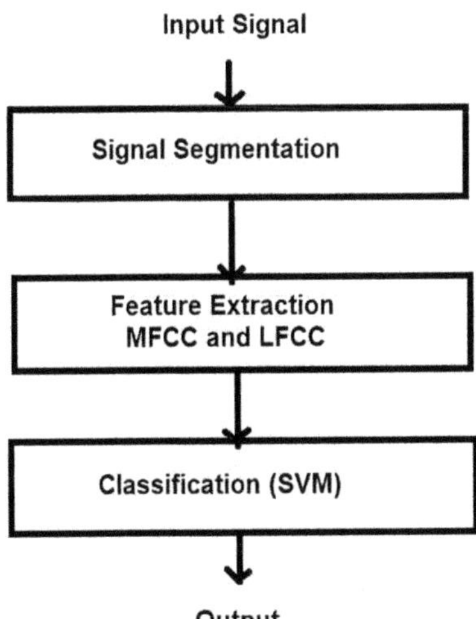

FIGURE 14.6 Steps involved in the classification process using a support vector machine (SVM). LFCC = linear frequency cepstral coefficient; MFCC = mel-frequency cepstral coefficient.

digital insect sound samples for the purpose of prediction. The data was taken from Canada and the United States. The samples consisted of 90 species and subspecies, from which only 88 were used in the experiment. These species were labeled with the help of biologists. The sound files were of an average of 15 seconds time span approximately. It was either at 44.1 kHz or 48 kHz sampling rate and in the MP3 format. To extract the insect calls from the background noise, automatic segmentation of each input audio file was done. The automatic signal segmentation was done using STFT. Each acoustic signal was divided into individual calls or syllables in this segmentation as was considered a unit for recognition purpose. Hamming window of 256 samples was considered to calculate the spectrum of the signal. There was an overlap of 35%.

The resulting spectrogram, represented by a time-frequency matrix M(f, t), is used in the following algorithm. First, the algorithm searches for the indices f_n and t_n in the matrix M, where $|M(f_n, t_n)|$ has the maximum magnitude among all elements. The position of the nth call is then set to t_n. The algorithm proceeds to calculate the amplitude of the first segment as follows:

Starting from t_n, it looks for the highest peaks of $|M(f, t)|$ for $t > t_n$ and $t < t_n$. Here, A_n represents the amplitude values, and t_0 is the initial time. The algorithm considers a condition where $(t_n - t_0)$ multiplied by a certain factor is less than the amplitude at time t_0 minus a specified threshold in decibels.

This process is repeated for each insect song in the database until the end of the spectrogram is reached. The resulting amplitude trajectories are stored as the nth call and are set for further analysis or processing.

$$M\left(f_n, [t_n - t_s, \dots, t_n + t_e]\right) = 0 \tag{14.2}$$

There were about 8,393 syllables derived from the database with 95 syllables by class (Noda et al. 2016).

14.7 METHODS OF FEATURE EXTRACTION

14.7.1 CONVOLUTIONAL NEURAL NETWORK

In past studies for the purpose of feature extraction, the mel-frequency cepstral coefficient (MFCC) technique has been used. MFCC was used to visualize the acoustic data that was preprocessed. The mel scale helps to represent the short-term power spectrum of a sound. A type of cepstral representation of an audio clip is used to derive the MFCCs. In the mel scale, the frequency bands are placed at equal intervals that are helpful in that they resemble the response of the human auditory system from a closer view. The mel filter bank is made use of to perform this task. The mel filter bank consists of triangular filters that are distributed on the mel scale. A filter bank was obtained for each speech signal window. The filter that was obtained first was very narrow. It indicated the energy present near 0 Hz. The filters keep expanding as the frequencies increase and the variations keep reducing. This study used 128 filter bands, with a time frame of 2,000. The sound data were allocated the value 0 if sound samples were less than 2,000. Most of the samples were recorded between the range of 5 s and 10 s. As an input to the CNN, 2,000 frames were considered to be enough, with a time equivalent to 20 s (Jung et al. 2021).

14.7.2 DECISION TREE

For the purpose of feature extraction, MFCC has been used in recent studies to get a representation of a new voice. This process has proved to be less redundant, more compact, and better suited for statistical modeling. The human ear's bandwidths which are critical yet known with reference to the frequency are the parameters on which the MFCC is based. In this the space is filtered at low frequencies and at high frequencies logarithmically. The MFCC helps for the important characteristics to be captured.

Steps used to implement MFCC are:

1. **Pre-emphasis**

   This filter is used so that before spectral analysis the speech spectrum can be flattened. It helps to make up for those speech signals that during the human sound production mechanism get overpowered. These are the high frequency parts of the speech signals. The filter that is mostly used is a high-pass finite impulse response (FIR) filter.

2. **Frame blocking**

   This is a framing process. It is the voice signal of the producer to which it is applied first. The signal is divided into N segments or frames, enabling the analysis of individual frames that can be represented by a single feature vector.

3. **Windowing**

   In the processing this is the third stage. Here each frame is windowed so that the signal discontinuities can be reduced. These are speech signals that are present in each frame at the edges. Hamming window is the window that is used the most.

4. **Fast Fourier transform**

   Fast Fourier transform (FFT) is the fourth stage in the process. In this the time domain is converted to frequency domain of each N samples. It is a computationally efficient algorithm of the discrete Fourier transform (DFT).

5. **Triangular bandpass filters**

   Here each triangular filter band's energy is calculated.

6. **Discrete cosine transform (DCT)**

   On the 20 log energy obtained from triangular bandpass filters the DCT is applied. It helps to get the L mel-scale cepstral coefficient (Chen et al. 2015).

14.7.3 LINEAR DISCRIMINANT ANALYSIS

In this case for the purpose of feature selection MFCC has been used according to studies. MFCC can represent in a compact form the signal spectrum, thus it has been used. In automatic speech recognition it is very useful. The input signal that is received is first partitioned into several frames. In each of these frames the MFCC is calculated, and they are labeled as the frame's features. The problem that arises here is that different syllables have a different number of frames. To solve this problem, the average of the MFCC of the frames in a syllable are calculated. This is used as a feature to represent the syllable. This ensures that the features are of a fixed number whatever might be the length of the acoustic syllable.

Steps to derive the MFCC:

Step 1: Pre-emphasis
   Apply a pre-emphasis equation to the input syllable signal, denoted as s[n], using a typical value of 0.95 for the pre-emphasis coefficient (â).
Step 2: Framing
   Divide each syllable into a series of overlapping frames, where the frame size is N samples and the overlap size is M samples between successive frames. This ensures that consecutive frames have minimal changes. In this case, N is 512 and M is 256.
Step 3: Windowing
   To reduce discontinuities at the ends of each frame, apply a Hamming window function to each frame by multiplying it element-wise with the frame samples.
Step 4: Spectral analysis
   Compute the Discrete Fourier Transform (DFT) of each frame using the Fast Fourier Transform (FFT) algorithm.
Step 5: Bandpass filtering
   Apply a set of triangular bandpass filters to the amplitude spectrum obtained from the previous step. Each filter corresponds to a specific frequency range, and the resulting amplitudes are calculated.
Step 6: DCT
   Compute the MFCCs for each frame by performing a DCT on the logarithm of the filtered amplitudes. The MFCCs are calculated using a formula that involves the number of filters (J), the filter coefficients, and the amplitude values. The resulting MFCCs form a feature vector of length L, which is 15 in this case.

In the proposed method, the filter bank comprises 25 triangular filters, and each frame is represented by a 15-dimensional MFCC feature vector (Lee et al. 2006).

14.7.4 Support Vector Machine

For the purpose of feature extraction, MFCCs and linear frequency cepstral coefficients (LFCCs) have been used in this study due to their good performance and easy implementation. To compare their features for each syllable, two vectors of features were extracted. With the application of 40 triangular bandpass filters MFCCs were mapped to the mel scale. DCT was used to get the MFCC features.

$$MFCC_n = \sum_{i=1}^{M}\left(\log|Y_i| \, .\cos\left[n(i-0.5)\frac{\pi}{M} \right] \right) \tag{14.3}$$

$$0 \leq n \leq N - 1$$

TABLE 14.3
Classification Results

Features	Accuracy Mean % ± std	F-Measure
MFCCs	98.01% ± 5.25	0.9878
LFCCs	99.08% ± 2.97	0.9926
MFCC+ΔMFCC	98.43% ± 4.63	0.9870
LFCC+ΔLFCC	98.94% ± 3.34	0.9909
MFCC+ΔMFCC+ΔΔMFCC	98.52% ± 4.43	0.9884
LFCC+ΔLFCC+ΔΔLFCC	98.71% ± 3.79	0.9887

In the context provided, *M* represents the number of triangular filters used in the feature extraction process. *N* refers to the number of cepstral coefficients, which are the resulting coefficients obtained after applying the triangular filters to the input signal. Each triangular filter output is denoted as Y_i, representing the output of the *i*th filter.

LFCCs are calculated similarly to MFCCs, but they utilize a triangular filter bank with a linear scale instead of mel-warped filters. The number of LFCC coefficients to extract has been determined through experimentation to achieve the highest identification rate, resulting in 18 coefficients extracted for both MFCCs and LFCCs.

In addition, first- and second-order regression coefficients, also known as deltas, are calculated for each feature element (C_n) using an equation that involves the neighbouring feature values. The specific equation for calculating deltas may vary, but it typically considers the adjacent feature elements to estimate the rate of change or variation within the sequence of feature values and the classification results are in Table 14.3 (Noda et al. 2016).

$$D_j = \frac{\sum_{n=1}^{N} \left(c_{n+i} - c_{n-i} \right)}{2\sum_{n=1}^{N} n^2} \tag{14.4}$$

14.8 DISCUSSION

In this chapter we review the implementation of four different types of algorithms that can be used in the field of bioacoustics. In all the algorithms it is observed that the results have improved after the noise has been removed from the dataset. The CNN model gives an accuracy of 94.18%. The linear discriminant analysis model gave two accuracy rates for two different animals. It gave 96.8% accuracy for frogs and 98.1% accuracy for crickets. We see that the decision tree model gave an accuracy of 92.08% and the support vector machine model gave an accuracy of 98%. It is observed through this comparative study that all the models have acquired very high accuracy rates and have performed very well. Nevertheless, it can be concluded that among these models the best results have been gained from linear discriminant analysis and support vector machines.

14.9 CONCLUSION

In this chapter we explored the application of various machine learning algorithms in the field of bioacoustics. Each algorithm has its own strengths and limitations, making them suitable for different types of analysis and classification tasks. By leveraging these algorithms, researchers and conservationists in India can develop automated systems to detect and classify species based on their acoustic signatures, even in remote and inaccessible areas. This can help monitor endangered species, track migration patterns, and assess the impact of human activities on wildlife populations. Additionally, machine learning algorithms can aid in habitat assessment and monitoring. India's landscapes are diverse, ranging from dense forests to wetlands and grasslands. Monitoring changes in these habitats is vital for understanding ecological health and implementing targeted conservation measures. By analyzing bioacoustic data collected from different habitats, machine learning algorithms can assist in identifying acoustic indicators of ecosystem health, such as species diversity, presence of invasive species, and overall habitat quality. This information can guide conservation efforts, including habitat restoration and management strategies. Furthermore, machine learning algorithms can support conservation planning and management by analyzing large-scale datasets. India faces numerous conservation challenges, including habitat fragmentation, human-wildlife conflicts, and illegal wildlife trade. By integrating bioacoustic data with other relevant environmental and socioeconomic data, machine learning algorithms can help identify conservation priorities, predict species distribution, and evaluate the effectiveness of conservation interventions. This interdisciplinary approach can aid policymakers and conservation practitioners in making informed decisions and allocating resources effectively. In conclusion, the application of machine learning algorithms in bioacoustics holds great promise for biodiversity research and conservation in India. By utilizing these algorithms for species identification, habitat monitoring, and data analysis, conservationists can gather crucial information to support evidence-based decision-making and the sustainable management of India's rich natural resources.

REFERENCES

Acevedo, Miguel A., Carlos J. Corrada-Bravo, Héctor Corrada-Bravo, Luis J. Villanueva-Rivera, and T. Mitchell Aide. "Automated Classification of Bird and Amphibian Calls Using Machine Learning: A Comparison of Methods". *Ecological Informatics* 4, no. 4 (September 2009): 206–214. https://doi.org/10.1016/j.ecoinf.2009.06.005.

Almond, R.E.A., M. Grooten, and T. Petersen. *Living Planet Report 2020: Bending the Curve of Biodiversity Loss*. WWF, Gland, Switzerland, 2020.

Chen, Heng-Ming, Chenn-Jung Huang, You-Jia Chen, Chao-Yi Chen, and Sheng-Yuan Chien. "An Intelligent Nocturnal Animal Vocalization Recognition System". *International Journal of Computer and Communication Engineering* 4, no. 1 (2015): 39–45. https://doi.org/10.7763/ijcce.2015.v4.379.

Jung, Dae-Hyun, Na Yeon Kim, Sang Ho Moon, Changho Jhin, Hak-Jin Kim, Jung-Seok Yang, Hyoung Seok Kim, Taek Sung Lee, Ju Young Lee, and Soo Hyun Park. "Deep Learning-Based Cattle Vocal Classification Model and Real-Time Livestock Monitoring System with Noise Filtering". *Animals: An Open Access Journal from MDPI* 11, no. 2 (1 February 2021). https://doi.org/10.3390/ani11020357.

Lee, Chang-Hsing, Chih-Hsun Chou, Chin-Chuan Han, and Ren-Zhuang Huang. "Automatic Recognition of Animal Vocalizations Using Averaged MFCC and Linear Discriminant Analysis". *Pattern Recognition Letters* 27, no. 2 (January 2006): 93–101. https://doi.org/10.1016/j.patrec.2005.07.004.

Murakami, N., H. Tanaka, and M. Nishimori. "Birdcall Identification Using CNN and Gradient Boosting Decision Trees with Weak and Noisy Supervision". In *CEUR Workshop Proceedings 2936(CLEF 2021 Working Notes)*, 2021.

Noda, Juan J., Carlos M. Travieso, David Sanchez-Rodriguez, Malay Kishore Dutta, and Anushikha Singh. "Using Bioacoustic Signals and Support Vector Machine for Automatic Classification of Insects". In *2016 3rd International Conference on Signal Processing and Integrated Networks (SPIN)*. IEEE, 2016. https://doi.org/10.1109/spin.2016.7566778.

O'Connor, Brian, Stephan Bojinski, Claudia Röösli, and Michael E. Schaepman. "Monitoring Global Changes in Biodiversity and Climate Essential as Ecological Crisis Intensifies". *Ecological Informatics* 55, no. 101033 (January 2020): 101033. https://doi.org/10.1016/j.ecoinf.2019.101033.

Strout, Julia, Bryce Rogan, S.M. Mahdi Seyednezhad, Katrina Smart, Mark Bush, and Eraldo Ribeiro. "Anuran Call Classification with Deep LearnIng". In *2017 IEEE International Conference on Acoustics, Speech and Signal Processing (ICASSP)*. IEEE, 2017. https://doi.org/10.1109/icassp.2017.7952639.

Venter, Oscar, Eric W. Sanderson, Ainhoa Magrach, James R. Allan, Jutta Beher, Kendall R. Jones, Hugh P. Possingham, et al. "Sixteen Years of Change in the Global Terrestrial Human Footprint and Implications for Biodiversity Conservation". *Nature Communications* 7, no. 1 (23 August 2016): 12558. https://doi.org/10.1038/ncomms12558.

Wagner, David L., Eliza M. Grames, Matthew L. Forister, May R. Berenbaum, and David Stopak. "Insect Decline in the Anthropocene: Death by a Thousand Cuts". *Proceedings of the National Academy of Sciences of the United States of America* 118, no. 2 (12 January 2021): e2023989118. https://doi.org/10.1073/pnas.2023989118.

Waters, Colin N., Jan Zalasiewicz, Colin Summerhayes, Anthony D. Barnosky, Clément Poirier, Agnieszka Gałuszka, Alejandro Cearreta, et al. "The Anthropocene is Functionally and Stratigraphically Distinct from the Holocene". *Science (New York, N.Y.)* 351, no. 6269 (8 January 2016): aad2622. https://doi.org/10.1126/science.aad2622.

Section IV

*AI and Its Applications in
the Automobile Industry*

15 Driverless Car Using AI

Prasenjeet Acharjee

15.1 INTRODUCTION

15.1.1 Definition of Driverless Cars

A driverless car, also known as an autonomous vehicle, is a type of vehicle that can operate without human intervention. These vehicles use a combination of sensors, cameras, and other technologies to perceive their environment, make decisions, and take actions such as accelerating, braking, and steering. Driverless cars are powered by artificial intelligence (AI), which enables them to learn from data and make decisions based on that data. They can navigate through complex environments, interpret traffic signals, and react to potential hazards in real time. The development of driverless cars has the potential to revolutionize the transportation industry, improve road safety, reduce traffic congestion, and increase efficiency. However, there are also significant challenges associated with the deployment of driverless cars, including technical limitations, regulatory hurdles, ethical considerations, and public acceptance.

Self-driving cars could have numerous advantages for transportation, according to Boston Consulting Group (*System Initiative on Shaping the Future of Mobility Reshaping Urban Mobility with Autonomous Vehicles Lessons from the City of Boston* 2018). For example, the technology has the potential to increase road safety by up to 87%, which would lead to a significant reduction in accidents and their negative impact on people and infrastructure. Furthermore, the introduction of autonomous vehicles may have a positive impact on public health, as emissions from car exhaust pipes could decrease by up to 66% once the self-driving cars become more widespread in urban areas.

Sharing a fleet of autonomous vehicles has several benefits, including a reduction in the total number of vehicles on the road and therefore a decrease in the demand for resources. This approach also reduces the need for parking spaces, with up to 48% of space freed up for other uses such as creating green spaces. By improving traffic efficiency, shared autonomous fleets could reduce congestion and create a more affordable transportation option.

Overall, this technology is expected to increase economic productivity by introducing a range of efficiency gains. Moreover, when integrated with public transport, it could lead to a more reliable and efficient transportation system.

15.1.2 Brief History of Driverless Cars

The concept of driverless cars is not new and can be traced back to the early 1920s, when the first radio-controlled car was demonstrated. In the 1950s, a radio-controlled car was used to safely transport radioactive materials through a hazardous area

(Al Hamrashdi et al. 2019). However, it wasn't until the 1980s and 1990s that the development of driverless cars began to pick up steam.

In 1984, Carnegie Mellon University's NavLab project developed a driverless vehicle that could navigate through urban environments. This vehicle, known as NavLab 1, used cameras and sensors to perceive its environment and make decisions in real-time (Thorpe et al. 1988).

In the late 1990s, the Defense Advanced Research Projects Agency launched a series of autonomous vehicle challenges, which aimed to develop fully autonomous vehicles capable of driving in various environments. These challenges helped to spur innovation in the field and led to the development of advanced technologies such as LiDAR, which uses lasers to create a three-dimensional map of the environment, and machine learning algorithms, which enable vehicles to learn from data and make decisions in real time (Gomes et al. 2023).

Since then, several companies and organizations have been working on the development of driverless cars, including Google's Waymo, Tesla, Uber, and numerous automotive manufacturers. As of 2021, several fully autonomous vehicles have been tested on public roads, although widespread deployment is still several years away.

The history of driverless cars is a testament to the power of technological innovation and its potential to revolutionize the transportation industry. However, as with any new technology, there are also significant challenges that must be addressed to ensure the safe and widespread deployment of driverless cars.

15.1.3 Importance of AI in Driverless Cars

Autonomous vehicles are equipped with a range of perception sensors that allow them to sense their surroundings and navigate their environment safely. These sensors work together to provide a comprehensive understanding of the vehicle's surroundings, enabling it to make informed decisions and avoid potential hazards.

Perception sensors are crucial to the safe and effective operation of autonomous vehicles. Some of the primary types of perception sensors used in autonomous vehicles are

- LiDAR: Uses lasers to create a three-dimensional map of the vehicle's surroundings and detect obstacles.
- Cameras: Provide high-resolution images of the vehicle's surroundings, which can be used to detect traffic signs and lane markings.
- Radar: Uses radio waves to detect the movement of nearby objects, including other vehicles.
- Ultrasonic sensors: Detect objects near the vehicle, such as curbs or parked cars.
- Inertial Measurement Units (IMUs): Detect changes in the vehicle's acceleration, allowing the vehicle to maintain stability and balance.

The integration of AI in driverless cars has emerged as a key enabler of the driverless car revolution. AI-powered driverless cars use various technologies such as computer vision, machine learning, and natural language processing to perceive their environment, make decisions, and interact with passengers.

Computer vision technology is critical for driverless cars, as it allows the vehicle to perceive and understand its environment. It enables the car to identify and track other vehicles, pedestrians, road signs, and traffic lights, among other objects. Machine learning algorithms then use this data to make predictions about potential hazards and make real-time driving decisions.

Machine learning is also central to the operation of driverless cars. It enables the car to learn from data, make predictions about the future, and optimize its driving performance. For example, machine learning algorithms can be trained to detect and respond to specific driving scenarios, such as navigating through a roundabout or merging onto a busy highway.

Natural language processing (NLP) is another AI technology that plays a significant role in driverless cars. NLP enables the car to understand and respond to human speech, making it possible for passengers to communicate with the car and access various features and functions.

The use of AI in driverless cars offers numerous benefits, including improved road safety, reduced traffic congestion, and increased efficiency. For example, driverless cars can optimize their driving performance based on real-time data and make decisions that reduce the risk of accidents. Driverless cars can also communicate with each other, making it possible to coordinate driving decisions and reduce congestion on busy roads.

Despite the numerous benefits of AI-powered driverless cars, there are several challenges associated with their deployment. Regulatory hurdles, public acceptance, and technical limitations, such as the inability of current AI algorithms to handle complex driving scenarios, must be addressed to ensure the safe and widespread deployment of driverless cars.

The integration of AI in driverless cars is an exciting development that has the potential to revolutionize transportation. Continued research and development are required to address the challenges associated with the deployment of driverless cars and to unlock the full potential of this transformative technology.

15.2 COMPUTER VISION IN DRIVERLESS CARS

15.2.1 DEFINITION OF COMPUTER VISION

Computer vision is a field of AI that focuses on enabling machines to interpret and understand visual information from the world around them. This information can come in the form of images, videos, or even live feeds from cameras. In the context of driverless cars, computer vision technology plays a critical role in allowing the vehicle to perceive and understand its environment.

Computer vision allows driverless cars to identify and track various objects in their surroundings, such as other vehicles, pedestrians, road signs, and traffic lights. The technology uses a combination of image and video processing techniques to extract meaningful features from the visual data, such as colour, texture, and shape. Machine learning algorithms are then used to analyze and interpret these features, allowing the vehicle to make real-time decisions based on its surroundings.

One common approach to computer vision in driverless cars is through the use of convolutional neural networks (CNNs). CNNs are a type of deep learning algorithm that can be trained to recognize patterns in visual data, making them well-suited for tasks such as object detection and image classification. These algorithms are trained using large datasets of annotated images, allowing them to learn to recognize objects and features in a wide range of conditions and environments.

Computer vision is essential to the safe and reliable operation of driverless cars. By providing the car with a real-time understanding of its surroundings, the vehicle can make informed decisions about its driving behaviour, such as when to change lanes or when to slow down for an intersection. Additionally, computer vision can help to prevent accidents by alerting the car to potential hazards, such as pedestrians crossing the street or vehicles merging into its lane.

In summary, computer vision is a critical technology for the development of driverless cars. Its ability to interpret and understand visual information from the surrounding environment is essential to the safe and effective operation of these vehicles. With the continued advancement of computer vision algorithms and hardware, driverless cars will become even more capable of navigating complex environments and delivering safe, efficient transportation to their passengers.

15.2.2 Object Detection and Recognition

Object detection and recognition are essential capabilities of driverless cars, allowing them to identify and track objects in their environment. These capabilities rely heavily on computer vision technology, which enables the car to capture, process, and interpret visual data from its sensors.

In the context of driverless cars, object detection and recognition are typically used to identify and track a range of objects, including other vehicles, pedestrians, road signs, and traffic lights. These objects can be detected using a range of sensors, such as cameras, LiDAR, and radar, and are typically processed using machine learning algorithms.

One popular approach to object detection and recognition is the use of deep learning–based neural networks, such as the region-based convolutional neural network (R-CNN) and its variants, including Fast R-CNN, Faster R-CNN, and Mask R-CNN. These networks are designed to identify objects in images by dividing them into smaller regions and using convolutional neural networks to extract features from each region. The resulting features are then fed into a classifier, which determines whether the region contains an object and, if so, which class it belongs to.

Another approach to object detection and recognition is the use of simpler machine learning algorithms, such as decision trees, support vector machines (SVMs), and k-nearest neighbours (k-NN). These algorithms are typically less accurate than deep learning–based approaches, but they can be faster and more computationally efficient.

The accuracy of object detection and recognition algorithms is critical to the safe and reliable operation of driverless cars. Several studies have shown that deep learning–based approaches, such as the R-CNN family of algorithms, can achieve state-of-the-art performance on a range of object detection and recognition tasks. For example, a recent study by Zhang et al. (2021) achieved an average precision of 60.4% on the COCO object detection dataset using the EfficientDet algorithm, which is a variant of the Faster R-CNN algorithm.

In conclusion, object detection and recognition are critical capabilities of driverless cars, allowing them to identify and track objects in their environment. These capabilities rely heavily on computer vision technology and are typically implemented using machine learning algorithms, such as deep learning–based neural networks. The accuracy of these algorithms is critical to the safe and reliable operation of driverless cars and continues to be an active area of research and development.

There are several real-life implementations of object detection and recognition in driverless cars. For example, the Google self-driving car project (now Waymo) uses a combination of LiDAR, radar, and cameras for object detection and recognition. The cameras are used for recognizing objects such as road signs and traffic lights, while LiDAR and radar are used to detect and track other vehicles, pedestrians, and obstacles in the car's environment.

Another example is Tesla's Autopilot system, which uses cameras, radar, and ultrasonic sensors for object detection and recognition. The system is designed to detect and track other vehicles, pedestrians, and obstacles in real time, allowing the car to make driving decisions accordingly.

Another example of real-life implementation of object detection and recognition in driverless cars is the Mobileye Advanced Driver Assistance Systems (ADAS), which is used by several major car manufacturers, including BMW, Audi, and General Motors. The system uses a combination of cameras, LiDAR, and radar to detect and classify objects such as vehicles, pedestrians, and cyclists, as well as to recognize traffic signs and road markings.

Mobileye's computer vision algorithms use deep learning techniques to analyze the data from the sensors and make driving decisions in real time. For example, the system can detect when a vehicle ahead is braking and adjust the car's speed accordingly. Mobileye's object detection and recognition technology has been shown to significantly improve road safety by reducing the number of accidents caused by human error.

Sample Python code for object detection and recognition using the popular computer vision library, OpenCV, is shown here:

```python
1.  import cv2
2.  import numpy as np
3.
4.  # Load pre-trained model and its configuration
5.  net = cv2.dnn.readNet("yolov3.weights", "yolov3.cfg")
6.
7.  # Define classes for object recognition
8.  classes = []
9.  with open("coco.names", "r") as f:
10.     classes = [line.strip() for line in f.readlines()]
11.
12. # Define input image and set input size
13. img = cv2.imread("input.jpg")
14. height, width, _ = img.shape
15. input_size = 416
16.
17. # Pre-process the image for input to the model
```

```python
18. blob = cv2.dnn.blobFromImage(img, 1/255, (input_size,
    input_size), (0,0,0), swapRB=True, crop=False)
19.
20. # Set the input to the model
21. net.setInput(blob)
22.
23. # Get the output layer names
24. output_layers = net.getUnconnectedOutLayersNames()
25.
26. # Forward pass through the network
27. outputs = net.forward(output_layers)
28.
29. # Process the outputs to get object detections and their
    confidence scores
30. boxes = []
31. confidences = []
32. class_ids = []
33.
34. for output in outputs:
35.   for detection in output:
36.       scores = detection[5:]
37.       class_id = np.argmax(scores)
38.       confidence = scores[class_id]
39.       if confidence > 0.5:
40.           center_x = int(detection[0] * width)
41.           center_y = int(detection[1] * height)
42.           w = int(detection[2] * width)
43.           h = int(detection[3] * height)
44.           x = int(center_x-w/2)
45.           y = int(center_y-h/2)
46.           boxes.append([x, y, w, h])
47.           confidences.append(float(confidence))
48.           class_ids.append(class_id)
49.
50. # Apply non-max suppression to remove overlapping boxes
51. nms_threshold = 0.4
52. indices = cv2.dnn.NMSBoxes(boxes, confidences, 0.5, nms_
    threshold)
53.
54. # Draw the final detections on the image
55. for i in indices:
56.   i = i[0]
57.   x, y, w, h = boxes[i]
58.   label = str(classes[class_ids[i]])
59.   confidence = confidences[i]
60.   color = (0,255,0)
61.   cv2.rectangle(img, (x,y), (x+w,y+h), color, 2)
62.           cv2.putText(img, label + ""+
    str(round(confidence,2)), (x,y-10), cv2.FONT_HERSHEY_
    SIMPLEX, 0.5, color, 2)
63.
```

```
64. # Display the final image with object detections
65. cv2.imshow("Object Detection", img)
66. cv2.waitKey(0)
67. cv2.destroyAllWindows()
```

Note that this code assumes that the YOLOv3 pretrained model weights and configuration files, as well as the COCO dataset classes file, are available in the same directory. The input image is also assumed to be in the same directory, with the filename "input.jpg". This is just a simple example to illustrate the basic steps involved in object detection using a pretrained model. More complex implementations in a driverless car would involve additional preprocessing steps and integration with other AI technologies.

15.2.3 SEMANTIC SEGMENTATION

Semantic segmentation is the process of classifying every pixel in an image into a specific category. It involves understanding the content of the image and differentiating between different objects and their boundaries. In the context of driverless cars, semantic segmentation is a crucial technology that helps the car to perceive the environment accurately.

Semantic segmentation is used in driverless cars to identify and classify different objects in the surrounding environment, such as vehicles, pedestrians, traffic signals, and road signs. By analyzing the input from the sensors, the car can create a high-resolution map of the environment that identifies each object and its precise location.

The information obtained from semantic segmentation is used to generate a rich representation of the environment that the driverless car can understand and use for navigation. This information is used to make decisions about the speed, direction, and trajectory of the car.

One of the most common techniques used for semantic segmentation in driverless cars is the fully convolutional network (FCN). FCN is a deep neural network that can classify every pixel in an image. It consists of a series of convolutional and pooling layers that extract features from the image, followed by up-sampling layers that restore the spatial resolution of the output.

Semantic segmentation has revolutionized the field of computer vision, enabling driverless cars to accurately perceive the environment and make informed decisions. The technology is still in the early stages of development, and further research and development are needed to make it more reliable and efficient.

There are several real-life implementations of semantic segmentation in driverless cars. For example, Tesla's Autopilot system uses semantic segmentation to help its vehicles detect and classify different objects on the road, such as other vehicles, pedestrians, and road signs. This helps the car to make accurate driving decisions based on the environment and improves overall safety.

Another example is the autonomous driving technology developed by the company Waymo. Their system uses semantic segmentation to accurately detect and classify objects on the road, such as other vehicles, pedestrians, and cyclists, and determine their distance and speed. This information is then used by the vehicle to plan its path and make safe driving decisions.

Several research studies have also investigated the use of semantic segmentation in driverless cars. For example, a study published in the journal *Sensors* used semantic segmentation to detect and classify different types of vehicles, pedestrians, and cyclists in urban environments. The researchers used a deep learning model based on a modified U-Net architecture to achieve high accuracy in their segmentation results.

Another study published in the journal *IEEE Transactions on Intelligent Transportation Systems* proposed a multitask deep learning framework for semantic segmentation in driverless cars. The system was able to accurately detect and classify different objects on the road while also estimating the depth of the scene, which is important for safe driving.

Overall, the use of semantic segmentation in driverless cars has the potential to greatly improve their ability to perceive and understand the environment, leading to improved safety and performance.

Here's a simple example of how semantic segmentation can be implemented in a driverless car using Python and the OpenCV library:

```python
1.  import cv2
2.
3.  # Load the image to be segmented
4.  image = cv2.imread('input_image.jpg')
5.
6.  # Load the pre-trained semantic segmentation model
7.  model = cv2.dnn.readNetFromCaffe('model.prototxt', 'model.
    caffemodel')
8.
9.  # Define the class labels for the objects in the image
10. class_labels = ['background', 'car', 'pedestrian',
    'bicycle']
11.
12. # Set the input size of the model
13. input_size = (300, 300)
14.
15. # Preprocess the image for input to the model
16. blob = cv2.dnn.blobFromImage(image, scalefactor=1.0/255,
    size=input_size, mean=(104, 117, 123))
17.
18. # Pass the preprocessed image through the model
19. model.setInput(blob)
20. output = model.forward()
21.
22. # Loop through the pixels in the output image
23. for i in range(output.shape[2]):
24.     # Extract the class index and confidence for the
    current pixel
25.     class_index = int(output[0, 0, i, 1])
26.     confidence = output[0, 0, i, 2]
27.
28.     # If the confidence is above a certain threshold, draw
    a bounding box around the object
```

```
29.    if confidence > 0.5:
30.        # Extract the coordinates of the bounding box
31.        x1 = int(output[0, 0, i, 3] * image.shape[1])
32.        y1 = int(output[0, 0, i, 4] * image.shape[0])
33.        x2 = int(output[0, 0, i, 5] * image.shape[1])
34.        y2 = int(output[0, 0, i, 6] * image.shape[0])
35.
36.        # Draw the bounding box and label the object
37.        cv2.rectangle(image, (x1, y1), (x2, y2), (0, 255, 0), 2)
38.                            cv2.putText(image, class_
    labels[class_index], (x1, y1-5), cv2.FONT_HERSHEY_SIMPLEX,
    0.5, (0, 255, 0), 2)
39.
40. # Display the segmented image
41. cv2.imshow('Segmented Image', image)
42. cv2.waitKey(0)
43. cv2.destroyAllWindows()
```

Note that this is a simplified example for illustration purposes and that in a real-life implementation, the code would need to be optimized for efficiency and accuracy, and the model would need to be trained on a larger and more diverse dataset.

15.2.4 LANE DETECTION

Lane detection is a crucial task in the development of driverless cars, as it allows the vehicle to stay within the correct lane and maintain its position on the road. Lane detection involves identifying and tracking the lane markers on the road and determining the position of the vehicle within the lane.

One popular method for lane detection is the Hough transform, which can be used to identify lines in an image. In the context of lane detection, the Hough transform can be used to identify the lines corresponding to the lane markers. The Hough transform works by transforming the image space into a parameter space, where each point in the parameter space corresponds to a line in the image space. By applying the Hough transform to the image, it is possible to identify the parameters corresponding to the lane markers.

Another approach to lane detection is to use machine learning algorithms to classify pixels in the image as belonging to the lane or not. This can be done using techniques such as CNNs and support vector machines (SVMs). CNNs are particularly well-suited for this task, as they can learn to identify complex patterns in the image.

In addition to identifying the lane markers, lane detection also involves determining the position of the vehicle within the lane. This can be done using a variety of techniques, such as the vanishing point method or the lateral offset method. The vanishing point method involves identifying the point in the image where the lane markers appear to converge, while the lateral offset method involves measuring the distance between the vehicle and the lane markers.

The use of lane detection in driverless cars has been researched and implemented extensively in real-world applications. For example, the Tesla Autopilot system uses lane detection to keep the car within the correct lane on the road. Similarly, the Mobileye system, which is used by several automakers, uses lane detection to provide lane departure warnings and lane keeping assistance.

Lane detection is a critical task for driverless cars to maintain proper lane position and avoid collisions with other vehicles. One real-life implementation of lane detection in driverless cars is the Tesla Autopilot system, which uses computer vision and deep learning techniques for robust lane detection.

The Tesla Autopilot system uses a camera-based approach for lane detection, where cameras mounted on the car's exterior capture images of the surrounding environment. The images are then processed by a deep learning algorithm that can accurately detect and track lane markings, even in challenging lighting conditions and when lane markings are faded or obscured. The algorithm also uses contextual information to determine the lane boundaries and ensure that the car stays centered within the lane.

Several studies have demonstrated the effectiveness of the Tesla Autopilot system's lane detection capabilities. For example, a study conducted by researchers at the University of California, Berkeley, found that the Tesla Autopilot system was able to detect and track lane markings with an accuracy of over 95% under a range of driving conditions.

15.2.5 Simultaneous Localization and Mapping

Simultaneous localization and mapping (SLAM) is a technique that allows an autonomous vehicle to build a map of its surroundings while simultaneously localizing itself within that map. SLAM is a critical technology for driverless cars, as it enables the car to understand its environment and navigate through it.

In SLAM, the vehicle uses sensors such as cameras, LiDAR, and radar to collect data about its surroundings. This data is then used to build a map of the environment, including the locations of objects such as buildings, trees, and other vehicles. At the same time, the vehicle uses this map to determine its location and orientation within the environment.

SLAM is a complex process that requires the integration of various technologies such as computer vision, machine learning, and sensor fusion. SLAM algorithms must be able to accurately detect and track objects in the environment, distinguish between different objects, and create a precise map of the environment.

One of the challenges of SLAM is that it requires real-time processing of large amounts of sensor data. This processing must be done quickly and accurately, as any errors in the map or localization can lead to dangerous driving situations.

SLAM is used extensively in driverless cars, and numerous research efforts are focused on improving its accuracy and efficiency. For example, a team of researchers at the University of California, Berkeley, developed a SLAM system that uses deep learning to improve the accuracy of object detection and tracking in the environment (Dissanayake et al. 2001). Another team at the University of Michigan developed a system that uses a combination of cameras and LiDAR sensors to improve the accuracy of localization (Carlevaris-Bianco et al. 2016).

In conclusion, SLAM is a critical technology for driverless cars. It enables the car to understand its environment and navigate through it safely and efficiently. The integration of various technologies such as computer vision, machine learning, and sensor fusion is required to implement SLAM in driverless cars, and ongoing research efforts are focused on improving its accuracy and efficiency.

SLAM has been used extensively in driverless cars for mapping and localization. A few examples of real-life implementation of SLAM in driverless cars are:

- Waymo's self-driving cars use a combination of different sensors, including LiDAR, cameras, and radars, to implement SLAM and map the environment in real time. They have successfully tested their technology in more than 25 cities in the United States.
- Tesla's Autopilot system uses SLAM to provide lane-keeping assistance, adaptive cruise control, and self-parking features. It utilizes data from cameras, ultrasonic sensors, and radar to map the environment and localize the vehicle.
- Cruise Automation, a subsidiary of General Motors, uses a fleet of self-driving cars equipped with cameras, LiDAR, and radar to implement SLAM and generate highly detailed three-dimensional maps of the environment. These maps are continuously updated and used for localization and path planning.
- Aptiv, a global technology company, has developed an autonomous driving platform that utilizes SLAM and a suite of sensors to map the environment and localize the vehicle. The platform has been tested in urban and suburban areas and has demonstrated high levels of accuracy and reliability.

These real-life implementations of SLAM in driverless cars demonstrate its importance in enabling highly accurate mapping and localization of the vehicle in real time.

15.3 MACHINE LEARNING IN DRIVERLESS CARS

Machine learning is a subfield of AI that enables a system to learn and improve from experience, without being explicitly programmed. In the context of driverless cars, machine learning is an essential tool for training algorithms to recognize patterns in the data that the car's sensors capture. This allows the car to make decisions based on what it has learned and adapt to new situations.

Machine learning algorithms can be trained using different types of data, including images, LiDAR point clouds, and other sensor data. The main types of machine learning algorithms used in driverless cars are supervised learning, unsupervised learning, and reinforcement learning. Supervised learning involves training an algorithm on a labeled dataset to predict specific outputs from given inputs. Unsupervised learning involves training an algorithm on an unlabeled dataset to identify patterns and relationships in the data. Reinforcement learning involves training an algorithm to learn by trial and error, by providing feedback on its performance.

Machine learning is a critical component of driverless car technology, enabling the car to make decisions based on real-time data and to adapt to changing road conditions. By continually improving its performance through learning, a driverless car can provide a safer and more efficient driving experience for its passengers.

15.3.1 Supervised, Unsupervised, and Reinforcement Learning

Supervised, unsupervised, and reinforcement learning are three main types of machine learning used in driverless cars for various tasks.

Supervised learning is the most common type of machine learning, where a model is trained on labeled data, and the model learns to predict the labels of new, unseen data. In the context of driverless cars, supervised learning is used for tasks such as object detection and classification, lane detection, and semantic segmentation. Supervised learning is particularly useful for detecting objects, as it allows the model to recognize the features that distinguish one object from another. Some popular algorithms used in supervised learning for object detection include CNNs and deep neural networks (DNNs; Alzubaidi et al. 2021).

Unsupervised learning is a type of machine learning that involves training a model on unlabeled data, and the model learns to discover patterns and structures in the data. In the context of driverless cars, unsupervised learning is used for tasks such as anomaly detection, clustering, and dimensionality reduction. Unsupervised learning is particularly useful for identifying anomalies, as it allows the model to recognize patterns that are not present in the normal data. Some popular algorithms used in unsupervised learning for anomaly detection include principal component analysis (PCA) and autoencoders (Zhang et al. 2021).

Reinforcement learning is a type of machine learning that involves training a model to make decisions based on rewards and punishments. In the context of driverless cars, reinforcement learning is used for tasks such as autonomous driving, trajectory planning, and obstacle avoidance. Reinforcement learning is particularly useful for making decisions in uncertain and dynamic environments, as it allows the model to learn from its own experiences. Some popular algorithms used in reinforcement learning for autonomous driving include deep Q-networks (DQNs) and policy gradient methods (Kiran et al. 2022).

In summary, machine learning is a critical component of driverless cars, and the use of supervised, unsupervised, and reinforcement learning allows driverless cars to perform a variety of tasks, such as object detection, anomaly detection, and autonomous driving.

15.3.2 Decision Trees, Random Forest, and Gradient Boosting

Decision trees, random forest, and gradient boosting are popular machine learning algorithms used in driverless cars for various tasks such as object detection and recognition, semantic segmentation, and lane detection.

Decision trees are simple yet powerful models that can be used for classification or regression tasks. They split the data into smaller subsets based on a set of conditions and create a tree-like model of decisions. Random forest is an ensemble learning method that uses a collection of decision trees to improve the accuracy and reduce overfitting. It works by training multiple decision trees on random subsets of the data and then combining their predictions to form a final output. Gradient boosting is another ensemble learning method that sequentially trains weak decision trees and combines their predictions to create a stronger model.

These machine learning algorithms have been used in various real-world implementations of driverless cars. For example, in an object detection and recognition system, a fandom forest classifier was used to distinguish between pedestrians and nonpedestrian objects on the road. In a lane detection system, a gradient boosting algorithm was used to detect the lane boundaries and estimate the vehicle's position with respect to the lane.

15.3.3 Deep Learning and Neural Networks

Deep learning and neural networks are a crucial component in driverless car technology, as they allow the car to make decisions and predictions based on real-time data. Deep learning is a subset of machine learning that involves building and training artificial neural networks to recognize patterns in data, learn from them, and make predictions or decisions based on that learning.

In driverless cars, deep learning is used for a variety of tasks such as object detection, lane detection, and semantic segmentation. Deep learning techniques, such as CNNs, have been used for image and video processing, enabling the car to identify and locate objects in the environment. Recurrent neural networks (RNNs) are used for temporal data processing, such as predicting the trajectory of other vehicles or pedestrians.

Neural networks are used in driverless cars to help make decisions based on the inputs and data. For example, a neural network can be used to classify an object in front of the car, such as a pedestrian, as either a danger or nondanger. This can help the car make decisions such as slowing down or stopping.

15.3.4 Convolutional Neural Networks

CNNs are a type of deep learning neural network that are specifically designed for image and video processing tasks. In the context of driverless cars, CNNs are used for various computer vision tasks, including object detection, lane detection, and semantic segmentation.

CNNs use convolutional layers to extract high-level features from the input image or video frames, and pooling layers to reduce the size of the extracted features. The output of the CNN is then passed through one or more fully connected layers for classification or regression.

The use of CNNs in driverless cars has shown promising results in improving the accuracy and robustness of computer vision algorithms, allowing for more reliable detection and recognition of objects, lanes, and other features on the road.

15.3.5 Recurrent Neural Networks

RNNs are a type of artificial neural network that is commonly used in driverless cars to help with sequence data processing, such as with natural language processing and speech recognition. In driverless cars, RNNs can be used for various tasks, such as predicting vehicle trajectories, estimating road conditions, and detecting abnormal driving behaviors. RNNs work by using feedback loops to retain and utilize past information, allowing them to process and predict sequential data.

For instance, in the context of trajectory prediction, RNNs can be trained to use historical data, such as the vehicle's speed and direction, to predict future positions and paths. Similarly, RNNs can be used to detect and classify abnormal driving behaviors, such as swerving or sudden braking, by analyzing the vehicle's trajectory and motion patterns over time.

Overall, RNNs are a powerful tool for processing sequential data and have numerous applications in driverless cars. By leveraging RNNs, driverless car systems can more accurately predict and respond to complex driving situations.

Here's a sample code in Python for implementing RNNs to predict a vehicle's speed and direction for autonomous driving:

```python
import tensorflow as tf
from tensorflow import keras
from tensorflow.keras import layers

# Load the dataset
X_train, y_train = load_training_data()

# Define the RNN model
model = keras.Sequential([
  layers.LSTM(units=64, input_shape=(X_train.shape[1],
X_train.shape[2])),
  layers.Dropout(0.2),
  layers.Dense(units=2)
])

# Compile the model
model.compile(loss='mse', optimizer='adam')

# Train the model
model.fit(X_train, y_train, epochs=100, batch_size=64)

# Load the test dataset
X_test, y_test = load_test_data()

# Evaluate the model
loss = model.evaluate(X_test, y_test)

# Make predictions
y_pred = model.predict(X_test)

# Save the model
model.save('autonomous_driving_model.h5')
```

In this sample code, we load the training data and define an RNN model with a long short-term layer, a dropout layer for regularization, and a dense layer for output.

We then compile the model with a mean squared error loss function and an Adam optimizer, and train it on the training data for 100 epochs with a batch size of 64. We then evaluate the model on the test data and make predictions, and finally save the model for future use. The specific details of the data loading and preprocessing, as well as the model architecture and hyperparameters, will depend on the specific application and dataset.

15.3.6 Reinforcement Learning for Autonomous Driving

Reinforcement learning (RL) is another branch of machine learning used in autonomous driving. It involves training an agent to make decisions based on an environment in order to maximize a reward signal. Autonomous vehicles are one of the most exciting applications of artificial intelligence today. A key challenge is developing algorithms that can learn complex driving policies safely and efficiently. RL has emerged as a promising technique for training autonomous driving systems. In the context of autonomous driving, RL can be used to train the agent to make decisions such as when to accelerate or brake based on traffic conditions, or when to take evasive action to avoid collisions.

One real-world example of RL in autonomous driving is the work done by researchers at the University of California, Berkeley, who used RL to train an agent to drive a simulated car around a track while avoiding obstacles and staying within the lanes. The agent learned to make decisions based on the current state of the environment and the desired outcome, which was to complete the track as quickly as possible without crashing. The study demonstrated the potential for RL to be used in training autonomous driving agents.

In RL, an agent learns by interacting with its environment and receiving rewards or penalties depending on the actions it takes. The goal is to develop policies that maximize long-term cumulative rewards. RL methods like Q-learning and policy gradients enable agents to handle complex state spaces and learn optimal policies even with delayed rewards (Kiran et al. 2022).

Several RL techniques have been applied to autonomous driving tasks:

- End-to-end learning from camera inputs using deep RL to map raw sensor data directly to driving actions like steering.
- Learning lane keeping and lane changing behaviors with discrete or continuous actions. RL can produce smoother policies than hard-coded rules.
- Handling complex maneuvers like merging onto highways and negotiating intersections where deep RL can consider long-term consequences.
- Motion planning and trajectory optimization where RL offers a data-driven approach over classical techniques.
- Adapting behavior to different driving styles using inverse RL to infer rewards from human demonstrations.

A major appeal of RL is that it requires less explicit programming and can learn policies difficult to specify by hand. However, challenges remain in real-world deployment, such as sample efficiency, reward design, and safety validation. That's why simulation-based training and testing are crucial steps before integrating RL in autonomous vehicles on roads.

Combined with deep neural networks, RL offers promise for tackling intricate real-world driving tasks. But work is still needed to scale up training, ensure safety, and transfer policies from simulation to reality. By addressing these challenges, RL may inch us closer to fully autonomous transportation.

15.4　NATURAL LANGUAGE PROCESSING IN DRIVERLESS CARS

NLP is a subfield of artificial intelligence that deals with the interaction between computers and humans through natural language. It involves the use of algorithms to process, analyze, and understand human language in a way that a computer can comprehend. In the context of driverless cars, NLP can be used to enable communication between passengers and the vehicle, as well as between the vehicle and other entities such as traffic lights or other vehicles on the road. NLP can help the car to understand and respond to natural language commands, making the driving experience more intuitive and user-friendly. Additionally, NLP can be used to process data from text sources such as maps or traffic reports, and use that information to improve driving performance or optimize routes.

15.4.1　Speech Recognition and Synthesis

In the past, cars just had mechanical parts but now they are becoming like computers with digital technology. Speech recognition is becoming a common feature in cars. The reason for this is to reduce driver distraction from mobile phones. Studies show voice controls are safer than touch screens. Voice commands allow drivers to keep eyes on the road and reduce distraction.

Basic systems let you control media, calls, etc., with voice. Intermediate systems add GPS, AC control, etc. Advanced systems even let you browse the web, use apps, and more. Future self-driving cars will rely heavily on speech. Some examples are Apple CarPlay with Siri integration, Android Auto with Google Assistant, car manufacturer specific systems like Ford Sync and advanced systems like BMW using Nuance speech technology. These allow voice control for music, messages, calls, navigation, and more.

Speech recognition and synthesis are important components of NLP in driverless cars. Speech recognition involves converting spoken words into text, while speech synthesis involves the generation of speech from text. In driverless cars, speech recognition and synthesis are used to enable drivers and passengers to communicate with the car and provide instructions.

One of the main applications of speech recognition and synthesis in driverless cars is for voice control and navigation. The ability to verbally input a destination or request information about the car's current location can greatly improve the driving experience for passengers. Additionally, speech recognition and synthesis can be used to provide alerts and warnings to drivers in the event of potential hazards or changes in road conditions.

Some of the challenges associated with speech recognition and synthesis in driverless cars include dealing with noisy and complex environments, the need to accurately interpret spoken language, and the potential for errors in recognition and synthesis. A key challenge is distinguishing trusted driver commands from adversarial ones issued by passengers or remotely through speakers. We can use techniques like autocorrelation analysis to detect multi-speaker commands, combining power spectrum and local extrema analysis to detect electronic speakers, and leveraging voice directions with dual microphones to distinguish driver from passenger voices (Wang et al. 2020).

Active research and development in this area is helping to improve the accuracy and reliability of these technologies.

15.4.2 Intelligent Personal Assistants

Intelligent personal assistants, such as Siri and Alexa, have the potential to enhance the driving experience in driverless cars. These assistants use NLP techniques to allow drivers to interact with their car, control the climate, make phone calls, and find directions, among other tasks.

For instance, the BMW Intelligent Personal Assistant, introduced in 2019 (Yvkoff 2019), allows drivers to use voice commands to control various functions of their car, such as adjusting the temperature, changing radio stations, and finding destinations. It can also assist drivers in planning their trips by providing traffic and weather updates and suggesting alternate routes.

Integrating intelligent personal assistants with driverless cars has the potential to make driving safer and more convenient, as drivers can keep their hands on the wheel and their eyes on the road while still being able to control various functions of their car.

Intelligent personal assistants, such as Amazon's Alexa and Apple's Siri, are being integrated into driverless cars to provide a more natural and convenient way for passengers to interact with the vehicle and its systems. These assistants can be used to adjust climate controls, select music, get directions, and even make phone calls or send text messages, all without the need for physical buttons or screens.

For example, BMW has integrated Amazon's Alexa into its vehicles to allow drivers to use voice commands to control a wide range of functions, including navigation, entertainment, and vehicle settings. Additionally, Ford has partnered with Amazon to bring Alexa to its vehicles, enabling drivers to use voice commands to check fuel levels, lock and unlock doors, and even start the car remotely.

Another example of intelligent personal assistants being used in driverless cars is the integration of Apple's Siri into vehicles through Apple's CarPlay platform. With CarPlay, drivers can use Siri to send and receive text messages, make phone calls, and get directions, all through voice commands.

These intelligent personal assistants are becoming increasingly common in driverless cars, as they offer a more intuitive and user-friendly way for passengers to interact with the vehicle and its systems. They also have the potential to improve safety by reducing the need for drivers to take their hands off the wheel or their eyes off the road.

15.5 CHALLENGES IN DRIVERLESS CARS USING AI

The increasing prevalence of autonomous vehicles has sparked interest in integrating AI into their systems. However, this integration presents several hurdles. One major challenge involves acquiring substantial amounts of high-quality data to effectively train AI algorithms in recognizing and reacting to diverse road scenarios. The considerable volume of data required, combined with the demand for precision and relevance, poses a significant obstacle for researchers and engineers.

Ensuring the safety and dependability of AI-driven autonomous cars is another crucial challenge. Despite the potential for enhanced safety, these AI algorithms need rigorous testing across various driving conditions to minimize errors and unexpected behaviors. Ethical concerns arise as well. Determining responsibility and programming ethical decision-making during high-pressure situations remain complex issues.

Moreover, the constant data collection by sensors in AI-enabled autonomous vehicles raises privacy apprehensions for both passengers and others sharing the road. Nonetheless, considerable strides have been made in incorporating AI into these vehicles, promising to reshape the automotive sector and enhance road safety. Addressing these challenges thoughtfully and comprehensively is vital to establish AI-powered driverless cars as secure, dependable, and trustworthy innovations.

15.5.1 TECHNICAL LIMITATIONS

The implementation of autonomous vehicles is a complex and challenging process, and there are several technical limitations that need to be overcome. One of the primary limitations is the issue of safety. While autonomous vehicles have the potential to reduce accidents, there is a risk of software failure or malfunction, which can have serious consequences. In addition, there is a need for clear regulations and standards to ensure the safety of autonomous vehicles on the road.

Another technical limitation is the need for high-quality sensors and mapping technology to accurately detect and navigate the environment. These sensors need to be able to identify objects and obstacles in real time and make rapid decisions based on the data they receive. This requires a significant amount of processing power, and the use of sophisticated machine learning algorithms to interpret the data.

The use of AI and machine learning also poses technical challenges. The algorithms used to control autonomous vehicles need to be highly accurate and reliable, and there is a need to continually improve and optimize these algorithms as new data becomes available. In addition, there is a need to ensure that the algorithms are transparent and explainable, so that stakeholders can understand how decisions are being made.

Another challenge is the need for infrastructure to support autonomous vehicles, such as charging stations, communication networks, and road infrastructure. This requires significant investment and coordination between different stakeholders, including governments, car manufacturers, and technology companies.

Overall, the implementation of autonomous vehicles is a complex process that requires significant investment, technical expertise, and collaboration between different stakeholders. While there are many technical limitations that need to be overcome, the potential benefits of autonomous vehicles are significant, including increased safety, improved efficiency, and reduced environmental impact.

15.5.2 REGULATORY HURDLES

The deployment of autonomous vehicles faces both technical and regulatory obstacles. Government approval is essential, involving complex safety regulations, testing, and certification. A clear legal framework is lacking, especially regarding liability in accidents. Potential job losses in sectors like transportation raise economic concerns. Collaboration between government, the automotive industry, and stakeholders is crucial to tackle these challenges and guarantee passenger and road user safety in the adoption of autonomous vehicles.

15.5.3 ETHICAL CONSIDERATIONS

The rise of autonomous vehicles brings ethical concerns. Safety is a prime worry due to potential hardware or software failures despite accident reduction. Job loss looms as automation expands, affecting industries like transport. Privacy issues arise from data collection by vehicle sensors. Unclear liability emerges from accidents, raising questions about responsibility. Ethical challenges also concern decision-making algorithms during unavoidable accidents. As self-driving vehicles become common, addressing these ethical considerations becomes crucial.

15.5.4 PUBLIC ACCEPTANCE AND PERCEPTION

Public acceptance and perception are pivotal for the successful integration of autonomous vehicles. Despite technological strides, doubts persist regarding their safety and reliability. Concerns encompass potential accidents due to software and hardware issues, job losses, and moral dilemmas in decision-making. To assuage these fears, transparent communication by manufacturers is key for trust-building. Safety must be paramount in development, aligning with robust testing and high standards.

Collaboration between policymakers and regulatory bodies is essential to establish updated safety norms. Public outreach and education are vital for dispelling misconceptions. A Pew Research study notes more worry than enthusiasm regarding automation, with potential benefits acknowledged alongside societal impacts. Addressing public sentiment is essential via manufacturer, policy, and public cooperation to ensure safe, dependable, and societally beneficial autonomous vehicles.

15.5.5 SAFETY CONCERNS

As with any technology that involves human lives, safety is a paramount concern when it comes to the implementation and adoption of autonomous vehicles. Despite the potential benefits, addressing safety issues is crucial. One major concern involves the vehicles' ability to accurately perceive and respond to their environment, including unexpected scenarios like pedestrians and complex traffic situations. Another worry is the vulnerability to cyberattacks, which could manipulate the vehicle's control systems. Human error, even in automated systems, remains a risk due to programming errors or malfunctions. Research emphasizes the potential for reducing accidents caused by human error but acknowledges challenges. Establishing safety standards and regulations is underway to ensure the safe use of autonomous vehicles on public roads.

Balancing benefits with safety concerns necessitates ongoing research, development, and regulatory efforts. One example of this is the National Highway Traffic Safety Administration in the United States, which has issued guidelines for the safe deployment of autonomous vehicles. These guidelines cover issues such as cybersecurity, the design of autonomous systems, and the testing and evaluation of these systems before they are deployed on public roads.

Another example is the European Union's General Safety Regulation, which mandates the inclusion of several advanced safety features in new vehicles, such as autonomous emergency braking and lane departure warning systems. This regulation also provides a framework for the testing and certification of autonomous vehicles.

15.6 CONCLUSION

AI technology has brought tremendous advancements in the field of driverless cars. With the use of machine learning techniques, driverless cars have come closer to becoming a reality. The development of intelligent personal assistants, speech recognition, and synthesis have improved the overall driving experience. However, there are still challenges to overcome in the implementation of autonomous vehicles, such as technical limitations, regulatory hurdles, ethical considerations, public acceptance and perception, and safety concerns. As the technology advances, it is important to address these challenges and work toward a future with safer and more efficient transportation.

Some of the key points discussed in the chapter are:

- Driverless cars are a significant application of AI and machine learning technologies.
- AI is used for various tasks in driverless cars, such as perception, decision-making, control, and communication.
- Supervised, unsupervised, and reinforcement learning algorithms are used in driverless cars for different purposes, such as object detection, path planning, and autonomous driving.
- Decision trees, random forests, and gradient boosting are machine learning models used in driverless cars for classification and regression tasks.
- Deep learning models such as convolutional neural networks (CNNs) and recurrent neural networks (RNNs) are used in driverless cars for tasks such as image and speech recognition, natural language processing, and predictive modelling.
- Natural language processing and speech recognition and synthesis are used in driverless cars for intelligent personal assistants and human-vehicle interaction.

- There are several technical, regulatory, ethical, public acceptance, and safety challenges that need to be addressed before fully implementing autonomous vehicles.
- Research is ongoing to improve the performance, safety, and acceptance of autonomous vehicles, including the use of advanced AI and machine learning technologies.

15.6.1 FUTURE PROSPECTS AND RESEARCH OPPORTUNITIES

AI in driverless cars is a dynamic research field with exciting opportunities. Current studies aim to enhance perception, decision-making, safety, and energy efficiency. Integration of AI could reduce congestion, boost accessibility, and enable new mobility forms. Future research might focus on fleet management, interaction, cybersecurity, and advanced techniques like generative adversarial networks (GANs) and transformers. The future holds promising prospects for innovative AI-driven research and advancements in driverless vehicles.

REFERENCES

Al Hamrashdi, H., Monk, S. D., & Cheneler, D. 2019. Passive Gamma-Ray and Neutron Imaging Systems for National Security and Nuclear Non-Proliferation in Controlled and Uncontrolled Detection Areas: Review of Past and Current Status. *Sensors*, *19*(11), 2638. https://doi.org/10.3390/s19112638

Alzubaidi, L., Zhang, J., Humaidi, A. J., Al-Dujaili, A., Duan, Y., Al-Shamma, O., Santamaría, J., Fadhel, M. A., Al-Amidie, M., & Farhan, L. 2021. Review of Deep Learning: Concepts, CNN Architectures, Challenges, Applications, Future Directions. *Journal of Big Data*, *8*(1), 53. https://doi.org/10.1186/s40537-021-00444-8

Carlevaris-Bianco, N., Ushani, A. K., & Eustice, R. M. 2016. University of Michigan North Campus Long-Term Vision and Lidar Dataset. *The International Journal of Robotics Research*, *35*(9), 1023–1035. https://doi.org/10.1177/0278364915614638

Dissanayake, G., Newman, P., Clark, S., Durrant-Whyte, H., & Csorba, M. 2001. A Solution to the Simultaneous Localization and Map Building (SLAM) Problem. *IEEE Transactions on Robotics and Automation*, *17*(3), 229–241. https://doi.org/10.1109/70.938381

Gomes, T., Matias, D., Campos, A., Cunha, L., & Roriz, R. 2023. A Survey on Ground Segmentation Methods for Automotive LiDAR Sensors. *Sensors*, *23*(2), 601. https://doi.org/10.3390/s23020601

Kiran, B. R., Sobh, I., Talpaert, V., Mannion, P., Sallab, A. a. A., Yogamani, S., & Pérez, P. 2022. Deep Reinforcement Learning for Autonomous Driving: A Survey. *IEEE Transactions on Intelligent Transportation Systems*, *23*(6), 4909–4926. https://doi.org/10.1109/tits.2021.3054625

System Initiative on Shaping the Future of Mobility Reshaping Urban Mobility with Autonomous Vehicles Lessons from the City of Boston. 2018. https://www3.weforum.org/docs/WEF_Reshaping_Urban_Mobility_with_Autonomous_Vehicles_2018.pdf

Thorpe, C., Hebert, M., Kanade, T., & Shafer, S. A. 1988. Vision and Navigation for the Carnegie-Mellon Navlab. *IEEE Transactions on Pattern Analysis and Machine Intelligence*, *10*(3), 362–373. https://doi.org/10.1109/34.3900.

Tunnell, J., Asher, Z. D., Pasricha, S., & Bradley, T. H. 2018. Toward Improving Vehicle Fuel Economy with ADAS. *SAE International Journal of Connected and Automated Vehicles*, *1*(2). https://doi.org/10.4271/12-01-02-0005

Wang, S., Cao, J., Sun, K., & Li, Q. 2020. *SIEVE: Secure In-Vehicle Automatic Speech Recognition Systems*. ResearchGate. https://www.researchgate.net/publication/357888553_SIEVE_Secure_In-Vehicle_Automatic_Speech_Recognition_Systems

Yvkoff, L. 2019. BMW Rolls-Out Its Intelligent Personal Assistant Feature Via Over-The-Air Update. *Forbes*.

Zhang, C., Liu, J., Chen, W., Shi, J., Yao, M., Yan, X., Xu, N., & Chen, D. 2021. Unsupervised Anomaly Detection Based on Deep Autoencoding and Clustering. *Security and Communication Networks*, *2021*, 1–8. https://doi.org/10.1155/2021/7389943

Section V

AI and Computer Vision

16 Comparative Analysis of Feature-Based Image Stitching Algorithms

Pawni Gupta, Namratha Betala,
and Sivakannan Subramani

16.1 INTRODUCTION

Image stitching is a commonly employed method in computer vision, encompassing the merging of multiple images depicting the same scene or object to generate a unified panoramic image. The objective of image stitching is to produce a cohesive and visually appealing composite image that captures a broader field of view than any individual image could achieve (Huang et al. 2019).

Image stitching finds various applications in different fields, such as photography (Wei et al. 2019), virtual reality (Ullah et al. 2022), and robotics (Chatterjee and Issac 2023). In these domains, it is utilized to craft high-resolution panoramic images, build immersive virtual environments, and create extensive maps of the surroundings (Wei et al. 2019). Additionally, image stitching plays a crucial role in industrial and scientific sectors, including medical imaging (Weibel et al. 2012), remote sensing (Meng et al. 2021; Manandhar et al. 2021), and surveillance.

The image stitching process typically encompasses several essential stages, including feature detection and matching, image alignment, and blending. Feature detection and matching algorithms are employed to recognize corresponding features or points of interest in different images, while image alignment techniques ensure the registration of images to a common coordinate system. Subsequently, blending techniques are utilized to merge the registered images into a unified composite image, with a focus on minimizing visual anomalies like seams and distortions.

The feature-based approach to image stitching has been extensively researched and proven effective across diverse imaging conditions and applications. However, challenges and limitations persist, necessitating attention. These include managing extensive image sets, addressing issues related to moving objects, and enhancing the precision and resilience of feature detection and matching algorithms (Bonny and Uddin 2016).

This chapter contributes significantly to image stitching by thoroughly exploring both traditional and modern feature-based methods. It covers a detailed study of these methods, classifying them into traditional and deep learning categories. The chapter also compares these methods based on their attributes, limitations, and advantages. Additionally, it highlights recent advancements in the field, aiding readers

DOI: 10.1201/9781003405436-21

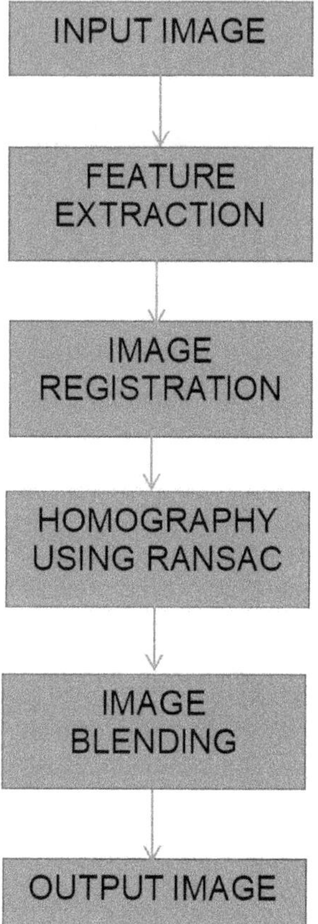

FIGURE 16.1   Feature-based image stitching process.

in selecting the most suitable algorithm for their specific applications. Figure 16.1 shows the feature-based image stitching process.

16.2 IMAGE STITCHING TECHNIQUES

Image stitching techniques can be broadly divided into two types: the direct method and the feature-based method, as shown in Figure 16.2.

16.2.1 DIRECT METHOD

These methods align multiple images by directly minimizing dissimilarities between pixels by comparing different pixel intensities in an image. This method can be complex since it compares each pixel with another one and is not reliable when it comes to image scaling and rotation (Adel et al. 2014). It takes advantage of information present in the image but lacks in taking adequate range of convergence (Kale and Singh 2015).

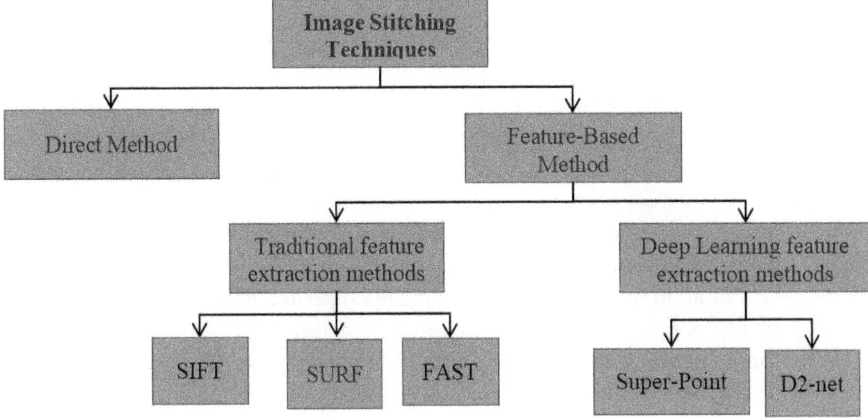

FIGURE 16.2   Image stitching techniques.

16.2.2 Feature-Based Methods

The most straightforward approach to identifying corresponding feature points in a pair of images involves comparing all features in one image to those in the other, using a local descriptor. Unfortunately, this method exhibits quadratic complexity with respect to the anticipated number of features, rendering it unfeasible for certain applications. When it comes to image stitching using feature-based techniques, it's important to recognize that feature extraction, registration, and blending are distinct steps integral to the stitching process. Feature-based methods initiate the procedure by establishing correspondences among various geometric entities such as points, lines, edges, and corners. Desirable attributes for robust detectors include resistance to image noise, scale, translation, and rotation invariance.

16.2.2.1 Terminologies in Feature Detection and Matching

a) **Feature.** A feature denotes a distinct and recognizable pattern or structure present in an image, identifiable through algorithms. Features often manifest as points or regions in an image, characterized by their geometric or photometric attributes, such as edges, corners, or blobs. Through the detection and matching of features across different images, computer vision algorithms can accomplish tasks such as object recognition, image alignment, and stereo vision.

b) **Interest point.** This term refers to a specific location within an image possessing unique visual properties, facilitating easy identification and tracking. Algorithms designed to detect interest points typically search for distinctive patterns or structures in the image, such as edges, corners, or blobs, which are unlikely to be replicated in other parts of the image. Once identified, these interest points serve as a foundation for subsequent analysis or processing.

c) Descriptor. A feature descriptor is a concise portrayal of the visual attributes of a specific point of interest within an image. Following the detection of interest points in an image, feature descriptors come into play to encapsulate the distinctive qualities of each point, encompassing aspects such as shape, texture, color, or gradient orientation. Typically, these descriptors are computed by analyzing a small neighborhood of pixels surrounding the identified interest point.

A descriptor can be divided into two sets of attributes:

 i) Local descriptors provide a description of specific regions or points within an image. Local descriptors are computed around interest points or key points detected in an image and capture the unique properties of those points, such as their shape, texture, or color. Examples of local descriptors include scale-invariant feature transform (SIFT), speeded up robust features (SURF), oriented FAST and rotated BRIEF (ORB), etc. Where BRIEF stands for binary robust independent elementary features and FAST for features from accelerated segment test.

 ii) Global descriptors provide a single representation of the entire image or object, summarizing its overall visual properties. Examples of global descriptors include color histograms, which count the number of pixels in an image with each color value, or bag-of-visual-words representations, which group similar local descriptors into visual "words" and count their occurrences in an image.

d) Matching. Refers to the process of finding corresponding features or interest points between two or more images. This process involves identifying the same physical objects or regions in different images by matching their unique visual properties or feature descriptors.

By finding corresponding features in different images, algorithms can determine how objects or scenes in the images are related to each other, despite variations in lighting, viewpoint, or scale.

16.2.2.2 Traditional Algorithms

Expanding on the success of image feature representation in various computer vision applications, numerous methods have emerged, aiming to extend their applicability to higher-dimensional data. The core concept involves capturing the essence of an image by describing its content through specific points or neighborhoods rather than considering the entire image as a whole.

Some of the widely used methods are SIFT, SURF, FAST

a) SIFT

Scale-invariant feature transform or SIFT was introduced by D. Lowe in 2004. In image mosaicking or image stitching, multiple features can be detected using this method. In particular, SIFT is a method that is used to detect key points in an image that are invariant to scale, orientation, and other changes. The common problem in image mosaicking is matching features. When the images are of the same scale and orientation then normal corner detection method can be used to extract the features (Lowe 2004). However, the same method will not work once there is a change in the scaling and orientation of the image. Therefore, SIFT is a popular method used in such cases because of its invariant nature towards the scale and orientation of the image. SIFT's distinctive feature descriptors enable effective matching of key points, making it well-suited for image registration and alignment in the stitching process. However, SIFT does have some disadvantages, including computational complexity and high memory requirements, which can be limiting for real-time applications on resource-constrained devices (Wang et al. 2013).

b) **SURF**

SURF (speeded up robust features) is a feature detection and description algorithm that was developed by Herbert Bay, Andreas Ess, Tinne Tuytelaars, and Luc Van Gool in 2006. It is an improvement over the earlier SIFT algorithm, with faster computation time and improved robustness to image noise and changes in viewpoint. SURF is a scale-invariant feature detector that can detect distinctive points in an image regardless of scale or orientation. SURF demonstrates robustness to variations in lighting, rotation, and various forms of image distortion (Bay et al. 2006). In the context of image stitching, the typical application of the SURF algorithm involves its implementation on individual images to detect and match features. Subsequently, the aligned and matched features are used to blend the images together. A notable strength of the SURF method lies in its swift computational capability, facilitating real-time applications such as tracking and object recognition. It improves upon the detection process employed in SIFT while maintaining the quality of identified points. However, it should be noted that while SURF excels in expediting the matching phase, it may exhibit limitations in handling changes in viewpoint and illumination.

c) **FAST**

The feature accelerated segment test (FAST) is a technique for detecting corners, which can be employed to obtain feature points for tracking. FAST is a corner detection method used in computer vision to extract feature points for tracking and mapping objects in various applications. It was developed by Edward Rosten and Tom Drummond and first published in 2006.

The FAST algorithm uses nonmaximum suppression in image stitching. In this, the algorithm checks each feature point in a neighborhood and retains only the one with the highest score (e.g., strongest corner response or highest gradient magnitude). All other nearby feature points with weaker responses are suppressed or discarded, assuming they are less relevant or redundant. The goal of nonmaximum suppression is to remove as many redundant feature points as possible while retaining the most salient ones, reducing the computational cost and improving the overall performance of feature detection and matching algorithms

One of its primary advantages is its high computational efficiency, making it ideal for real-time video processing. Additionally, when used in conjunction with machine learning techniques, it can achieve even better performance in terms of computation time and resource usage. Therefore, the FAST corner detector is well-suited for real-time video processing applications that require high-speed performance (Viswanathan 2011).

16.2.2.3 Deep Learning Feature Extraction Techniques

Switching to convolutional neural networks (CNNs) can outperform traditional feature extractors. CNNs excel at capturing intricate image details, learning task-specific features, and doing so with greater efficiency.

a) **SuperPoint: Self-supervised interest point detection and description**
   SuperPoint represents a fully convolutional neural network architecture designed for SIFT-like interest point detection and description. The model is trained in a single forward pass using a self-supervised domain adaptation framework known as homographic adaptation. The architecture incorporates a VGG-style (visual geometry group) encoding mechanism for feature extraction, and it includes two distinct decoders. The interest point decoder is specialized in point detection, while the descriptor decoder is dedicated to point description (DeTone et al. 2018).

b) **D2-Net: A trainable CNN for joint description and detection of local features**
   A single CNN serves a dual purpose by simultaneously functioning as both a dense feature descriptor and a feature detector. Unlike traditional methods, the detection process is not carried out on low-level image structures; instead, it is deferred until more reliable information becomes available and is performed concurrently with the image description. Although the features may not be as precisely localized as those detected by conventional feature methods, they prove highly effective for tasks such as three-dimensional reconstruction (Dusmanu et al. 2019).

16.3 COMPARISON BETWEEN ALGORITHMS

Table 16.1 provides a comparison of different feature-based methods based on various image transformation and deformations.

16.4 RECENT WORKS

Recent advancements in feature-based methods have concentrated on creating more effective and precise techniques for extracting significant characteristics or attributes from data. The primary goal is to develop methods that can capture the most relevant information while eliminating redundant or irrelevant data. The reason for this is that effective feature extraction methods play a crucial role in enhancing the performance of machine learning algorithms. An optimized SIFT algorithm over the traditional SIFT was proposed by Huang et al. (2019). The traditional SIFT algorithm involves many calculations leading to mismatching and an unsatisfactory feature vector. The proposed optimized SIFT algorithm uses a Laplacian operator to sharpen the edges

TABLE 16.1

Comparison between Different Feature-Based Algorithms

Feature Descriptor	Algorithm	Feature Detection	Speed[1]	Invariate To[2]			Robustness to Noise/Occulation[3]
				Scale	Rotation	Affine Distortion	
SIFT	Scale-invariant feature transform	Yes	Slow	Yes	Yes	Yes	Good
SURF	Speeded up robust features	Yes	Fast	Yes	Yes	Yes	Good
FAST	Features from accelerated segment test	Yes	Fastest	Yes	No	No	Poor
SuperPoint		Yes	Fast	Yes	Yes	No	Good
D2-Net	Detect and describe network	Yes	Fast	Yes	Yes	Yes	Good

[1] Speed refers to the computational speed of the algorithm.

[2] Rotation and scale refer to how well the algorithm can detect and match features in an image even when the object is rotated or scaled; affine distortion refers to how well the algorithm can detect and match features in an image even when the object is distorted in a nonuniform way.

[3] Robustness to noise/occlusion refers to the ability of a computer vision algorithm or system to maintain accurate results despite the presence of noisy or obstructed data.

of the images and a bidirectional matching algorithm for matching the feature points. The proposed algorithm gives 93.79 % accuracy for outdoor scenes and 90.77% accuracy for indoor scenes. The RSIFT (rapid scale-invariant feature transform) operator (Rui et al. 2021) was proposed to reduce feature extraction time. In this algorithm the level of octave and levels of difference of Gaussians (DoG) are reduced at the same time, resulting in improved accuracy and stability.

Further, the adaptive as natural as possible warping algorithm (AANAP) was used for image registration for an unmanned aerial vehicles (UAV) image mosaic for obtaining a natural-looking mosaicking result with minimal registration errors. The results demonstrated that the proposed algorithm is improved by 78% compared to SIFT, and extraction time was 0.28 s. Yang et al. (2013) solved the problems of matching speed and ghosting in an image mosaic on SIFT. A fast registration algorithm is used to remove invalid key points, thereby decreasing registration computation and improving the registration accuracy. An improved frame difference method was proposed for detecting and eliminating ghosting in the image. The algorithm resulted in 90% accuracy in 0.81 s. Various algorithms (Kumar and Singh 2019), such as SIFT, SURF, binary robust invariant scalable keypoints (BRISK), and ORB, were compared to find the most appropriate algorithm for dog muzzle recognition and resulted in ORB being the most appropriate algorithm for the given problem statement. Guo et al. (2018) analyzed the theory of SIFT and used it on a poker image dataset to study SIFT attributes, that is, its invariant nature on rotation, translation, and scaling of an image.

The results indicate that SIFT performs well in certain aspects, but it has limitations, particularly in dealing with objects in complex backgrounds. An enhanced version of SIFT was proposed by Yang et al. (2018), aiming to expedite feature point extraction and reduce redundant information by employing the Hellinger kernel function instead of the Euclidean distance function. Devia et al. (2019) applied a swift technique for creating UAV image mosaics, relying on SIFT, particularly for monitoring crop growth. SURF and ORB algorithms were utilized in the context of rice crop monitoring using multispectral aerial imagery, machine learning, and image mosaicking processing techniques.

Fujinaga et al. (2018) proposed a method for generating a greenhouse map, the initial step in developing a robot for tomato cultivation. The approach involved using SURF with multimodal images. Jadhav and Singh (2018) demonstrated the success of deep residual networks in classifying remote sensing images, achieving an overall accuracy of 85% or higher for major crops. The study highlighted the superior performance of CNN over existing algorithms in terms of image classification and segmentation accuracy, though it didn't explore improving feature extraction or matching algorithms in image mosaicking.

In Zhang et al. (2018), a SURF-based image technique was employed in low-altitude UAV remote sensing images for rapid image slicing. Moreover, experiments were conducted to compare and analyze the performance of SURF and SIFT in detecting feature points. The results indicated that SURF has a higher speed in feature matching compared to SIFT, while maintaining a higher level of accuracy in image registration. To solve the problem of quality of stitched image and time cost, Adel et al. (2015) compared different feature detectors, such as the Harris corner detector, SIFT, SURF, FAST, GoodFeaturesToTrack, MSER, and ORB techniques, and manipulated the common feature used in implementation of these techniques to get a high quality mosaicked image. The result shows that SIFT takes the highest matching time whereas ORB takes the least matching time. The report concluded that ORB is best for real-time applications. Even though ORB detects fewer key points, it is more accurate than SIFT and SURF. Ro and Kim (2021) analyzed various problems such expensive equipment making in situ analysis impossible. Therefore, to overcome the limited field of view of the microscope and enhance observation, the image stitching technique was implemented. The experimental findings indicated that for the three sets of mineral images, all characteristic points were identified within a search region of 30%, with no unnecessary expansion required. This resulted in an enhanced matching speed. Weibel et al. (2012) aimed to address the limitations of the endoscopic camera field of view in minimally invasive surgery by focusing on tracking and extracting features in soft-tissue operation mosaicking. The researchers were motivated by the necessity to enhance surgical navigation, particularly for surgeons dealing with challenges such as deformed organs and patient movements, regardless of their experience level. To address the issue of soft-tissue video mosaicking, they conducted experimental tests employing various techniques such as SIFT, SURF, good features to track (GFTT), and the FAST pyramid approach. Their algorithm incorporated 128-dimensional SURF and 64-bit BRIEF descriptors, with a specific focus on the analysis of BRIEF descriptors concerning local tissue deformation.

In a study conducted by Karami et al. (2017), the authors compared different matching techniques, including SIFT, SURF, BRIEF, and ORB, against a range of image transformations and deformations such as scaling, rotation, noise, fish-eye distortion, and shearing. The study aimed to assess matching parameters, detected features, execution time, and matching rates. The results indicated that ORB exhibited the fastest performance, while SIFT consistently performed the best across various scenarios. However, it was noted that in situations where the rotation angle approached 90 degrees, both ORB and SURF outperformed SIFT.

When dealing with noisy images, ORB and SIFT showed similar performances. The researchers observed that ORB features were primarily concentrated in objects located at the center of the image, whereas features in SURF, SIFT, and FAST key point detectors were distributed throughout the entire image. The paper proposes an unsupervised deep image stitching framework consisting of two stages—unsupervised coarse image alignment and unsupervised image reconstruction for unlabeled data. This model provides better quality when compared with other feature-based methods and supervised solutions (Nie et al. 2021). A novel approach was introduced, leveraging CNN-inspired semantic feature extraction. This method involves computing and quantifying the semantic features of individual pixels in an image using a neural network. These features serve to represent the contribution of each pixel to the overall semantics of the image. The results of the experiments indicate that this method can decrease reliance on traditional feature extraction techniques, specifically those based on shallow features like edges and corners (Shi et al. 2020).

The insights presented showcase the versatility of various feature extraction and detection methods across diverse applications (Table 16.2). The modifications and adaptations made to traditional algorithms, such as SIFT and SURF, highlight the continuous efforts to enhance their efficiency and applicability. From improved accuracy in object recognition to optimized computational efficiency in agriculture and defense applications, these methods contribute significantly to the evolving landscape of computer vision. However, the choice of algorithm often depends on the specific requirements of the task and the characteristics of the data involved.

16.5 CONCLUSION

In summary, the chapter presents a comparative analysis of three traditional and two deep learning state-of-the-art feature-based image stitching algorithms. The chapter evaluates the algorithms based on their ability to produce high-quality panoramic images while minimizing distortion and processing time. The study concludes that SIFT and SURF algorithms produce the highest quality panoramic images with minimal distortion. Deep learning methods such as D2-Net and SuperPoint have shown improvement in automating feature extraction and matching tasks, making image stitching faster and more accurate. These methods continue to be the subjects of ongoing research and development. The spectrum of image feature extraction algorithms, such as improved SIFT, RSIFT + AANAP, ORB, CNN, and SURF, among others, demonstrates their versatile applications, from military defence and

TABLE 16.2

Analysis of Different Feature-Based Algorithms

Citations	Application	Algorithm	Methodology	Results	Dataset
Huang et al. (2019)	General	Improved SIFT	Uses Laplacian operator in SIFT instead of Gaussian to reduce mismatched and unsatisfactory feature vector	93.79% accuracy for outdoor image and 90.77% accuracy for indoor image	116 indoor images, 236 outdoor images
Rui et al. (2021)	Military and defense	RSIFT + AANAP	Reduces the level of octave and DoG at the same time to reduce extraction time	Result is improved by 78% when compared with SIFT and extraction time was 0.28 s	635 images, mainly including forest, mountain, farmland, water, urban construction, etc.
Yang et al. (2013)	General	Improved SIFT	Reducing the invalid key points, hence improving registration accuracy	Gives 90% accuracy and featured extraction time is 0.81s	NS
Kumar and Singh (2019)	Animal welfare	SIFT, SURF, BRISK, ORB	All the algorithms are compared to find the most appropriate for dog muzzle reorganization	ORB resulted in being the most appropriate	55 dog muzzle pattern images acquired from 11 dogs
Guo et al. (2018)	General	SIFT	Studied theory of SIFT and its attributes i.e., scale invariance, orientation invariance, etc.	SIFT performs well for certain features, but struggles with objects in complex backgrounds	Poker images
Yang et al. (2018)	General	Improved SIFT	Uses Hellinger function to improve feature extraction efficiency and reduce invalid key points	High accuracy and efficiency	NS
Devia et al. (2019)	Agriculture	SURF, ORB	Using aerial imagery with multiple spectral bands to monitor rice crops	NS	600 rice images
Fujinaga et al. (2018)	Agriculture	SURF	A map creation that shows the growth stages of tomatoes	NS	70 tomato images
Jadhav and Singh (2018)	Agriculture	Harris corner detection, CNN	Optimizing legacy feature-domain algorithms by improving computational efficiency, processing capacity, and storage requirements	CNN attained an accuracy of at least 85% for all the main crops and achieved higher detection rate and better repeatability using Harris corner	Landsat public dataset

Zhang et al. (2018)	Agriculture	SURF	Addressing the issue of splicing in low-altitude remote sensing images captured by UAVs	Eliminates invalid points from nonoverlapping region	150 images of rice
Adel et al. (2015)	General	Harris corner detector, SIFT, SURF, FAST, ORB	Evaluate different algorithms to identify the optimal one for varying contexts	ORB takes the least time for matching features	NS
Ro and Kim (2021)	Mineralogical analysis	SURF	Overcome the limited field of view of the microscope and enhance observation	Enhanced matching speed	Thin sections of sandstone images, using a polarizing microscope
Weibel et al. (2012)	Medical	SURF and BRIEF descriptors	Better understanding of feature extraction and tracking in image mosaicking	Robust matches were not found in the sequence	In vivo soft tissue
Karami et al. (2017)	General	SIFT, SURF, BRIEF, ORB	Compare various algorithms to determine the most suitable one for different scenarios	ORB was the fastest, whereas SIFT performed best in most of the scenarios	Images with varying intensity and color composition values
Nie et al. (2021)	General	Unsupervised homography	Unsupervised deep image stitching with two stages: coarse alignment and image reconstruction for unlabeled data	Provides better quality when compared with other feature-based methods and supervised solutions	Synthetic dataset
Meng et al. (2021)	Satellite	SIFT, SURF, ORB	Evaluation of different algorithms for visible camera and satellite images in UAV multispectral image registration	Sift performs better than the other algorithms	>10,000 images under different weather conditions
Shi et al. (2020)	General	CNN	Computing and quantifying the semantic features of individual pixels in an image using a neural network	Decrease reliance on traditional feature extraction techniques	Live image database

AANAP = adaptive as natural as possible warping algorithm; BRIEF = binary robust independent elementary features; CNN = convolutional neural network; NS = not specified; ORB = oriented FAST and rotated BRIEF; RSIFT = rapid scale-invariant feature transform; SIFT = scale-invariant feature transform; SURF = speeded up robust features; UAV = unmanned aerial vehicles. BRISK = binary robust invariant scalable keypoints; DoG = difference of Gaussians; FAST = features from accelerated segment test.

agriculture to mineralogical analysis and medical imaging, reflecting ongoing efforts to enhance accuracy and efficiency across diverse domains and scenarios. In summary, the research provides valuable insights into the strengths and limitations of different image stitching algorithms and their suitability for various applications.

REFERENCES

Adel, Ebtsam, Elmogy, Mohammed and Elbakry, Hazem. 2014. Image Stitching Based on Feature Extraction Techniques: A Survey, *International Journal of Computer Applications*, Volume 99.

Adel, Ebtsam, Elmogy, Mohammed and Elbakry, Hazem. 2015. Image Stitching System Based on ORB Feature-Based Technique and Compensation Blending, *International Journal of Advanced Computer Science and Applications*, Volume 6.

Bay, Herbert, Tuytelaars, Tinne and Van Gool, Luc. 2006. SURF: Speeded Up Robust Features, *Computer Vision-ECCV*, Volume 3951.

Bonny, M. Z. and Uddin, M. S. 2016. Feature-Based Image Stitching Algorithms, *International Workshop on Computational Intelligence (IWCI)*, Dhaka, Bangladesh, pp. 198–203. DOI: 10.1109/IWCI.2016.7860365

Chatterjee, Saurabh and Issac, K. Kurien. 2023. Viewpoint Planning and 3D Image Stitching Algorithms for Inspection of Panels, *NDT & E International*, Volume 137.

DeTone, D., Malisiewicz, T. and Rabinovich, A. 2018. SuperPoint: Self-Supervised Interest Point Detection and Description, *IEEE/CVF Conference on Computer Vision and Pattern Recognition Workshops (CVPRW)*, Salt Lake City, UT, USA, pp. 337–33712. DOI: 10.1109/CVPRW.2018.00060.

Devia, C., Rojas, J., Petro, E., Martinez, C., Mondragon, I., Patino, D., Rebolledo, C. and Colorado, J. 2019. Aerial Monitoring of Rice Crop Variables Using an UAV Robotic System, pp. 97–103. DOI: 10.5220/0007909900970103.

Dusmanu, M., Rocco, I., Pajdla, T., Pollefeys, M., Sivic, J., Torii, A. and Sattler, T. 2019. D2-Net: A Trainable CNN for Joint Description and Detection of Local Features, *IEEE/CVF Conference on Computer Vision and Pattern Recognition (CVPR)*, pp. 8084–8093.

Fujinaga, Takuya, Yasukawa, Shinsuke, Li, Binghe and Ishii, Kazuo. 2018. Image Mosaicing Using Multi-Modal Images for Generation of Tomato Growth State Map, *Journal of Robotics and Mechatronics*, Volume 30.

Guo, F., Yang, J., Chen, Y. and Yao, B. 2018. Research on Image Detection and Matching Based on SIFT Features, *3rd International Conference on Control and Robotics Engineering (ICCRE)*, Nagoya, Japan, pp. 130–134. DOI: 10.1109/ICCRE.2018.8376448.

Huang, Z., Wang, H. and Li, Y. 2019. The Research of Image Mosaic Techniques Based on Optimized SIFT Algorithm, *RSVT '19: Proceedings of the International Conference on Robotics Systems and Vehicle Technology*, pp. 80–86. DOI: 10.1145/3366715.3366737.

Jadhav, Jagannath K. and Singh, R. P. 2018. Automatic Semantic Segmentation and Classification of Remote Sensing Data for Agriculture, *Mathematical Models in Engineering*, Volume 4.

Kale, P. and Singh, K. R. 2015. A Technical Analysis of Image Stitching Algorithm Using Different Corner Detection Methods, *International Journal of Innovative Research in Computer and Communication Engineering*, pp. 3300–3309.

Karami, E., Prasad, S. and Shehata, M. S. 2017. Image Matching Using SIFT, SURF, BRIEF and ORB: Performance Comparison for Distorted Images. ArXiv, abs/1710.02726.

Kumar, Santosh and Singh, Sanjay. 2019. Cattle Recognition: A New Frontier in Visual Animal Biometrics Research, *Proceedings of the National Academy of Sciences, India Section A: Physical Sciences*, Volume 90.

Lowe, David G. 2004. Distinctive Image Features from Scale-Invariant Key Points, *International Journal of Computer Vision*, Volume 60.

Manandhar, P., Jalil, A., AlHashmi, K. and Marpu, P. R. 2021. Automatic Generation of Seamless Mosaics Using Invariant Features, *Remote Sensing*, Volume 13, Issue 16, p. 3094. https://doi.org/10.3390/rs13163094

Meng, Lingxuan, Zhou, Ji, Liu, Shaomin, Ding, Lirong, Zhang, Jirong, Wang, Shaofei and Lei, Tianjie. 2021. Investigation and Evaluation of Algorithms for Unmanned Aerial Vehicle Multispectral Image Registration, *International Journal of Applied Earth Observation and Geoinformation*, Volume 102.

Nie, Lang, Lin, Chunyu, Liao, Kang, Liu, Shuaicheng and Zhao, Yao. 2021. Unsupervised Deep Image Stitching: Reconstructing Stitched Features to Images, *IEEE Transactions on Image Processing*, Volume 30.

Ro, Sung-Hyok and Kim, Se-Hun. 2021. An Image Stitching Algorithm for the Mineralogical Analysis, *Minerals Engineering*, Volume 169.

Rui, Ting, Hu, Yucheng, Yang, Chengsong, Wang, Dong and Liu, Xun. 2021. Research on Fast Natural Aerial Image Mosaic, *Computers & Electrical Engineering*, Volume 90.

Shi, Zaifeng, Li, Hui, Cao, Qingjie, Ren, Huizheng and Fan, Boyu. 2020. An Image Mosaic Method Based on Convolutional Neural Network Semantic Features Extraction, *Journal of Signal Processing Systems*, Volume 92.

Ullah, Hayat, Afzal, Sitara and Ullah Khan, Imran. 2022. Perceptual Quality Assessment of Panoramic Stitched Contents for Immersive Applications: A Prospective Survey, *Virtual Reality & Intelligent Hardware*, Volume 4.

Viswanathan, Deepa. 2011. Features from Accelerated Segment Test (FAST). https://www.semantic scholar.org/paper/Features-from-Accelerated-Segment-Test-(-FAST-)-Viswanathan/ cd267a4b04d835dbecf01d47fc69ed3a38c23055

Wang, Yan, Lin, Bing and Sun, Hua. 2013. Image Mosaic Based on Improved SIFT, *Advanced Materials Research*, Volume 760–762.

Wei, Lyu, Zhou, Zhong, Lang, Chen and Zhou, Yi. 2019. A Survey on Image and Video Stitching, *Virtual Reality & Intelligent Hardware*, Volume 1.

Weibel, Thomas, Daul, Christian, Wolf, Didier, Rosch, Ronald and Guillemin, Francois. 2012. Graph Based Construction of Textured Large Field of View Mosaics for Bladder Cancer Diagnosis, *Pattern Recognition*, Volume 45.

Yang, Chunde, Wu, Ge and Shi, Jing. 2018. The Algorithm of Fast Image Stitching Based on Multi-Feature Extraction, *American Institute of Physics Conference Series*, Volume 1967.

Yang, Y., Xu, Q., Gan, Z. and Liu, F. 2013. A Research on Registration and De-ghosting Algorithm in Image Mosaic, *International Conference on Wireless Communications and Signal Processing*, Hangzhou, pp. 1–5, DOI: 10.1109/WCSP.2013.6677232.

Zhang, Weiping, Li, Xiujuan, Yu, Junfeng, Kumar, Mohit and Mao, Yihua. 2018. Remote Sensing Image Mosaic Technology Based on SURF Algorithm in Agriculture, *EURASIP Journal on Image and Video Processing*, Volume 2018.

Section VI

AI and Its Applications in Material Science

17 Revolutionizing Concrete Engineering
Predicting Material Properties with AI Insights

Khushi Sabarad and Sivakannan Subramani

17.1 INTRODUCTION

The construction industry has experienced a growing demand for new concrete mixtures, which has led to the need for research on the physical characteristics of concrete. This research aims to address the industry's difficulties in generating new concrete types and ultimately improve design standards. In the field of concrete prediction models, empirical relationships derived from statistical analysis, particularly, there has been a wide use of the regression model (Cheng et al. 2012; Yuan et al. 2014). These regression models utilize mathematical expressions to find the factors that influence the correlation between output variable concrete strength (y) and multiple input variables (x_i), resulting in a generic equation in the form:

$$y = f(b_i, x_i). \tag{17.1}$$

The work of several researchers suggests the use of various regression models for estimating the physical characteristics of concrete, for example, elastic modulus, tensile strength, compressive strength, and shear strength (Popovics and Ujhelyi 2008; Slater et al. 2012; Silva and Dhir 2015; Namyong et al. 2004). Although these models have benefits in specific cases, their development is time consuming, they lack reliability in complex scenarios, and can be expensive. Chou and Tsai (2012) have specifically pointed out the challenges of analyzing the mechanical strength of high-performance concrete (HPC) because of the intricate connection between components of the concrete and the resultant mechanical strength. Additionally, the existing statistical equations included in design standards and codes often neglect the impacts of new mixture components, raising concerns about their capability to foresee the mechanical strength of novel types of concrete with multiple attributes.

To overcome the limitations of standard regression models, machine learning (ML) methods are proving to be a promising and reliable alternative for estimating concrete strength (Yaseen et al. 2018). Based on the type of learning the model requires, ML methods can be categorized into three major types as follows: supervised learning, unsupervised learning, and reinforced learning. Supervised learning,

DOI: 10.1201/9781003405436-23

in particular, has demonstrated reliability in determining the mechanical properties of concrete by utilizing computer algorithms to generate hypotheses and patterns from data, which can then be used to anticipate future values (DeRousseau et al. 2018).

The four major ML approaches predominantly used for predicting concrete strength are artificial neural networks (ANNs), support vector machines (SVMs), decision trees, and search algorithms. The selection of a suitable ML model is dependent primarily on the dimensional parameters, such as the number of input variables and the size of the dataset. It is critical to measure the performance of these ML models and this is done using various metrics but all of them essentially compare the actual and predicted values.

This study aims to provide an analytical comparison of published research that covers different ML models to predict the physical characteristics of concrete. It involves an in-depth literature review, identifying and utilizing relevant references for each ML approach. ML approaches with limited information or rarely observed data were excluded from consideration, and the focus was placed on techniques that were beneficial to the study.

17.2 PREDICTION OF MECHANICAL PROPERTIES OF CONCRETE

ML models utilize large datasets, which are divided for the purposes of training, testing, and validation of the model. Among the first steps in developing a model is providing exposure to the data through the training set, while the validation dataset aids in regularization. If the error rate in the validation dataset increases, the training process is halted to prevent overfitting. Subsequently, the fitted model leverages the validation dataset to predict outputs for new observations. The performance of a model is indicated by its accuracy, which is obtained by exposing the model to the test set. The underlying principle of accuracy measure of various types of ML models is to assess the degree of conformity between estimated values and actual data. Furthermore, they facilitate sensitivity analysis by considering the weight of input parameters during evaluation (Xu et al. 2019a; Belalia Douma et al. 2017; Van Dao et al. 2019).

17.2.1 Artificial Neural Network

The basic framework of the human brain influenced the creation of ANNs (Nazemi et al. 2019; Sharifzadeh et al. 2019). ANN is a network of nodes (neurons) with connections (synapses) that transmit signals, taking the output obtained from one node as the input value of another. These connections are called edges. When a node receives information, it merges it with the information gained from other nodes. The nodes and edges carry a weight that changes as the model learns. The weight increases or decreases based on how much value the input parameters hold to the output variable (DeRousseau et al. 2018). Merged information is then transmitted to other nodes. This iterative procedure stops when the data is fitted, which can be seen by the merge in the rate of error, or when we hit the iteration limit (Bourdeau et al. 2019). Nodes are aggregated into 1 input layer, hidden layer(s), and output layer.

Activation functions are essential to produce output and make sure that the data is transmitted through the different layers (Esfe et al. 2015).

17.2.1.1 Extreme Learning Machine

Feed-forward neural networks consisting of a single hidden layer is trained by a neural network technique called an extreme learning machine (ELM; Yaseen et al. 2018). ELM overcomes slow training and overfitting problems. ELM initializes the hidden nodes randomly and applies rectification without iterative tuning, which is a distinctive feature of the algorithm. Yaseen et al. (2018) used an ELM model to predict the compressive strength (FCU) of foamed concrete. Findings disclosed that ELM was better in prediction in terms of accuracy than the other three ML algorithm models. Al-Shamiri et al. (2019) steadily raised the model's hidden neuron count from 10 to 200. It was found that 110, by far, gave the best possible result to predict the FCU of high strength concrete (HSC). The correlation coefficient value revealed that ELM has a great predictive ability.

17.2.1.2 Hybrid ANN-Based Model

The concept of a "hybrid approach" refers to the combination of two distinct methodologies to create a novel and enhanced model. Researchers have shown great interest in this approach due to its ability to provide a nuanced perspective and capitalize on the strengths of both models. Consequently, there has been extensive research conducted on the reliability of models such as the adaptive neuro-fuzzy inference system (ANFIS; Yuan et al. 2014; Van Dao et al. 2019; Khademi et al. 2016, 2017; Amani and Moeini 2012; Ahmadi-Nedushan 2012). According to the universal approximation theorem, neural networks possess the capability to approximate any function. ANFIS models serve as universal approximators by integrating ANN and fuzzy logic (FL) parameters. In this setup, ANN optimizes the membership capacity to minimize error rates, while FL rules contribute expert knowledge (Van Dao et al. 2019). By employing fuzzy "if-then" rules, this model generates specific input-output sets.

The ANFIS model was deployed to assess the FCU of geopolymer concrete, as well as a sustainable variety of concrete that incorporated industrial byproducts such as fly ash and slag from blast furnaces. Research findings have revealed that ANFIS outperforms back-propagation neural networks (BPNNs) in terms of prediction ability. Additionally, this model has successfully evaluated the shear strength of RC and HSC beams (Amani and Moeini 2012; Bashir and Ashour 2012), demonstrating accurate predictions exceeding currently existing design codes used by several developed countries having good infrastructure.

Yuan et al. introduced the genetic algorithm (GA) model that further improves upon the performance of BPNN by optimizing the weights and thresholds of the network (Yuan et al. 2014). This optimization technique is a resemblance to the natural selection process observed in organisms, where fit candidates are chosen for procreation to ensure the next generation produced is resilient (Kramer 2017). GA proves to be a strong candidate for BPNN optimization, as it can identify a global optimum solution while avoiding local optima. The GA-ANN approach was found to be more effective than BPNN in predicting FCU containing slag and fly ash.

To determine the most effective ANN structure, Behnood and Golafshani (2018) proposed a multi-objective gray wolves optimization (MOGWO) algorithm. This hybrid MOGWO-ANN model was used to predict the FCU of silica fume concrete, yielding satisfactory accuracy. Sensitivity analysis indicated that the maximum aggregate size significantly influenced the FCU of concrete.

Furthermore, Bui et al. (2018) predicted the tensile and compressive strength of HPC utilizing an ANN model whose weights and biases were optimized by a modified firefly algorithm (MFA). The findings revealed that the MFA-ANN model offers improved speed and accurate predictions.

17.2.2 Back-Propagation Neural Network

Back propagation (BP) in neural networks is short for 'backward propagation of errors'. It is a standard approach to train ANN (Xu et al. 2019a). The weights of neural networks can be fine-tuned to minimize the error rates acquired during the iteration and increase its generalization making it more reliable. In previous studies, researchers utilized BPNN to make predictions regarding the FCU of HPC (Deepa et al. 2010; Chou et al. 2011; Chithra et al. 2016). The variables considered as inputs in these studies included the constituents of the concrete mixture and the age of testing. A thorough analysis disclosed that BPNN is very good at prediction and is more accurate than the regression models. The FCU of recycled aggregate concrete (RAC) was predicted by Topçu and Saridemir (2008), who created a BPNN using an FL model and gradient descent algorithm. Both models were accurate, but in terms of R^2, root mean squared error, and mean absolute percentage error (MAPE), BPNN showed better performance than FL. Asteris et al. (2016) created an ANN model using the Levenberg-Marquardt algorithm. The findings showed that BPNN can accurately predict the FCU of self-compacting concrete (SCC). Sensitivity analysis disclosed that the compressive strength was mostly affected by viscosity-modifying admixtures added into SCC. Through sensitivity analysis, Naderpour et al. studied the BPNN performance and investigated how individual parameters affected the compressive strength of RAC (Naderpour et al. 2018). Six input parameters and 18 hidden nodes were incorporated in their model. The findings disclosed that BPNN predicted RAC's compressive strength accurately, and the concrete strength was strongly affected by aggregate water absorption and the water-to-total-material ratio. Behnood et al. (2015b) created a model to forecast the tensile strength of steel fiber–reinforced concrete (SFRC). Taking concrete compressive strength as an input, the model outperformed SVM in terms of accuracy. Golafshani and Behnood (2018a) created a BPNN model using the Levenberg-Marquartt learning algorithm to predict the RAC elastic modulus and compared it with a radial basis function neural network (RBFNN). Findings disclosed that BPNN was better in prediction than RBFNN. Table 17.1 shows an overview of ANN-based models.

17.2.3 Support Vector Machine

Support vectors establish a hyperplane, which is a critical element in SVM. SVM needs a labelled dataset, which makes it a supervised ML algorithm. SVM aims to

TABLE 17.1

Overview of Artificial Neural Network Models

Reference	Size of the Dataset	Type of Concrete	Methods	Output Variable
Gupta et al. (2019)	324	Rubberized concrete	BPNN	Compressive strength
Bachir et al. (2018)	112	Rubberized concrete	BPNN	Compressive strength
Naderpour et al. (2018a)	177	FRP reinforced concrete	BPNN	Shear strength
Golafshani and Behnood (2018a)	400	RAC	BPNN, RBFNN	Elastic modulus
Asteris et al. (2016)	205	SCC	BPNN	Compressive strength
Deng et al. (2018)	74	RAC	BPNN, CNN	Compressive strength
Naderpour et al. (2018b)	139	RAC	BPNN	Compressive strength
Van Dao et al. (2019)	210	GPC	ANFIS, BPNN	Compressive strength
Xu et al. (2019b)	421	RAC	BPNN	Elastic modulus
Xu et al. (2019b)	346	RAC	BPNN	Tensile strength

ANFIS = adaptive neuro-fuzzy inference system; BPNN = back-propagation neural network; CNN = convolutional neural network; FRP = fiber-reinforced polymer; GPC = geopolymer concrete; RAC = recycled aggregate concrete; RBFNN = radial basis function neural network.

identify the hyperplane that maximizes the separation margin among two categories. By maximizing this margin, SVM results in better accuracy when evaluated on the test set. The margin is a measure of the direct distance between the nearest point in each of the classes in the data and the hyperplane (DeRousseau et al. 2018). But not all classes can be separated by a linear hyperplane. To overcome this, a kernel function is utilized to convert the data into a higher-dimensional space. Different SVM algorithms make use of different kernel functions, such as radial basis function (RBF), linear, sigmoid, polynomial, and exponential. It improves the efficiency of the SVM by making it simpler to arrive at the idealistic parameter values of the hyperplane. Its solution can be defined as a weighted blend of kernel function values at support vectors. Support vector regression (SVR) is built on the SVM concept (Chou et al. 2011). It is used for regression analysis and better accuracy is obtained in a faster training process (Yaseen et al. 2018). SVM-based models are used in many studies; some studies used them as standalone models whereas some used meta-heuristic algorithms.

17.2.3.1 Standalone SVM Models

Much research has been carried out to better understand the practical use of standalone SVM models in this domain of concrete strength prediction. Chou et al. (2011) created an SVM model utilizing a radial basis kernel function to estimate the FCU of HPC. The findings disclosed that, based on the MAPE value, SVM exhibited high predictive accuracy. The work of Omran et al. (2016) focuses on implementing an SVM model built on sequential minimal optimization to estimate the FCU of a type of concrete whose mixture contained Portland limestone cement and an environmentally friendly Haydite lightweight cement. Deng et al. (2018) successfully predicted the FCU of RAC. Behnood et al. (2015a) predicted the tensile strength of SFRC. The findings disclosed that the SVM model had better predictive accuracy than nonlinear regression analysis. The model obtained acceptable results in terms of predicted values of the tensile modulus of HSC and RAC (Golafshani and Behnood 2018b; Yan and Shi 2010). The least square support vector machine (LS-SVM), is a set of associated supervised ML methods that recognize patterns by studying the data, using which optimization problems are created and accuracy is improved (Kaytez et al. 2015). Yaseen et al. (2018) used a least square support vector regression (LS-SVR) which is based on LS-SVM to predict the FCU values of foamed concrete with great accuracy.

17.2.3.2 Hybrid SVM-Based Models

Hybrid SVM-based models exceed the standalone SVM models in terms of performance. Pham et al. (2016) predicted the FCU of HPC by incorporating firefly algorithm (FA) into least square support vector regression (LSSVR), i.e., FA-LS-SVR hybrid model. In this model, FA is used primarily to evaluate the LS-SVR hyperparameters. Two tests were carried out; the dataset used in the study was divided into training and test sets in the first one while the second test used tenfold cross-validation. The findings disclosed that FA-LS-SVR has a strong prediction ability, indicated by MAPE. This algorithm was employed in predicting the shear strength of a type of concrete that is usually stronger than just plain concrete as it is reinforced with steel rods, commonly called reinforced concrete (RC). The algorithm was also used for predicting the shear strength of SFRC beams, as well as FRP-reinforced slabs. The input parameters consisted of mechanical properties of reinforcing components and geometrical characteristics of slabs and beams. The model predicted accurately. Yu et al. (2018) predicted the FCU of HPC using enhanced cat swarm optimization (ECSO). The principal components of SVM were optimized using an ESCO model. Statistical measures showed that the ESCO-SVM model has strong predictive accuracy. Keshtegar et al. (2019) combined SVM and the response surface method (RSM) to forecast the shear strength of SFRC beams and obtained reasonable accuracy. RSM-SVR models have the best predictive accuracy compared to many empirical formulae and models. Cheng et al. (2012) created a model for predicting the FCU of HPC using an evolutionary fuzzy SVM inference model. This model utilized SVM, FL, and genetic algorithms (GAs) to analyze time series data. The scatter plot displaying real and estimated values disclosed that the model outperformed SVM and BPNN.

17.2.4 Decision Trees

Decision trees are a type of supervised learning method used to predict the target variable value by learning the data patterns and generating formal decision rules (DeRousseau et al. 2018). It is used for classification and regression.

17.2.4.1 M5P-Tree

The M5P algorithm is an adaptation of Quinlan's M5 algorithm that integrates linear regression functions at the nodes of a traditional decision tree. It is done in a series of

TABLE 17.2

Overview of Implemented Support Vector Machine Models

Reference	Size of the Dataset	Type of Concrete	Methods	Output Variable
Gholampour et al. (2018)	650	Concrete containing coarse recycled concrete aggregates	LS-SVR	Compressive strength
Gholampour et al. (2018)	421	Concrete containing coarse recycled concrete aggregates	LS-SVR	Elastic modulus
Gholampour et al. (2018)	346	Concrete containing coarse recycled concrete aggregates	LS-SVR	Tensile strength
Keshtegar et al. (2019)	139	Steel fiber–reinforced concrete	Response Surface Method Combined with SVR	Shear strength
Yu et al. (2018)	1,761	High-performance concrete	ECSO-SVM	Compressive strength
Al-Musawi et al. (2018)	139	Steel fiber–reinforced concrete	Firefly algorithm combined with SVR	Shear strength

ECSO = enhanced cat swarm optimization; LS = least squares; SVM = support vector machine; SVR = support vector regression.

three major steps (Zhan et al. 2011; Behnood et al. 2015a). The initial step involves creating a regression tree by utilizing the training dataset. For each node, a linear model is calculated. The second step tries to overcome the data overfitting by pruning that deletes the unwanted nodes. The third step is the smoothing process, where the abrupt disruption that takes place between the consecutive linear models at the pruned trees are removed. In this step, the size of the tree is minimized while maintaining its accuracy. To increase the accuracy, the last two steps that require pruning are done at the same time. The process of dividing the input space and constructing the regression tree involves using the standard deviation ratio (SDR), representing the maximum decrease in output errors (Behnood et al. 2015a, 2017). To address the handling of missing values and enumerated attributes, an enhancement was made to the M5 algorithm, resulting in the development of M5P. Prior to the construction of the tree, the enumerated attributes are transformed into binary variables in the M5P algorithm (Almasi et al. 2017). Table 17.2 shows a summary of SVM-based models that have been implemented.

17.2.4.2 Multiple Additive Regression Trees

Multiple additive regression trees (MART) is a resilient meta-classifier that integrates stochastic gradient boosting with classification and regression trees (CART). It combines and fits a succession of low-error-rate models to produce an ensemble model that performs better (Elish 2009). Chou et al. estimated the FCU of HPC by employing MART. In terms of R^2, the model outperformed both ANN and SVM and obtained reasonable prediction accuracy (Chou et al. 2011).

17.2.4.3 Random Forest

Random forest (RF) uses ensemble learning, an approach to combine classifiers which then finds ways to deal with difficult problems. This model comprises multiple decision trees and is trained using the technique of bagging or bootstrap aggregating (Han et al. 2019). Bagging is an algorithm that improves the level of accuracy of machine learning algorithms via two steps. The original dataset is resampled randomly, by which independent, identically distributed datasets are formed, in the first step. In step two, a different dataset is used in training of the base predictors. The average of the predicted values of each decision tree predictor is aggregated to obtain the final result. Han et al. (2019) estimated the FCU of HPC and Mangalathu and Jeon (2018) predicted the shear strength of joints at the intersection of a beam and column using RF. The two studies concluded that RF generates accurate predictions. The one-dimensional compressive strength of self-compacting concrete (SCC) was estimated using an RF model by Zhang et al. (2019). The model was upgraded using a beetle antennae search (BAS) algorithm. BAS mirrors the antenna's functions and unpredictable flight patterns of the beetle, which flies to a site that has a stronger odour. The process of searching and detecting is employed (Jiang and Li 2018). Findings of a study disclosed that BAS has excellent ability to determine if the RF hyperparameters were optimal and that the hybrid BAS-RF algorithm has a great prediction ability.

17.2.5   Evolutionary Computation Algorithms

Evolutionary computation (EC) algorithms is a generic population-based meta-heuristic optimization algorithm that utilizes biological evolution–inspired mechanisms to locate a solution (Eiben and Smit 2011). A population reflecting a candidate solution set is created randomly and then the fitness function regulates the quality of the solutions. Through recombination and mutation, the following generation containing a superior collection of candidates is produced. Recombination uses binary operators on the previous generations to generate new candidates (parents). Mutation is a genetic operator that is used to preserve and modify the genes of a population, from one generation to the next. A better collection of candidates of the next generation is formed. When the desired value of fitness function or the highest possible number of generations is attained, the iteration is ceased. Gandomi et al. 2014; 2017 utilized gene expression programming (GEP) to make predictions regarding the shear strength of slender RC beams in their study. The researchers achieved the most effective GEP algorithm by minimizing the objective function, which encompassed statistical indexes linked to the testing, validation, and training data. Findings disclosed that the GEP model outperformed design codes like ACI and Eurocode 2 in terms of accuracy. In another study, linear genetic programming (LGP) was used to predict the FCU of a type of concrete reinforced with carbon fiber–reinforced polymer (CFRP) (Gandomi et al. 2010). The LGP model was used to create four

TABLE 17.3
Summary of Decision Tree Models

Reference	Size of the Dataset	Type of Concrete	Methods	Output Variable
Gholampour et al. (2018)	650	Concrete containing coarse recycled concrete aggregates	M5	Compressive strength
Gholampour et al. (2018)	650	Concrete containing coarse recycled concrete aggregates	M5	Compressive strength
Gholampour et al. (2018)	421	Concrete containing coarse recycled concrete aggregates	M5	Elastic modulus
Gholampour et al. (2018)	346	Concrete containing coarse recycled concrete aggregates	M5	Tensile strength
Behnood and Golafshani (2020)	470	Concrete containing waste foundry sand	M5P-Tree	Compressive strength
Behnood and Golafshani (2020)	172	Concrete containing waste foundry sand	M5P-Tree	Elastic Modulus
Behnood and Golafshani (2020)	295	Concrete containing waste foundry sand	M5P-Tree	Tensile Strength

distinct formulations. The findings disclosed that the formulations had a high level of accuracy. The extent to which the parameters influenced the final result was obtained through sensitivity analysis. Golafshani and Behnood (2018a) could accurately predict the tensile modulus of RAC using the nature-inspired concept of artificial bee colony programming, biogeography-based programming, and genetic programming (GP). The elastic modulus of RAC was affected by the compressive strength of concrete, water absorption, and the ratio of fine aggregates to total aggregates. Table 17.3 is a compilation of decision tree models.

17.2.6 Selection of Model Inputs

Expertise and skill from human professionals, in addition to computational efforts, are essential in selecting optimal parameters for performing training and carry out the testing of various ML models. This selection process is crucial for simplifying the models by eliminating parameters that have low impact, improving overall performance, and saving computational time. Several studies (Cheng et al. 2012; Deepa et al. 2010; Chou et al. 2011; Asteris et al. 2016; Asteris and Kolovos 2017; Bui et al. 2018; Yu et al. 2018; Han et al. 2019) have highlighted the significance of common parameters such as aggregates, binder content, fly ash, and blast furnace slag in predicting the mechanical properties of concrete.

Parameters related to aggregates, including hardness, granular size distribution, and cleanliness, significantly influence the physical properties of concrete materials, with strength being an important property. Supplementary cementitious materials (SCMs) like fly ash, blast furnace slag, and silica fume are frequently incorporated into concrete due to their beneficial pozzolanic characteristics and micro-filler effect, which positively impact the compressive strength (Ayaz et al. 2015). Fly ash, for example, can reduce the required water content for mixing, while silica fume particles can enhance the density of the zone that lies in the interaction between the cement paste and aggregate.

The water-to-binder (w/b) ratio and chemical admixtures are also crucial input factors that determine concrete strength. Higher w/b ratios tend to cause a dip in the mechanical strength of concrete by increasing the porosity in the structure and reducing the ratio of hydrated products.

17.3   DISCUSSION AND CRITICAL ANALYSIS

Statistical metrics obtained from notable research studies demonstrate the substantial benefits of using ML approaches over empirical formulas. Many researchers adopted ML methodologies to effectively predict the mechanical strength of concrete. These methodologies considered the constituents of the concrete mixture and their curvilinear correlation with the resulting FCU. The quantity of training data used to develop the ML models are way greater than the data used in empirical models. Training using a smaller dataset leads to a high error rate when predicting unseen data, which then limits the use of empirical models where the concrete has a simpler structure, as traditional approaches are more fitted in providing detailed statistical formulae. Empirical models have limited capability in terms of the accuracy in predicting the impact of novel ingredients in the concrete mixture. ML models can alter the number of features and ingredients included. Sensitivity analysis enables the evaluation of the effect of various inputs on one of the important parameters of concrete, which is its strength.

Numerous studies use ANN. ANN models have a well-defined structure composed of well-defined sets of weights and biases, few hidden layers and neurons gained after many trials. But this trial-and-error method gets very time consuming. In BPNN, a gradient descent algorithm to minimize the errors that arise, is used for the training process (Wang et al. 2015). The issue arises with back-propagation's tendency to converge towards local minima and avoid global solutions. This can be minimized using an extreme learning machine (ELM) and more simplicity can be provided as stopping criteria and learning rate are not required (Christou et al. 2019). Al-Shamiri et al. (2019) used this model and observed that the ELM model has a better performance rate than BPNN. But it requires relatively more hidden neurons compared to BPNN, because of the randomness involved in determining the value of input weights and biases (Zhu et al. 2005). A situation of model overfit may arise in such a case where more hidden neurons are used. It can be concluded that ANN overestimates the complexity of concrete properties (Behnood and Golafshani 2018). To enhance ANN performance, many ensemble and metaheuristic models have been put forward.

The ANFIS model is a fusion of ANN and fuzzy learning (Yuan et al. 2014). Şahin and Erol (2017) and Van Dao et al. (2019) uncovered that ANFIS, with its ability to learn quickly, is capable of identifying the nonlinear structure processes. But ANFIS may suffer from FL selection as it cannot produce more than one output

TABLE 17.4

Overview of Evolutionary Algorithms

Reference	Size of the Dataset	Type of Concrete	Methods	Output Variable
Sarveghadi et al. (2015)	208	Steel fiber–reinforced concrete	Multi-expression programming	Shear strength
Gandomi et al. (2011)	1,938	Reinforced concrete	Linear genetic programming	Shear strength
Golafshani and Behnood (2018b)	400	Recycled aggregate concrete	Genetic programming, artificial bee colony programming, biogeography-based programming	Elastic modulus
Gandomi et al. (2010)	101	Concrete fiber-reinforced polymer confined concrete	Linear genetic programming	Compressive strength
Gandomi et al. (2014)	1,942	Reinforced concrete	Gene expression programming	Shear strength
Gandomi et al. (2017)	466	Reinforced concrete	Gene expression programming	Shear strength

variable (Yu et al. 2018). Yuan et al. used the GA model to optimize the accuracy of ANN. The GA-ANN model increases computation time and model complexity (Yuan et al. 2014).

SVM models have the capability to do nonlinear mapping and generalization (Yu et al. 2018). SVM integrates support vectors while training, which stops the nonsupport vectors from causing any effect on the model's performance. Choosing a suitable kernel function is a heuristic approach and it takes a lot of time as it is determined by an experimental process based on the error in each scenario. The mapping of a nonlinear input into high-dimensional feature space is a complex process due to which the SVR performance cannot be easily interpreted.

ANN and SVM have been regarded as 'black-box' models because they contain huge node sizes and internal connections (Yadav and Chandel 2014). Because of this, it is difficult to produce statistical formulas that report the underlying correlation between the input and output variables through this model. As a solution, decision trees and evolutionary computation (EC) algorithms are utilized. They can produce the mathematical theory that gives a detailed account of the correlations between variables and the matching outputs. Decision trees sometimes led to model overfit. The disadvantages of decision tree models are reduced by using RF models and MART. Chou et al. obtained better results from MART than ANN and SVM (Chou et al. 2011). The accuracy of hybrid and standalone SVM and ANN models are higher than decision trees and evolutionary algorithm (EA) (Behnood et al. 2015a; Bui et al. 2018; Yu et al. 2018). Table 17.4 shows a summary of SVM-based models that have been implemented and Table 17.5 illustrates a comparative assessment of various ML models using the same testing data.

TABLE 17.5

An Evaluation of Multiple Machine Learning Models Using Identical Testing Data

Reference	Size of the Dataset	Output Variable	Models	Evaluation Metrics	
Yu et al. (2018)	1,761	Compressive strength	BPNN	R^2	0.96
			ELM		0.934
			ANFIS		0.906
			SVM		0.793
			M5		0.937
			ECSO-SVM		0.942
Pham et al. (2016)	239	Compressive strength	ANN	R^2	0.76
			SVM		0.79
			FA-LS-SVR		0.87
Chou et al. (2011)	1,030	Compressive strength	ANN	R^2	0.909
			SVM		0.885
			MART		0.911
Van Dao et al. (2019)	210	Compressive strength	ANN	R^2	0.851
			ANFIS		0.879
Yuan et al. (2014)	180	Compressive strength	ANN	R^2	0.68
			GA-ANN		0.813
			ANFIS		0.95

ANFIS = adaptive neuro-fuzzy inference system; ANN = artificial neural network; BPNN = back-propagation neural network; ELM = extreme learning machine; ECSO = enhanced cat swarm optimization; FA = firefly algorithm; GA = generic algorithm; LS = least squares; MART = multiple additive regression trees; SVM = support vector machine; SVR = support vector regression.

17.4 APPLICABLE SUGGESTIONS AND AREAS OF INFORMATION DEFICIENCY

The best model selection for prediction is based on different criteria as all ML approaches have their own list of pros and cons. The relationship between the components of the concrete mixture and their mechanical strength significantly influences the choice of a model. In the case of nonlinear correlation that is influenced by multiple factors, adopting models such as ANN and SVM can be a smart option as they excel in solving nonlinear problems with lower error. Hybrid ANN-based and SVM-based models outperform standalone models and offer the highest accuracy and process in predicting the concrete strength. Decision trees and EA can be used when model transparency is necessary, as they have the ability to create explicit statistical formulae which describe the dependency of results on the parameters. Ensemble models improve the accuracy of decision trees, with a downside of increased computing time and model complexity.

Using the hybrid ANN and SVM models on big data with optimal feature selection could produce the best result, despite the increased computing time. Yet, the selection of the optimal metaheuristic model can only be done by trial and error. As a result, there is no exact method to select the best optimization algorithm as diverse outcomes can be obtained depending on the problem.

Many growing technologies are merging the physical, digital, and biological worlds in this industrial revolutionary world but the construction industry is far behind in seizing the opportunities. ML predictions of mechanical properties of construction materials and structures contribute to generative intelligent design. However, mechanical engineers must still overcome a wide range of knowledge gaps before they can replicate methods employed in other emerging domains.

17.5 CONCLUSIONS

Much research has been undertaken to predict the strength of complex concrete mixtures and identify their pros and cons. Generally, the traditional statistical and empirical models are expensive, time consuming, and imprecise. To resolve these cons, machine learning models have been proposed by researchers.

In this chapter, ML models are categorized into four groups: ANN, SVM, decision trees, and EA. The mechanical properties of concrete, such as shear strength, tensile strength, compression strength, and elastic modulus were predicted using these models. The pros and cons of the approaches have been analyzed and compared. How big the training dataset is, the correlation between the composition of the concrete mixture and its constituent ingredients, its strength and number of features included in the model have been found to have a direct effect on the accuracy of the model. Civil engineers should be able to determine the best-suited ML model to carry out prediction of the mechanical properties of concrete based on the evaluation of the different approaches put forward in this study. Considering the concrete strength is accurately predicted, further research to examine the ML models' reliability to predict the properties of novel concrete types will be undertaken.

REFERENCES

Ahmadi-Nedushan, Behrouz. 2012. Prediction of Elastic Modulus of Normal and High Strength Concrete Using ANFIS and Optimal Nonlinear Regression Models. *Construction and Building Materials* 36 (November): 665–673. https://doi.org/10.1016/j.conbuildmat.2012.06.002.

Almasi, S. Najmedin, Raheb Bagherpour, Reza Mikaeil, Yılmaz Özçelik, and Hamid Kalhori. 2017. Predicting the Building Stone Cutting Rate Based on Rock Properties and Device Pullback Amperage in Quarries Using M5P Model Tree. *Geotechnical and Geological Engineering* 35 (4): 1311–1326. https://doi.org/10.1007/s10706-017-0177-0.

Al-Musawi, Abeer A., Afrah Abdulelah Hamzah Alwanas, Sinan Q. Salih, Zainab Hasan Ali, Tran Minh Tung, and Zaher Mundher Yaseen. 2018. Shear Strength of SFRCB without Stirrups Simulation: Implementation of Hybrid Artificial Intelligence Model. *Engineering With Computers* 36 (1): 1–11. https://doi.org/10.1007/s00366-018-0681-8.

Al-Shamiri, Abobakr Khalil, Joong Hoon Kim, Tian Yuan, and Young Soo Yoon. 2019. Modeling the Compressive Strength of High-Strength Concrete: An Extreme Learning Approach.

Construction and Building Materials 208 (May): 204–219. https://doi.org/10.1016/j. conbuildmat.2019.02.165.

Amani, J., and Ramtin Moeini. 2012. Prediction of Shear Strength of Reinforced Concrete Beams Using Adaptive Neuro-Fuzzy Inference System and Artificial Neural Network. *Scientia Iranica* 19 (2): 242–248. https://doi.org/10.1016/j.scient.2012.02.009.

Asteris, Panagiotis G., and Konstantinos G. Kolovos. 2017. Self-Compacting Concrete Strength Prediction Using Surrogate Models. *Neural Computing and Applications* 31 (S1): 409–424. https://doi.org/10.1007/s00521-017-3007-7.

Asteris, Panagiotis G., Konstantinos G. Kolovos, Maria G. Douvika, and Konstantinos Roinos. 2016. Prediction of Self-Compacting Concrete Strength Using Artificial Neural Networks. *European Journal of Environmental and Civil Engineering* 20 (sup1): s102–s122. https://doi.org/10.1080/19648189.2016.1246693.

Ayaz, Yaşar, Adnan Fatih Kocamaz, and Mehmet Burhan Karakoç. 2015. Modeling of Compressive Strength and UPV of High-Volume Mineral-Admixtured Concrete Using Rule-Based M5 Rule and Tree Model M5P Classifiers. *Construction and Building Materials* 94 (September): 235–240. https://doi.org/10.1016/j.conbuildmat.2015.06.029.

Bachir, Rahali, Sidi Mohamed Aissa Mamoune, and Habib Trouzine. 2018. Using Artificial Neural Networks Approach to Estimate Compressive Strength for Rubberized Concrete. *Periodica Polytechnica-Civil Engineering* (May). https://doi.org/10.3311/ppci.11928.

Bashir, Rizwan, and Ashraf Ashour. 2012. Neural Network Modelling for Shear Strength of Concrete Members Reinforced with FRP Bars. *Composites Part B: Engineering* 43 (8): 3198–3207. https://doi.org/10.1016/j.compositesb.2012.04.011.

Behnood, Ali, Venous Behnood, Mahsa Modiri Gharehveran, and Kürşat Esat Alyamaç. 2017. Prediction of the Compressive Strength of Normal and High-Performance Concretes Using M5P Model Tree Algorithm. *Construction and Building Materials* 142 (July): 199–207. https://doi.org/10.1016/j.conbuildmat.2017.03.061.

Behnood, Ali, and Emadaldin Mohammadi Golafshani. 2018. Predicting the Compressive Strength of Silica Fume Concrete Using Hybrid Artificial Neural Network with Multi-Objective Grey Wolves. *Journal of Cleaner Production* 202 (November): 54–64. https://doi.org/10.1016/j.jclepro.2018.08.065.

Behnood, Ali, and Emadaldin Mohammadi Golafshani. 2020. Machine Learning Study of the Mechanical Properties of Concretes Containing Waste Foundry Sand. *Construction and Building Materials* 243: 118152.

Behnood, Ali, Jan Olek, and Michał A. Glinicki. 2015a. Predicting Modulus Elasticity of Recycled Aggregate Concrete Using M5' Model Tree Algorithm. *Construction and Building Materials* 94 (September): 137–147. https://doi.org/10.1016/j.conbuildmat.2015.06.055.

Behnood, Ali, Kho Pin Verian, and Mahsa Modiri Gharehveran. 2015b. Evaluation of the Splitting Tensile Strength in Plain and Steel Fiber-Reinforced Concrete Based on the Compressive Strength. *Construction and Building Materials* 98 (November): 519–529. https://doi.org/10.1016/j.conbuildmat.2015.08.124.

Belalia Douma, O., B. Boukhatem, M. Ghrici, and A. Tagnit-Hamou. 2017. Prediction of Properties of Self-Compacting Concrete Containing Fly Ash Using Artificial. *Neural Computing and Applications* 28: 707–718.

Bourdeau, Mathieu, Xiao Qiang Zhai, Elyes Nefzaoui, Xiao-Feng Guo, and Patrice Chatellier. 2019. Modeling and Forecasting Building Energy Consumption: A Review of Data-Driven Techniques. *Sustainable Cities and Society* 48 (July): 101533. https://doi.org/10.1016/j.scs.2019.101533.

Bui, Dac-Khuong, Tuấn Ngọc Nguyễn, Jui-Sheng Chou, H. Nguyen-Xuan, and Tuan Ngo. 2018. A Modified Firefly Algorithm-Artificial Neural Network Expert System for Predicting Compressive and Tensile Strength of High-Performance Concrete. *Construction and Building Materials* 180 (August): 320–333. https://doi.org/10.1016/j.conbuildmat.2018.05.201.

Cheng, Min-Yuan, Jui-Sheng Chou, Andreas F. V. Roy, and Yuwei Wu. 2012. High-Performance Concrete Compressive Strength Prediction Using Time-Weighted Evolutionary Fuzzy Support Vector Machines Inference Model. *Automation in Construction* 28 (December): 106–115. https://doi.org/10.1016/j.autcon.2012.07.004.

Chithra, S., Sanjeev Kumar, K. Chinnaraju, and F. Alfin Ashmita. 2016. A Comparative Study on the Compressive Strength Prediction Models for High Performance Concrete Containing Nano Silica and Copper Slag Using Regression Analysis and Artificial Neural Networks. *Construction and Building Materials* 114 (July): 528–535. https://doi.org/10.1016/j.conbuildmat.2016.03.214.

Chou, Jui-Sheng, C. Chiu, M. Farfoura, and I. Al-Taharwa. 2011. Optimizing the Prediction Accuracy of Concrete Compressive Strength Based on a Comparison of Data-Mining. *Techniques* 25: 242–253. https://doi.org/10.1061/(ASCE)CP.1943-5487.

Chou, Jui-Sheng, and Chih-Fong Tsai. 2012. Concrete Compressive Strength Analysis Using a Combined Classification and Regression Technique. *Automation in Construction* 24 (July): 52–60. https://doi.org/10.1016/j.autcon.2012.02.001.

Christou, Vasileios, Markos G. Tsipouras, Νικόλαος Γιαννακέας, Alexandros T. Tzallas, and Gavin Brown. 2019. Hybrid Extreme Learning Machine Approach for Heterogeneous Neural Networks. *Neurocomputing* 361 (October): 137–150. https://doi.org/10.1016/j.neucom.2019.04.092.

Deepa, C., K. Sathiyakumari, and V. Pream Sudha. 2010. Prediction of the Compressive Strength of High Performance Concrete Mix Using Tree Based Modeling. *International Journal of Computer Applications* 6 (5): 18–24. https://doi.org/10.5120/1076-1406.

Deng, Fang-Ming, Yigang He, Shan Zhou, Yun Yu, Haigen Cheng, and Xiang Wu. 2018. Compressive Strength Prediction of Recycled Concrete Based on Deep Learning. *Construction and Building Materials* 175 (June): 562–569. https://doi.org/10.1016/j.conbuildmat.2018.04.169.

DeRousseau, M. A., J. R. Kasprzyk, and Wil V. Srubar. 2018. Computational Design Optimization of Concrete Mixtures: A Review. *Cement and Concrete Research* 109 (July): 42–53. https://doi.org/10.1016/j.cemconres.2018.04.007.

Eiben, A. E., and Selmar K. Smit. 2011. Parameter Tuning for Configuring and Analyzing Evolutionary Algorithms. *Swarm and Evolutionary Computation* 1 (1): 19–31. https://doi.org/10.1016/j.swevo.2011.02.001.

Elish, Mahmoud O. 2009. Improved Estimation of Software Project Effort Using Multiple Additive Regression Trees. *Expert Systems With Applications* 36 (7): 10774–10778. https://doi.org/10.1016/j.eswa.2009.02.013.

Esfe, Mohammad Hemmat, Somchai Wongwises, Ali Naderi, Amin Asadi, Mohammad Reza Safaei, Hadi Rostamian, Mahidzal Dahari, and Arash Karimipour. 2015. Thermal Conductivity of Cu/TiO_2–Water/EG Hybrid Nanofluid: Experimental Data and Modeling Using Artificial Neural Network and Correlation. *International Communications in Heat and Mass Transfer* 66 (August): 100–104. https://doi.org/10.1016/j.icheatmasstransfer.2015.05.014.

Gandomi, Amir H., Amir H. Alavi, Mostafa Gandomi, and Sadegh Kazemi. 2017. Formulation of Shear Strength of Slender RC Beams Using Gene Expression Programming, Part II: With Shear Reinforcement. *Measurement* 95 (January): 367–376. https://doi.org/10.1016/j.measurement.2016.10.024.

Gandomi, Amir H., Amir H. Alavi, Sadegh Kazemi, and Mostafa Gandomi. 2014. Formulation of Shear Strength of Slender RC Beams Using Gene Expression Programming, Part I: Without Shear Reinforcement. *Automation in Construction* 42 (June): 112–121. https://doi.org/10.1016/j.autcon.2014.02.007.

Gandomi, Amir H., Amir H. Alavi, and Mohammad Ghasem Sahab. 2010. New Formulation for Compressive Strength of CFRP Confined Concrete Cylinders Using Linear Genetic Programming. *Materials and Structures* 43: 963–983.

Gandomi, Amir H., Amir H. Alavi, and Gun Jin Yun. 2011. Nonlinear Modeling of Shear Strength of SFRC Beams Using Linear Genetic Programming. *Structural Engineering and Mechanics* 38 (1): 1–25. https://doi.org/10.12989/sem.2011.38.1.001.

Gholampour, Aliakbar, Iman Mansouri, Özgür Kişi, and Togay Ozbakkaloglu. 2018. Evaluation of Mechanical Properties of Concretes Containing Coarse Recycled Concrete Aggregates Using Multivariate Adaptive Regression Splines (MARS), M5 Model Tree (M5Tree), and Least Squares Support Vector Regression (LSSVR) Models. *Neural Computing and Applications* 32 (1): 295–308. https://doi.org/10.1007/s00521-018-3630-y.

Golafshani, E. M., and A. Behnood. 2018a. Application of Soft Computing Methods for Predicting the Elastic Modulus of Recycled Aggregate Concrete. *Journal of Cleaner Production* 176 (March): 1163–1176. https://doi.org/10.1016/j.jclepro.2017.11.186.

Golafshani, E. M., and A. Behnood. 2018b. Automatic Regression Methods for Formulation of Elastic Modulus of Recycled Aggregate Concrete. *Applied Soft Computing* 64 (March): 377–400. https://doi.org/10.1016/j.asoc.2017.12.030.

Gupta, Trilok, K. A. Patel, Salman Siddique, Ravi Kumar Sharma, and Sandeep Chaudhary. 2019. Prediction of Mechanical Properties of Rubberised Concrete Exposed to Elevated Temperature Using ANN. *Measurement* 147 (December): 106870. https://doi.org/10.1016/j.measurement.2019.106870.

Han, Qinghua, Changqing Gui, Jie Xu, and Giuseppe Lacidogna. 2019. A Generalized Method to Predict the Compressive Strength of High-Performance Concrete by Improved Random Forest Algorithm. *Construction and Building Materials* 226 (November): 734–742. https://doi.org/10.1016/j.conbuildmat.2019.07.315.

Jiang, Xiangyuan, and Shuai Li. 2018. BAS: Beetle Antennae Search Algorithm for Optimization Problems. *International Journal of Robotics and Control* 1 (1): 1. https://doi.org/10.5430/ijrc.v1n1p1.

Kaytez, Fazıl, M. Cengiz Taplamacıoğlu, Ertuğrul Çam, and Fırat Hardalaç. 2015. Forecasting Electricity Consumption: A Comparison of Regression Analysis, Neural Networks and Least Squares Support Vector Machines. *International Journal of Electrical Power & Energy Systems* 67 (May): 431–438. https://doi.org/10.1016/j.ijepes.2014.12.036.

Keshtegar, B., M. Bagheri, and Z. Mundher Yaseen. 2019. Shear Strength of Steel Fiber-Unconfined Reinforced Concrete Beam Simulation: Application of Novel Intelligent Model. *Composite Structures* 212 (March): 230–242. https://doi.org/10.1016/j.compstruct.2019.01.004.

Khademi, Faezehossadat, Mahmoud Akbari, Sayed Mohammadmehdi Jamal, and Mehdi Nikoo. 2017. Multiple Linear Regression, Artificial Neural Network, and Fuzzy Logic Prediction of 28 Days Compressive Strength of Concrete. *Frontiers of Structural and Civil Engineering* 11 (1): 90–99. https://doi.org/10.1007/s11709-016-0363-9.

Khademi, Faezehossadat, Sayed Mohammadmehdi Jamal, Neela Deshpande, and Shreenivas Londhe. 2016. Predicting Strength of Recycled Aggregate Concrete Using Artificial Neural Network, Adaptive Neuro-Fuzzy Inference System and Multiple Linear Regression. *International Journal of Sustainable Built Environment* 5 (2): 355–369. https://doi.org/10.1016/j.ijsbe.2016.09.003.

Kramer, Oliver. 2017. Genetic Algorithm Essentials. *Studies in Computational Intelligence.* https://doi.org/10.1007/978-3-319-52156-5.

Mangalathu, Sujith, and Jong-Su Jeon. 2018. Classification of Failure Mode and Prediction of Shear Strength for Reinforced Concrete Beam-Column Joints Using Machine Learning Techniques. *Engineering Structures* 160 (April): 85–94. https://doi.org/10.1016/j.engstruct.2018.01.008.

Naderpour, Hosein, Omid Poursaeidi, and Masoud Ahmadi. 2018a. Shear Resistance Prediction of Concrete Beams Reinforced by FRP Bars Using Artificial Neural Networks. *Measurement* 126 (October): 299–308. https://doi.org/10.1016/j.measurement.2018.05.051.

Naderpour, Hosein, Amir Hossein Rafiean, and Pouyan Fakharian. 2018b. Compressive Strength Prediction of Environmentally Friendly Concrete Using Artificial Neural Networks. *Journal of Building Engineering* 16 (March): 213–219. https://doi.org/10.1016/j.jobe.2018.01.007.

Namyong, Jee, Yoon Sangchun, and Cho Hongbum. 2004. Prediction of Compressive Strength of In-Situ Concrete Based on Mixture Proportions. *Journal of Asian Architecture and Building Engineering* 3 (1): 9–16. https://doi.org/10.3130/jaabe.3.9.

Nazemi, Ehsan, M. Dinca, Mohammad Amir, B. Rokrok, and M. H. Choopan Dastjerdi. 2019. Estimation of Volumetric Water Content during Imbibition in Porous Building Material Using Real Time Neutron Radiography and Artificial Neural Network. *Nuclear Instruments and Methods in Physics Research Section A: Accelerators, Spectrometers, Detectors and Associated Equipment* 940 (October): 344–350. https://doi.org/10.1016/j.nima.2019.06.052.

Omran, Behzad Abounia, Qian Chen, and Ruoyu Jin. 2016. Comparison of Data Mining Techniques for Predicting Compressive Strength of Environmentally Friendly Concrete. *Journal of Computing in Civil Engineering* 30 (6). https://doi.org/10.1061/(asce)cp.1943-5487.0000596.

Pham, Anh-Duc, Nhat-Duc Hoang, and Quang-Trung Nguyen. 2016. Predicting Compressive Strength of High-Performance Concrete Using Metaheuristic-Optimized Least Squares Support Vector Regression. *Journal of Computing in Civil Engineering* 30 (3). https://doi.org/10.1061/(asce)cp.1943-5487.0000506.

Popovics, S., and J. Ujhelyi. 2008. Contribution to the Concrete Strength Versus Water-Cement Ratio Relationship. *Journal of Materials in Civil Engineering* 20: 459–463. https://doi.org/10.1061/(ASCE)0899-1561(2008)20:7(459).

Şahin, Mehmet, and Rızvan Erol. 2017. A Comparative Study of Neural Networks and ANFIS for Forecasting Attendance Rate of Soccer Games. *Mathematical and Computational Applications* 22 (4): 43. https://doi.org/10.3390/mca22040043.

Sarveghadi, Masoud, Amir H. Gandomi, Hamed Bolandi, and Amir H. Alavi. 2015. Development of Prediction Models for Shear Strength of SFRCB Using a Machine Learning Approach. *Neural Computing and Applications* 31 (7): 2085–2094. https://doi.org/10.1007/s00521-015-1997-6.

Sharifzadeh, Mahdi, Alexandra Sikinioti-Lock, and Nilay Shah. 2019. Machine-Learning Methods for Integrated Renewable Power Generation: A Comparative Study of Artificial Neural Networks, Support Vector Regression, and Gaussian Process Regression. *Renewable & Sustainable Energy Reviews* 108 (July): 513–538. https://doi.org/10.1016/j.rser.2019.03.040.

Silva, Rui V., and Ravindra K. Dhir. 2015. Tensile Strength Behaviour of Recycled Aggregate Concrete. *Construction and Building Materials* 83 (May): 108–118. https://doi.org/10.1016/j.conbuildmat.2015.03.034.

Slater, Emma, Moniruzzaman Moni, and M. Shahria Alam. 2012. Predicting the Shear Strength of Steel Fiber Reinforced Concrete Beams. *Construction and Building Materials* 26 (1): 423–436. https://doi.org/10.1016/j.conbuildmat.2011.06.042.

Topçu, İLker Bekir, and Mustafa Sarıdemir. 2008. Prediction of Mechanical Properties of Recycled Aggregate Concretes Containing Silica Fume Using Artificial Neural Networks and Fuzzy Logic. *Computational Materials Science* 42 (1): 74–82. https://doi.org/10.1016/j.commatsci.2007.06.011.

Van Dao, Dong, Haï-Bang Ly, Son Hoang Trinh, Tien-Thinh Le, and Binh Thai Pham. 2019. Artificial Intelligence Approaches for Prediction of Compressive Strength of Geopolymer Concrete. *Materials* 12 (6): 983. https://doi.org/10.3390/ma12060983.

Wang, Lin, Yi Zeng, and Tao Chen. 2015. Back Propagation Neural Network with Adaptive Differential Evolution Algorithm for Time Series Forecasting. *Expert Systems with Applications* 42 (2): 855–863. https://doi.org/10.1016/j.eswa.2014.08.018.

Xu, Jinjun, Yuliang Chen, Tianyu Xie, Xinyu Zhao, Baiqing Xiong, and Zongping Chen. 2019a. Prediction of Triaxial Behavior of Recycled Aggregate Concrete Using Multivariable Regression and Artificial Neural Network Techniques. *Construction and Building Materials* 226 (November): 534–554. https://doi.org/10.1016/j.conbuildmat.2019.07.155.

Xu, Jinjun, Xinyu Zhao, Yu Yong, Tianyu Xie, Guosong Yang, and Jianyang Xue. 2019b. Parametric Sensitivity Analysis and Modelling of Mechanical Properties of Normal- and High-Strength Recycled Aggregate Concrete Using Grey Theory, Multiple Nonlinear Regression and Artificial Neural Networks. *Construction and Building Materials* 211 (June): 479–491. https://doi.org/10.1016/j.conbuildmat.2019.03.234.

Yadav, Amit Kumar, and S. S. Chandel. 2014. Solar Radiation Prediction Using Artificial Neural Network Techniques: A Review. *Renewable & Sustainable Energy Reviews* 33 (May): 772–781. https://doi.org/10.1016/j.rser.2013.08.055.

Yan, Kezhen, and Caijun Shi. 2010. Prediction of Elastic Modulus of Normal and High Strength Concrete by Support Vector Machine. *Construction and Building Materials* 24 (8): 1479–1485. https://doi.org/10.1016/j.conbuildmat.2010.01.006.

Yaseen, Zaher Mundher, Ravinesh C. Deo, Ameer A. Hilal, M. Abbas, Laura Cornejo Bueno, Sancho Salcedo–Sanz, and Moncef L. Nehdi. 2018. Predicting Compressive Strength of Lightweight Foamed Concrete Using Extreme Learning Machine Model. *Advances in Engineering Software* 115 (January): 112–125. https://doi.org/10.1016/j.advengsoft.2017.09.004.

Yu, Yang, Wengui Li, Jianchun Li, and Thuc N. Nguyen. 2018. A Novel Optimised Self-Learning Method for Compressive Strength Prediction of High Performance Concrete. *Construction and Building Materials* 184 (September): 229–247. https://doi.org/10.1016/j.conbuildmat.2018.06.219.

Yuan, Zhe, Linna Wang, and Xu Ji. 2014. Prediction of Concrete Compressive Strength: Research on Hybrid Models Genetic Based Algorithms and ANFIS. *Advances in Engineering Software* 67 (January): 156–163. https://doi.org/10.1016/j.advengsoft.2013.09.004.

Zhan, Chengjun, Albert Gan, and Mohammed Hadi. 2011. Prediction of Lane Clearance Time of Freeway Incidents Using the M5P Tree Algorithm. *IEEE Transactions on Intelligent Transportation Systems* 12 (4): 1549–1557. https://doi.org/10.1109/tits.2011.2161634.

Zhang, Junfei, Guowei Ma, Yimiao Huang, Junbo Sun, Farhad Aslani, and Brett Nener. 2019. Modelling Uniaxial Compressive Strength of Lightweight Self-Compacting Concrete Using Random Forest Regression. *Construction and Building Materials* 210 (June): 713–719. https://doi.org/10.1016/j.conbuildmat.2019.03.189.

Zhu, Qin-Yu, A. K. Qin, Ponnuthurai Nagaratnam Suganthan, and Guang-Bin Huang. 2005. Evolutionary Extreme Learning Machine. *Pattern Recognition* 38 (10): 1759–1763. https://doi.org/10.1016/j.patcog.2005.03.028.

Index